NUMERICAL SIMULATIONS AND CASE STUDIES USING VISUAL C++.NET

NUMERICAL SIMULATIONS AND CASE STUDIES USING VISUAL C++.NET

SHAHARUDDIN SALLEH
Universiti Teknologi Malaysia

ALBERT Y. ZOMAYA
University of Sydney

STEPHAN OLARIU
Old Dominion University

BAHROM SANUGI
Universiti Teknologi Malaysia

A JOHN WILEY & SONS, INC., PUBLICATION

Published by John Wiley & Sons, Inc., Hoboken, New Jersey.
Published simultaneously in Canada.

For general information on our other products and services please contact our Customer Care Department within the U.S. at 877-762-2974, outside the U.S. at 317-572-3993 or fax 317-572-4002.

Wiley also publishes its books in a variety of electronic formats. Some content that appears in print, however, may not be available in electronic format. For more information about Wiley products, visit our web site at www.wiley.com.

Library of Congress Cataloging-in-Publication Data is available.

Numerical simulations and case studies using Visual C++.Net / Shaharuddin Salleh . . . [et al].
p. cm.
Includes bibliographical references.
ISBN-10 0-471-69461-4 (cloth : alk. paper)
ISBN-13 978 0-471-69461-8
1. C++ (Computer program language) 2. Computer simulation. 3. Microsoft .NET. 4. Microsoft Visual C++. I. Salleh Shaharuddin. 1956–
QA76.76.C153N86 2005
005.13′3--dc22 2004025688

Printed in the United States of America.

10 9 8 7 6 5 4 3 2 1

CONTENTS

PREFACE

Numerical computing has come a long way since the discovery of the first computer in the early 1940s. Computers have brought changes to the world through their capability to handle and solve problems that were previously not solvable. Because of the improvements in information and communications technology that computers have brought, the world looks smaller. There have been vast improvements in the way computers are used for solving numerical problems in which hardware and software together form the building blocks.

This book has been written to discuss both problems in numerical methods and simulations, and their solutions using Visual C++. There are several objectives for doing this. First, there is a gap between a problem and its computing elements. A problem normally comes from a practitioner, whereas the computing elements are the work of a programmer. A practitioner understands the problem and its manual solution well but may encounter problems in implementing the idea through programming. A programmer, on the other hand, has strong analytical skills for programming but may be lacking in providing the expected solution to the problem as it is not his or her area of expertise. As a result, students end up buying some books on Visual C++ and another few books on numerical methods, but still have problems in combining both. A bridge is needed to close this gap. Most books on the market discuss either Visual C++ or the problem exclusively, not both of them at the same time. There are many books specializing in numerical methods and simulations but almost none of them connect to the resources in Visual C++, particularly the Microsoft Foundation Class, or MFC, library. Only a handful of them discuss the problems using the standard C++. For example, the problem of solving a system of linear equations is a fundamental problem in numerical methods. Unfortunately, there are no known books on the market that discuss this problem in depth, especially using the rich resources in MFC. MFC has a large collection of library functions for serving many requirements in programming, but the absence of books in the numerical simulations area may reduce its audience. Today, C++ is facing stiff competition from other languages such as Visual Basic and Java. There is also a relatively new language, C#, which has been mentioned to take over from C++. The issue here is no longer an option, it is survival. Therefore, to remain competitive,

MFC needs to be promoted so that it takes care of areas such as numerical simulations. We have looked at this deficiency and present this book as a solution.

Our second objective is to discuss a problem and its solution and present the solution in a friendly manner. Visualization is the keyword here. A problem will remain a problem as long as its friendliness form is not there. It takes time for people to appreciate a given problem if there is no tool to present its solution in a friendly manner. Today's requirements are very challenging to a problem solver. A numerical solution that displays only a series of numbers will not be appreciated anymore, not like in the time when FORTRAN dominated the programming world in the 1960s and 1970s. Instead, the solution must be presented in the form of charts, graphs, animation, and, sometimes, multimedia. Not only that, the interface on the computer should be friendly to the user when the use of windows, dialog boxes, mouse, menus, and images are necessary.

Our third objective is to put more emphasis on the problem and try to minimize the coding using MFC. This is necessary since an approach that involves too many codes often distracts the reader from understanding the method for solving the problem. We embark on this idea by implementing the nonwizard approach in MFC for most problems. The wizard, or guided approach, is presented in one chapter to compliment the nonwizard approach. Only one application class is involved in the program design in most examples. This is necessary in order to reduce the complexity in coding. The advantage of our approach can be seen from the small number of codes required in each application. This benefits the reader, as small lines of codes make the solution easier to understand. The interface for each problem has also been designed to be as simple as possible for this purpose. We do not add things like animation and sound to a typical numerical problem as this approach may be overreacting. Instead, some relevant things such as edit boxes and a list view window will be more practical for this application. The problem itself may require a lengthy discussion and coding. Also, the idea behind this approach is to have the readers understand the solution to each problem and use the method in their work. The minimum coding provided in each example will serve as a good beginning for the reader. It is expected that the readers will pick up the code and expand it in their real work.

The book is not intended for use as a beginning text book for learning MFC. The concepts of MFC are not discussed in depth in this book because this is not our main objective. There are dozens of books on the market today that provide lengthy discussions on MFC, and we do not wish to compete against them. However, the MFC concepts related to the topics discussed are explained. Also, it is assumed that the reader has acquired some programming skills using C++ and understands the object-oriented approach to programming prior to using this book. This is necessary since MFC requires some understanding of concepts such as inheritance, polymorphism, and overloading.

The programming work in this book has been developed wholly using Microsoft Visual C++.Net version 2002. The code is also compatible with version 2003 of the software and Microsoft Visual C++ version 6. The topics discussed in this book consist of several selected numerical methods and simulation problems. We chose problems that are fundamental in nature, and ones that will benefit a large audience. The topics range from trivial problems such as the fourth-order Runge–Kutta problem to something quite tricky such as the multiprocessor scheduling problem. We anticipate that the audience for this book will mostly be third-year undergraduate and beginning graduate students. The topics discussed are intended to help students develope their projects at the final-year undergraduate level, intermediate Masters, and beginning Ph.D. degree courses. The book is also suitable for use by practitioners, working professionals, researchers, and lecturers working in the simulation areas.

The work in this book is the result of some years of collaborative research and teaching between Universiti Teknologi Malaysia, University of Sydney, and Old Dominion University. Many materials in this book were developed by the first author for the SSM 3323 and MSM 5023 classes at the Department of Mathematics, Universiti Teknologi Malaysia. The authors would like to thank Professor Ariffin Samsuri, Dean of the Research Management Center at Universiti Teknologi Malaysia, for his support in completing this book. Special thanks also to Michael Till and his group at the CISCO Internetworking Unit, School of Information Technologies at the University of Sydney, Australia; and Kurt Maly, head of the Computer Science Department at Old Dominion University in the United States.

SHAHARUDDIN SALLEH
ALBERT Y. ZOMAYA
STEPHAN OLARIU
BAHROM SANUGI

April 2005

CHAPTER 1

DEVELOPING APPLICATIONS USING VISUAL C++.NET

1.1 OBJECT-ORIENTED APPROACH TO VISUAL C++.NET

An *object* is an instance of a class. A class is a set of entities that share the same parent. *Object-oriented programming* is a programming approach based on objects. C++ is one of the most popular object-oriented programming languages in the world. The main reason for its popularity is due to the fact that it is a high-level language but, at the same time, it runs as powerfully as the assembly language. In addition, C++ has its roots in ANSI C, which has been a very nicely crafted procedural language, popular in the 1970s and 1980s. But the real strength of C++ lies in its takeover from C to move to the era of object-oriented programming in the late 1980s. This conquest provides C++ with the powerful features of the procedural C and an added flavor for object-oriented programming.

The original product from Microsoft consists of the C compiler that runs under the Microsoft DOS (disk operating system), and it has been designed to compete against Turbo C, which was produced by the Borland Corp. In 1988, C++ was added to C and the compiler was renamed Microsoft C++. In early 1989, Microsoft launched the Microsoft Windows operating system, which includes the Windows API (Application Programming Interface). This interface is based on 16 bits and supports the procedural mode of programming using C. Improvements were made over the following years that include the Windows Software Development Kit (SDK). This development took advantage of the API for the graphical user interface (GUI) applications with the release of the Microsoft C compiler. As this language is procedural, the demands in the applications required an upgrade to the object-oriented language design approach, and this contributed to the release of the Microsoft

Numerical Simulations and Case Studies Using Visual C++.Net by Salleh, Zomaya, Olariu, and Sanugi

C++ compiler. With the appearance of the 32-bit Windows API (or Win32 API) in early 1990s, C++ was reshaped to tackle the extensive demands on Windows programming and this brought about the release of the Microsoft Foundation Classes (MFC) library. The library is based on C++ and it has been tailored with the object-oriented methodology for supporting the application architecture and implementation.

The Net platform refers to a huge collection of library functions and objects for creating full-featured applications both on the desktop and the enterprise Web. The classes and objects provide support for friendly user interface functions like multiple windows, menus, dialog boxes, message boxes, buttons, scroll bars, and labels. Besides, the platform also includes several tedious task-handling jobs like file management, error handling, and multiple threading. This platform also supports advanced frameworks and environments such as Passport, Windows XP, and Tablet PC. The strength of the Net platform is obvious in providing the Internet and web enterprise solutions. Web services include information sharing, e-commerce, HTTP, XML, and SOAP. XML, or Extensible Markup Language, is a platform-independent approach for creating markup languages needed in a web application.

Managed Extension Features

A new approach in Visual C++.Net is the Managed Extension, which performs automatic garbage collection for optimizing the code. Garbage collection involves the removal of memory and resources not used any more in the application, which is often neglected by the programmer. The managed extension is a more structured way of programming, and it is now the default in Visual C++.Net. Central to the .Net platform is the Visual Studio integrated development environment (IDE). It is in this platform that applications are built from a choice of several powerful programming languages that include Visual Basic, Visual C++, Visual C#, and Visual J++.

In addition, IDE also provides the integration of these languages in tackling a particular problem under the .Net banner. Visual C++.Net is one of the high-performance compilers that make up the .NET platform. This highly popular language has its roots in C, improved to include the object-oriented elements, and now, with the .Net extension, it is capable of creating solutions for Web enterprise requirements. A relatively new language called Visual C# in the .Net family was developed by combining the best features of Visual Basic visual tools with the programming power of Visual C++.

In addition to its single-machine prowess, Visual C++.Net presents a powerful approach to building applications that interact with databases through ADO.NET. This product evolved from the earlier ActiveX Data Objects (ADO) technology, and it encompasses XML and other tools for accessing and manipulating databases for several large-scale applications. This feature makes Visual C++.Net an ideal tool for several Web-based database applications.

1.2 MFC FUNDAMENTAL FEATURES

MFC is a library that consists of more than 200 classes. Each class has more than a dozen member functions that handle tasks ranging from a simple text display to the more challenging web data manipulation. The MFC library is arranged in a hierarchical manner, as shown in part in Figure 1.1. This hierarchy makes possible a class to derive common member functions from its predecessor classes, thus eliminating redundancies in the classes. The hierarchy also identifies the ranking of each class with respect to other classes, where a class in a higher level is a base class to the given class. In addition, the hierarchy system makes possible further extensions to several new functions of a class, and for the addition of new classes or removal of some obsolete classes. In other words, MFC has been designed in a very modular form so that its future releases will cater to the programming needs of the time.

One of the highest-ranking classes in MFC is `CObject`. This class is responsible for several general duties, particularly for supporting handle runtime, serializa-

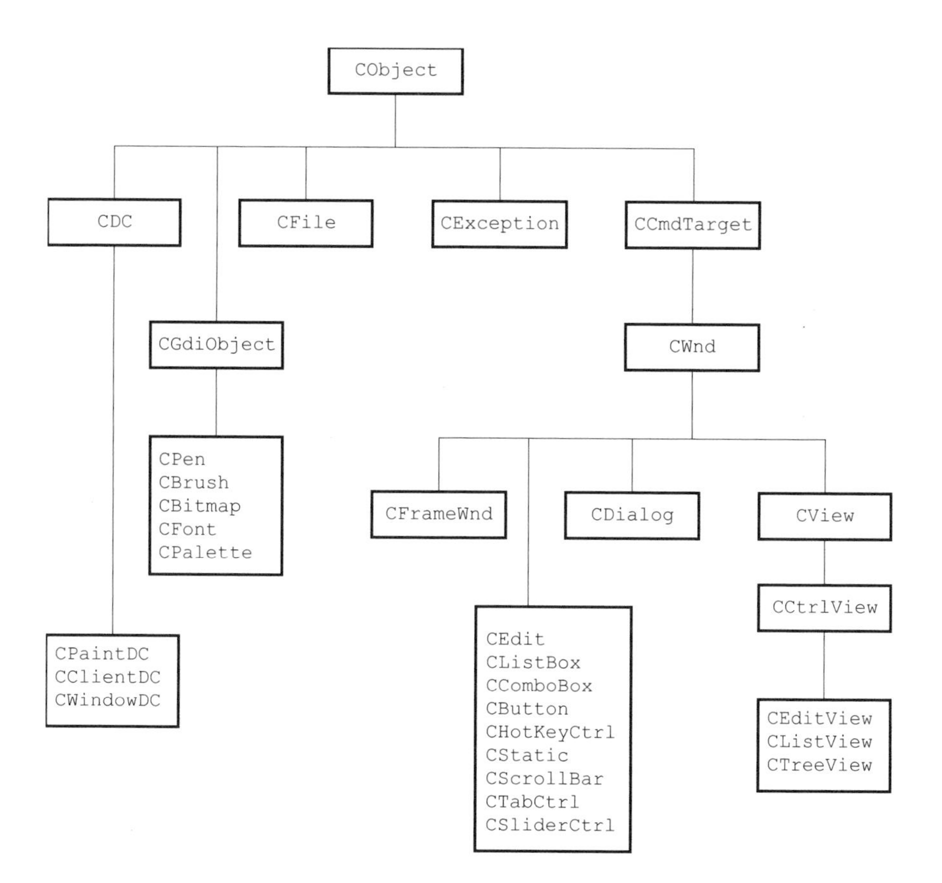

Figure 1.1 Hierarchy of some selected classes in MFC.

tion, and performing diagnostic output for several derived objects. It is from this base class that several other classes are derived, as shown in Figure 1.1.

Graphics Device Interface

The GDI, or *Graphics Device Interface,* is a layer in the Windows architecture that insulates the application from direct interaction with the hardware. A class that is commonly derived from `CObject` is the `CDC` class. In MFC, the `CDC` class is the base class for providing an interface with other classes, including `CPaintDC`, `CClientDC`, and `CWindowDC`. Each of these classes makes use of GDI to provide all the basic graphical and drawing functions for an application on Windows through an object abstraction called device context.

A *device context* is a data structure that is responsible for displaying text and graphics as output on Windows. The tools in the data structure are represented as graphic objects such as pens, brushes, fonts, and bitmaps. In reality, a device context is a logical device that acts as an interface between a physical device (such as the monitor and printer) and the application. A device context is a set of tools or attributes for putting text and drawing graphics on the screen using GDI functions.

There are four types of device contexts in GDI: display context, memory context, information context, and printer context. A *display context* supports operations for displaying text and graphics on a video display. Before displaying text and graphics, a display context links with MFC functions for creating a pen, brush, font, color palette, and other devices. A *memory context* supports graphics operations on a bitmap and interfaces with the display context by making it compatible before displaying the image on the window. An *information context* supports the retrieval of device data. A *printer context* provides an interface for supporting printer operations on a printer or plotter.

In Windows, everything including text is drawn as a graphics object. This is made possible as every text character and symbol is formed from pixels that may vary in shapes and sizes. This facility allows flexibility in the shape of the text by allowing it to be displayed from a selection of dozens of different typefaces, styles, and sizes. Text and graphics are managed by GDI functions that are called on every time a graphic needs to be displayed on the screen.

A device context object is created from one of the classes as listed in Table 1.1. For example, the device context (`dc`) in the main window is obtained by deriving this object from `CPaintDC`, as follows:

```
CPaintDC dc(this);
```

Table 1.1 Server/client classes derived from the CDC class

Class	Description
`CPaintDC`	Device context for the server area in Windows
`CClientDC`	Device context for the client area in Windows
`CWindowDC`	Device context for the whole window
`CMetaFileDC`	Device context for representing a Windows metafile, or a device-independent file for reproducing an image

Table 1.2 GDI objects for text and graphics

GDI object	Class	Description
Pen	`CPen`	To draw a line, rectangle, circle, polyline, etc.
Brush	`CBrush`	To brush a region with a color
Color palette	`CPalette`	Color palettes for pens and brushes
Font	`CFont`	To create a font for the text
Bitmap	`CBitmap`	To store a bitmap object

This object can then be linked with the available functions for displaying text, drawing lines, circles, rectangles, and so on. Some of the primitive objects for displaying text and graphics in Windows are pen, brush, font, bitmap, and color palette; these are described briefly in Table 1.2.

GDI Functions

There are dozens of GDI functions for displaying text and graphics. Table 1.3 describes some of the most commonly used GDI functions for displaying text and graphics. These functions are derived from the `CDC` class. Basically, a pen is a GDI device for drawing a line. The object is created from the class `CPen`. The default pen consists of a solid black line with a thickness of one pixel. This shape can be modified by changing the parameters in the class's constructor, `CPen()`. A brush is another GDI device for painting and filling a region using the current color. By

Table 1.3 Commonly used functions in the CDC class

Function	Description
`Arc()`	Draws an arc
`BitBlt()`	Copies a bitmap to the current device context
`llipse()`	Draws an ellipse (and a circle)
`FillRect()`	Fills a rectangular region with the indicated color
`FillSolidRect()`	Creates a rectangle using the specified fill color
`GetPixel()`	Gets the pixel value at the current position
`LineTo()`	Draws a line to the given coordinates
`MoveTo()`	Sets the current pen position to the indicated coordinates
`Polyline()`	Draws a series of lines passing through the given points
`Rectangle()`	Draws a rectangle according to the given coordinates
`RGB()`	Creates color from the combination of red, green, and blue palettes
`SelectObject()`	Selects the indicated GDI drawing object
`SetBkColor()`	Sets the background color of the text
`SetPixel()`	Draws a pixel according to the chosen color
`SetTextColor()`	Sets the color for the text
`TextOut()`	Displays a text message at the indicated coordinates

default, white is the color of the brush. This setting can be changed by modifying the parameters in the brush's constructor, `CBrush()`.

Text is displayed using the function `TextOut()`. By default, text is displayed using black color with a font of size of 12. These default settings can be changed by calling the appropriate functions. Color is created using the function `RGB()`. The color of the text and its background can be changed using the function `SetTextColor()` and `SetBkColor()`, respectively. For example, the following statements change the text color to green and the background to black:

```
dc.SetTextColor(RGB(0,255,0));
dc.SetBkColor(RGB(0,0,0));
```

Numerical Functions

MFC does not have special numerical functions for performing scientific computations, as the functions in the standard C++ library are sufficient for most applications. Some of the most common functions in the C++ library are listed and described in Table 1.4. These functions are available for use in an application by inserting their prototype file, `math.h`, in the preprocessing area. In using the functions, care must be taken in considering their domain and range correctly. For example, `log(-2)` will result in a crude answer as this function supports only positive numbers in its argument.

1.3 WRITING APPLICATIONS USING MFC

MFC provides powerful support for creating desktop or Web applications. This feature is observed through the rich features in MFC that allow the application to include a lot of advanced routines. Applications using Microsoft Visual C++ can be developed either on a guided or nonguided basis. In a guided approach, a tool called a *wizard* is provided to help in writing the code for the application. The wizard provides the initial skeleton of the program, and, therefore, the programmer does not have to worry about the detail steps in Windows such as calling functions for serializing objects and registering the application on the Windows interface. The programmer can concentrate on writing the code for the application. Many tedious jobs, ranging from a simple task like declaring a variable to something more difficult like creating a dialog box for an application, are done using the friendly menus provided by the wizard. Programming looks easier and more appealing this way.

However, a full guided tour has its drawbacks. The programmer may not benefit too much from the "free ride." It is important for us to know how to walk the stairs instead of using the elevator all the time. Climbing stairs is a generic skill every human being must possess. Taking a ride on the elevator is a luxury in the sense that elevators may not be available in many places. Therefore, persons lacking in this generic skill may not survive under certain conditions. The person may also not be

Table 1.4 Some of the most common numerical functions available in the standard C++ library

Function	Description
`double exp(double x)`	Returns the *exponent* value of its argument. For example, `exp(-1)` returns 0.3679.
`double log(double x)`	Returns the *logarithm* value of its argument. For example, `log(4.5)` returns 0.6532.
`double sin(double x)`	Returns the *sine* value of its argument. For example, `sin(4.5)` returns –0.9775.
`double cos(double x)`	Returns the *cosine* value of its argument. For example, `cos(4.5)` returns –0.2108.
`double tan(double x)`	Returns the *tangent* value of its argument. For example, `tan(4.5)` returns 4.6373.
`double sinh(double x)`	Returns the *hyperbolic sine* value of its argument. For example, `sinh(-1)` returns –1.1752.
`double cosh(double x)`	Returns the *hyperbolic cosine* value of its argument. For example, `cosh(-1)` returns 1.5431.
`double tanh(double x)`	Returns the *hyperbolic tangent* value of its argument. For example, `tanh(-1)` returns –0.7616.
`double asin(double x)`	Returns the *arc sine* value of its argument. For example, `asin(0.4)` returns 0.4115.
`double acos(double x)`	Returns the *arc cosine* value of its argument. For example, `acos(0.4)` returns 1.1592.
`double atan(double x)`	Returns the *arc tangent* value of its argument. For example, `atan(0.4)` returns 0.3805.
`int abs(int x)`	Returns the *absolute* value of its integer argument. For example, `abs(4)` and `abs(-4)` both return 4.
`double fabs(double x)`	Returns the *absolute* value of its double argument. For example, `fabs(4.5)` and `fabs(-4.5)` both return 4.5.

flexible enough to exercise several different options for meeting new challenges. Many fundamental steps such as creating a variable and declaring a class are considered the basic attributes of a language that a programmer should know.

Our approach in this book will be mostly to use the nonwizard option as we would like to concentrate on discussing the problems by writing small programs and keeping code writing to the minimum. The wizard approach involves some massive handling of the dialog windows and menus, which generate many files that are not related to the applications. A discussion on wizards in Chapter 3 should provide some relationship between the two options.

Creating a New Project

We start by discussing some basic ideas in creating an application on Windows. Microsoft Visual C++.Net provides an interface called Visual Studio for developing an application. Besides C++, this interface is shared by other languages in the family including Visual Basic and Visual C#. In order to develop an application using MFC, a person must know the C++ language very well. A good knowledge of C++ is a prerequisite to developing applications on Windows. This is necessary since MFC has classes and objects defined in a manner that can only be understood if one knows the language well.

A C++ project can be created in many ways, depending on user requirements. Table 1.5 lists some of the most common ways to create an application with Visual Studio. In its simplest form, a standard C++ project that runs without the support of any Windows functions is a console application. This option is necessary to a beginner in C++ or a person who does not wish to use the Windows facilities. The console option is available by choosing *New Project, Win32 Application* and by choosing *Console Application* in *Application Type.*

A *Win32 Project* is an option for creating an empty application with or without the support of MFC. This option does not provide a guide for creating an application, as the person must know all the details. One advantage to this option is the small amount of code required to generate an application. The option allows the application to exist as an executable file (EXE) or as a dynamic-link library (DLL).

The *MFC Application* option is a guided approach for creating an application using a tool known as the *wizard.* With this option, the details about Windows are prepared by Visual Studio through a series of menus and dialog windows in the wizard. Therefore, the user can concentrate on writing the code for an application. The wizard does not provide the whole solution for the application as it only assists by generating the code related to the Windows management.

The *Managed Extension* option is a structured way of writing an application. This new option provides an opportunity to integrate the application with .Net frameworks such as Passport, .Net My Services, Windows XP, and Tablet PC.

Creating a Window

The easiest way to create a nonwizard application using Windows is to use `CFrameWnd` as the framework. `CFrameWnd` is a class derived from `CWnd`. `CFrameWnd` also has rich ancestry in other classes such as `CCmdTarget`, `CObject`, and `CWnd`, which allows access to many functions and variables for creating applications. In a nonwizard application, a window is created by deriving the class from `CFrameWnd` using the function `Create()` (see Table 1.6).

Table 1.5 Some of the available new project options

Item	Description
Console Application	Native C++ project that supports no Windows
Win32 Project	Empty project with or without MFC
MFC Application	Wizard approach to creating a Windows application
Managed C++ Application	Managed C++ project with or without Windows support

Table 1.6 The function `Create()`

Function	Description
`Create()`	Creates the main window for the application

The main window is created using the function `Create()` from the `CFrameWnd` class. In its simplest form, the main window is created as follows:

```
Create(NULL,"My Main Window")
```

`Create()` has several parameters but only the first two need to be stated, as shown above. The first parameter indicates the default class used, whereas the second is the title of the application. Leaving the other parameters as is means we agree with the default settings of the window.

Several *child windows* are also created using the function `Create()`. A child window is a window derived from the main window. A child window can exist in the form of a push button, a list view window, an edit box, or a full window similar to the main window. To create a child window, an object must be derived from its class. For example, to create a push-button window on the main window an object is derived from the class `CButton`, and applied as follows:

```
CButton MyPushbutton;
MyPushbutton.Create("My Button",WS_CHILD | WS_VISIBLE |
    BS_DEFPUSH BUTTON,CRect(30,325,250,355),this,
    IDC_MYBUTTON);
```

The above statements create an object called `MyPushbutton` derived from the class `CButton`. We will discuss the parameters inside `Create()` in later in the chapters. The `MyPushbutton` object is a push-button window shown as a three-dimensional rectangle in the main window. Several other types of child windows are created and displayed in a similar manner.

A Windows application consists of a client/server process represented as objects. The server occupies the main window using using the functions `OnPaint()` or `OnDraw()`. These two special functions are the message handling functions that respond to the event detected by `WM_PAINT`. The device context object for these functions are created from the class `CPaintDC`. A client area is created using any function other than `OnPaint()`. In this function, a device context object is created by deriving its object from the class `CClientDC`.

1.4 WRITING THE FIRST NONWIZARD APPLICATION

In this section, we discuss a nonwizard approach for creating a simple application. The option can be started by choosing the appropriate icons and answering a series of questions, as follows:

Step 1: Start Visual C++. From the menu choose *File,* followed by *New* and *Project,* as shown in Figure 1.2. This step creates a new project.

Step 2: The screen, as shown in Figure 1.3, appears. Click the *Win32 Project* icon, for nonwizard MFC applications. Name the filename Code1 and choose a suitable folder for storing the project. Click OK to confirm the selection.

Step 3: The dialog window in Figure 1.4 appears. Choose *Application Settings* to set up the nonwizard features into the application. Choose *Windows Application* and *Empty Project* for developing an application using the nonwizard Windows option. Click the *Finish* button to complete the selection.

Step 4: The menu as shown in Figure 1.5 appears. Choose *Properties* to embed the features into the application.

Step 5: The dialog window as shown in Figure 1.6 appears. Highlight *Use of MFC* and choose the *Use MFC in a Static Library* option. This option embeds MFC into the application to produce an EXE file.

Step 6: Highlight *Source Files* in the Solution Explorer and right-click. The menu as shown in Figure 1.7 appears. Choose *Add* then *Add New Item* to insert two new files into the project.

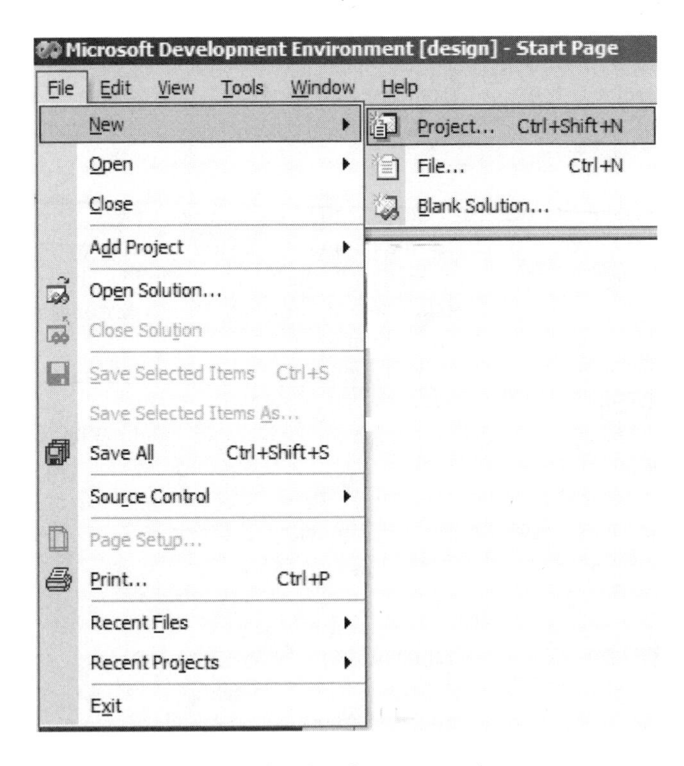

Figure 1.2 Creating a new project.

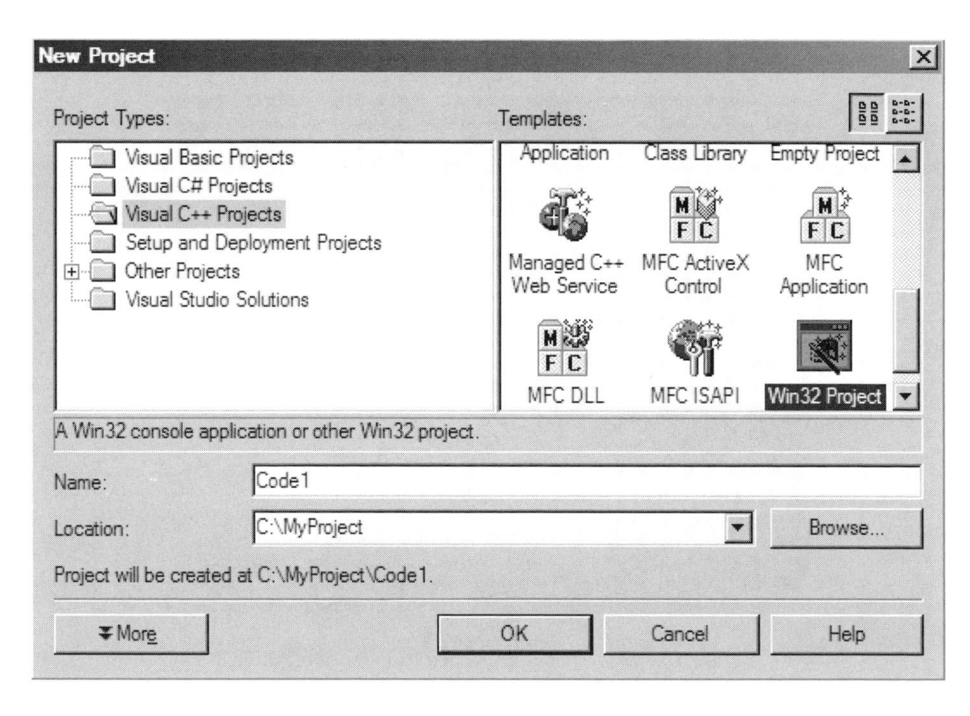

Figure 1.3 Win32 project option for MFC applications.

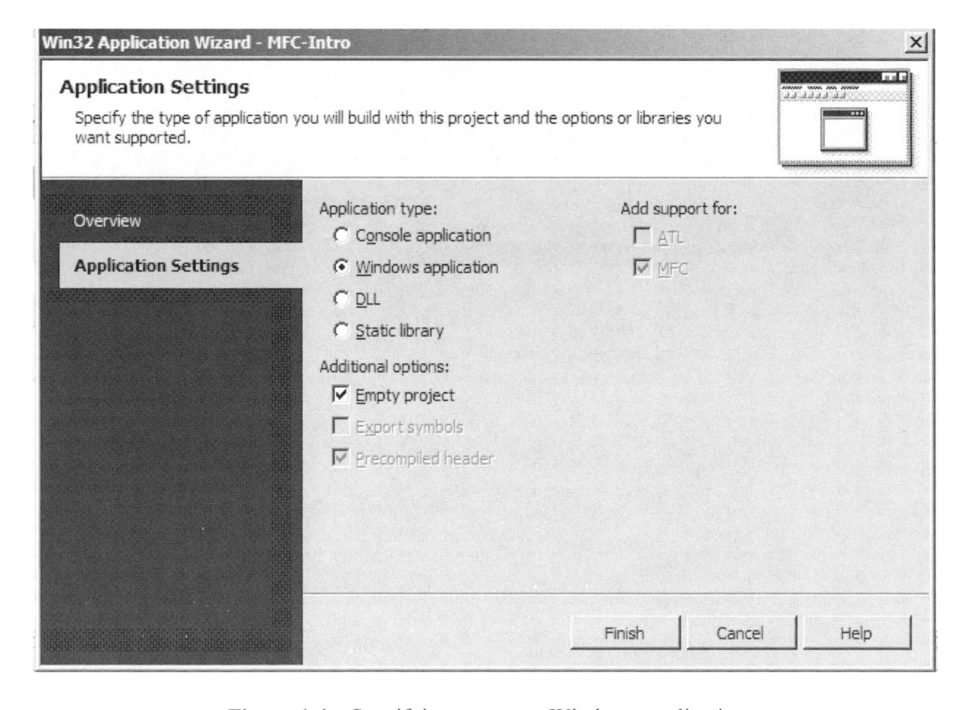

Figure 1.4 Specifying an empty Windows application.

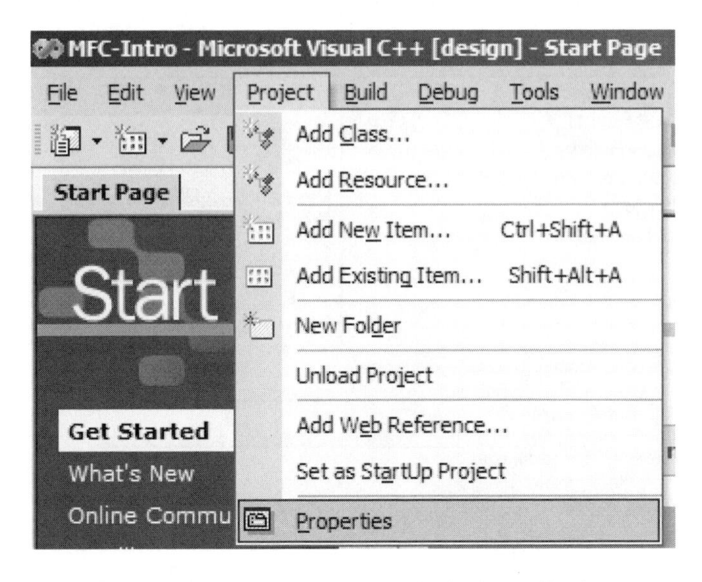

Figure 1.5 Specifying the properties in the application.

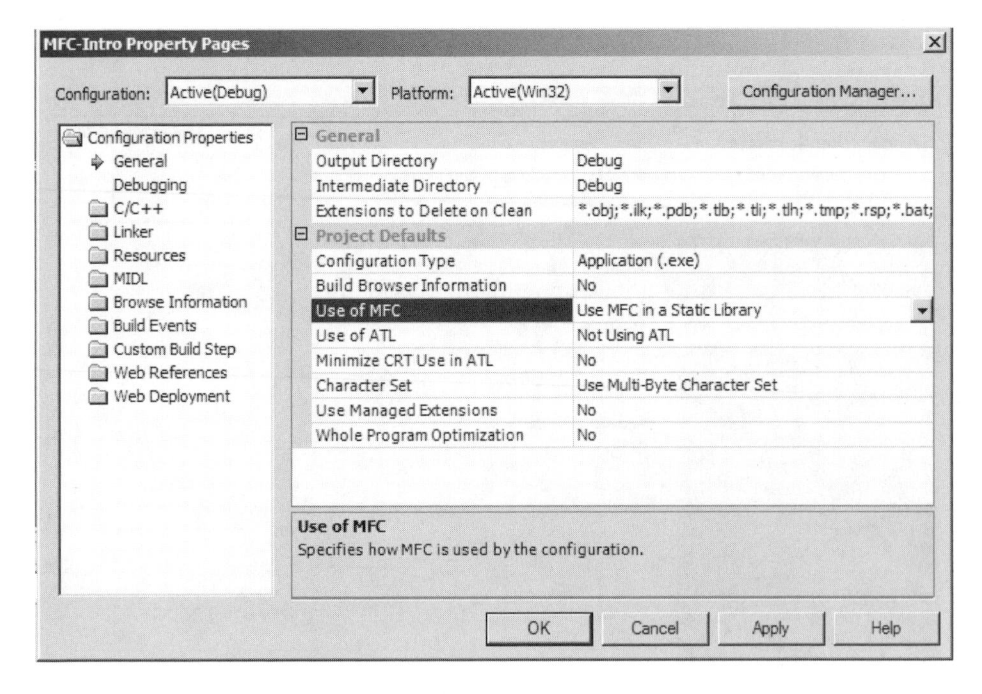

Figure 1.6 Specifying the static MFC library.

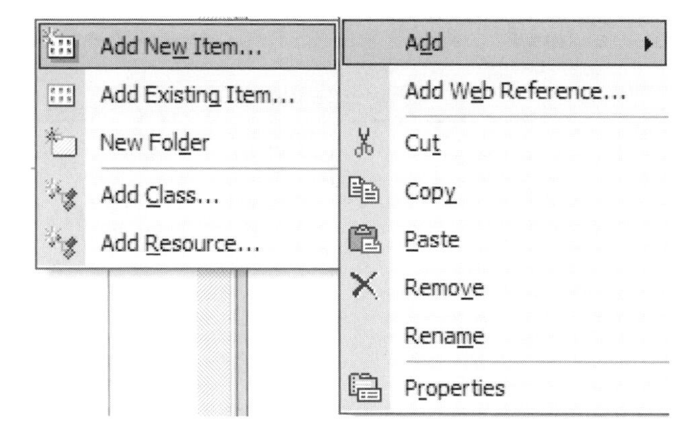

Figure 1.7 Specifying the static MFC library.

Step 7: A dialog window as shown in Figure 1.8 appears. Choose *Header File* and name the new file as Code1.h.

Type the following code into Code1.h:

```
// Code1.h
#include <afxwin.h>

class CCode1 : public CFrameWnd
{
public:
      CCode1();
      ~CCode1()          {}
      afx_msg void OnPaint();
      DECLARE_MESSAGE_MAP();
};

class CMyAppClass : public CWinApp
{
public:
      virtual BOOL InitInstance();
};
```

Step 8: Repeat Steps 6 and 7 to add the file Code1.cpp. Insert the following code:

```
// Code1.cpp
#include "Code1.h"
```

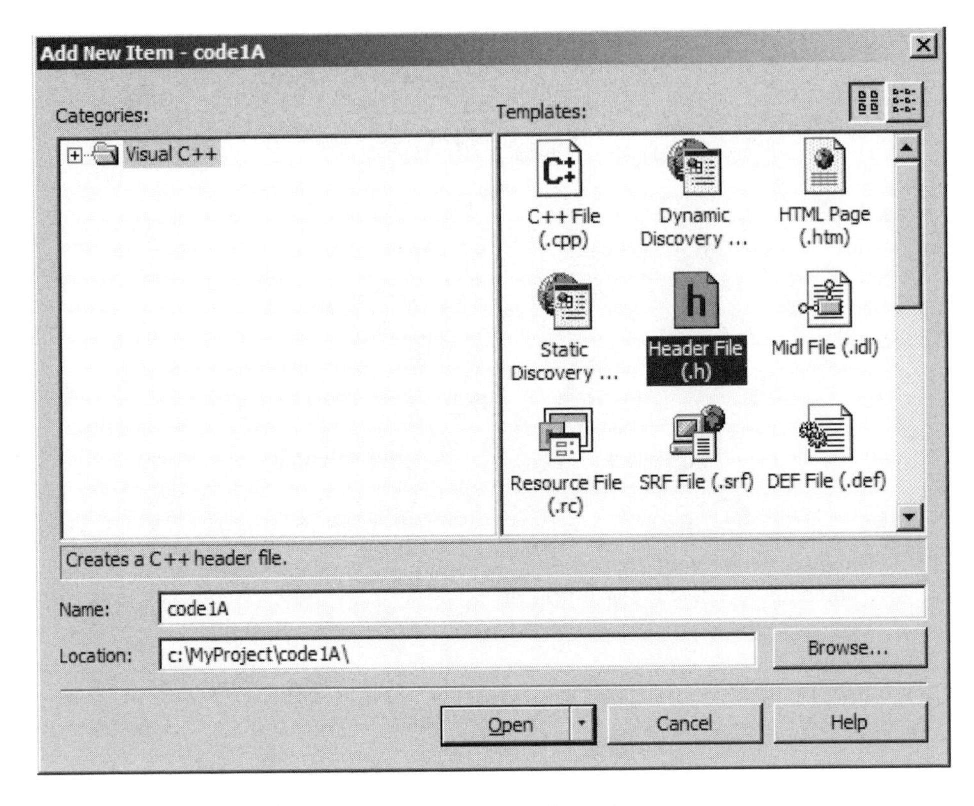

Figure 1.8 Inserting a new file into the project.

```
CMyAppClass MyApplication;
BOOL CMyAppClass::InitInstance()
{
      CCode1* pFrame = new CCode1;
      m_pMainWnd = pFrame;
      pFrame->ShowWindow(SW_SHOW);
      pFrame->UpdateWindow();
      return TRUE;
}

BEGIN_MESSAGE_MAP(CCode1, CFrameWnd)
      ON_WM_PAINT()
END_MESSAGE_MAP()

CCode1::CCode1()
{
      Create(NULL,"My First MFC Program");
}
```

```
void CCode1::OnPaint()
{
      CPaintDC dc(this);

      // draw the border
      CPen penBlue(PS_SOLID,5,RGB(0,0,255));
      dc.SelectObject(&penBlue);
      CRect rc;
      rc=CRect(CPoint(10,30),CPoint(300,200));
      dc.Rectangle(rc);

      // display the text messages
      dc.SetBkColor(RGB(255,255,255));         // white
                                               background
      dc.SetTextColor(RGB(255,0,0));           // red text
      dc.TextOut(50,80, "This is MFC");        // first
                                               message
      dc.SetTextColor(RGB(0,255,0));           // green
                                               text
      dc.TextOut(50,115, "Enjoy the ride!");   // second
                                               message
}
```

Step 9: Finally, compile and run to get the output as shown in Figure 1.9.

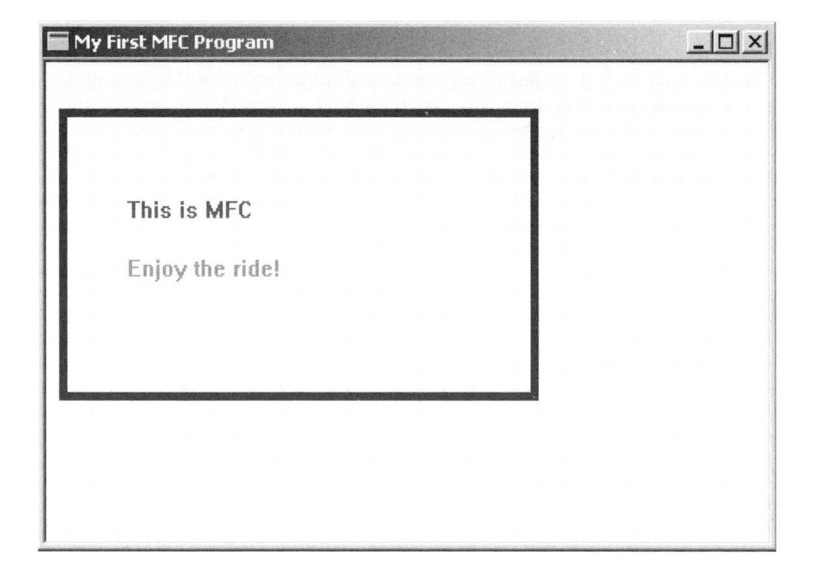

Figure 1.9 Output from Code1.

1.5 DISCUSSION

The above project, codenamed Code1, illustrates the nonwizard way of creating an application. The project is based on the files Code1.h and Code1.cpp. The output, as shown in Figure 1.9, consists of two lines of text and a blue rectangle displayed in a simple window. We discuss the implementation of the project below.

Window Creation Process

The process of creating a window involves several steps, namely, creating the objects for accessing the relevant classes in MFC, registering the class, showing the window, and updating. A function called afxwin.h needs to be included in the header file as it contains the declarations and prototypes of the member functions in the application class. The steps begin by creating an application class called `CMyAppClass`, which is derived from the MFC class `CWinApp`. This class has a member function called `InitInstance()` from the type `BOOL` (Boolean). This is done in the header file Code1.h, as follows:

```
#include <afxwin.h>
class CMyAppClass : public CWinApp
{
public:
      virtual BOOL InitInstance();
};
```

In creating the window, the class `CCode1` is derived from `CFrameWnd`. Since CFrameWnd is derived from `Wnd`, this means our window inherits all the class members from `Wnd`. Once the application class has been formed, an object is derived from this application class. This is done in the file Code1.cpp, as follows:

```
CMyAppClass MyApplication;
```

The object can have any name. In our case it is called `MyApplication`. The next step is to write the contents of the function `InitInstance()` into Code1.cpp, as follows:

```
BOOL CMyAppClass::InitInstance()
{
     CCode1* pFrame = new CCode1;
     m_pMainWnd = pFrame;
     pFrame->ShowWindow(SW_SHOW);
     pFrame->UpdateWindow();
     return TRUE;
}
```

InitInstance() is called up by a constructor from the MFC class CWinApp for constructing the window. It begins with the construction of CCode1 object:

```
CCode1* pFrame = new CCode1;
```

The above code helps in the creation of the window through Create() in the constructor. The next step is to attach the CCode1 object to the MyApplication object, as follows:

```
m_pMainWnd = pFrame;
```

Since m_pMainWnd is a member of CWinApp, this assignment allows the CCode1 object to be linked to MyApplication since CMyAppClass is derived from CWinApp. The next step is to show the window, as follows:

```
pFrame->ShowWindow(SW_SHOW);
```

And, finally, the update on the window:

```
pFrame->UpdateWindow();
```

1.6 SUMMARY AND CONCLUSION

This chapter describes the establishment of MFC and some of its fundamental components. The MFC library has more than 200 classes with thousands of member functions available for developing applications. These rich resources include many graphical user interface tools for developing applications on the desktop and Web, in databases, and in communications.

We discussed some fundamental features of MFC for desktop applications, including a MFC class hierarchical organization, graphics device interface (GDI), and device context. We also discussed the steps for creating the first Windows program using the nonwizard option. The nonwizard approach to programming is illustrated with a simple interface for displaying some text and graphical messages. The approach does not involve massive coding or the use of too many resources. This approach will be further discussed in most applications in this book.

BIBLIOGRAPHY

1. R. Jones, *Introduction to MFC Programming with Visual C++,* Prentice-Hall, 2000.
2. G. Shepherd, *Programming with Microsoft Visual C++.Net,* Microsoft Press, 2003.
3. D. J. Kruglinski, S. Wingo, and G. Shepherd, *Programming Visual C++,* Microsoft Press, 1998.

CHAPTER 2

INTERFACES FOR NUMERICAL PROBLEMS

2.1 VISUALIZING A NUMERICAL PROBLEM

Mathematics plays a pivotal role in generating the theoretical foundations for several forms of scientific and technological innovations and inventions. Ideas in the form of analytical reasoning and formulation in mathematics contribute in solving many problems that lead to many breakthroughs in technological innovations and inventions. The contributions of mathematics come in many forms. First, mathematics provides a set of structured steps that lead to an effective algorithm design. The algorithm, in turn, is key to the design of the pseudocode, and, eventually, the program code. Also, mathematics goes in line with the physical and logical properties of a given problem. In this case, mathematics equips its users with precious analytical and technical skills needed for solving problems of this nature.

One problem with the integration of mathematics in society is its "unfriendly" look. A mathematician normally has some idea of what a problem looks like and the steps that lead to its solution. However, he or she always has the trouble in explaining this idea to other people. To a layman, a solution to a problem in mathematics may be an abstraction that is not easy to understand. The difficulty arises from the fact that the understanding of the problem between a mathematician and a nonmathematician is not at the same level. Therefore, a bridge is needed to reduce this misunderstanding. It is also necessary to connect a problem involving mathematics with its solution in a more acceptable manner. A proper presentation of the problem and its solution in the form of an attractive graphical user interfaces will definitely help in closing the gap.

The main objective of this chapter is to illustrate the idea of visualizing a given mathematical problem, solve this problem, and visualize its solution as well. MFC has a very comprehensive set of tools in the form of graphical user interface library functions for achieving this objective. This chapter describes a few topics concerning the use of MFC tools in some selected mathematical problems that contribute to the overall development of solutions to a particular problem. We select four examples having tedious development for discussion. The first is an iterative method for finding the root of an equation that is easily implemented using the tools in MFC. This is followed by an example in solving a system of linear equations that makes use of dialog boxes to allow better interaction between the user and the problem. Next is the improvement in the graphical user interface presentation of the dialog window through the use of resource files for solving the same problem. Finally, the last section describes a simple modular approach for solving a problem involving matrix operations, which demonstrates an effective data-passing mechanism between functions.

The Art of Visualization

Visualization is a practical way of looking at a problem and its solution. Some useful pointers for good visualization include the use of graphs, charts, diagrams, and friendly interfaces such as menus, dialog boxes, and buttons. A good visualization model on the computer describes the problem and its solution well. Good visualization also makes it possible for the solution to be understood even by an end user who is not necessarily a technical person. Several forms of visualization have been produced depending on the nature of the problem. The most practical is the use of the computer as today's machines are capable of handling the required number of calculations. A computer provides the textual and graphical means of describing a problem and the series of steps that lead to its solution.

A typical problem may be difficult to understand at first. For example, the thermal convection in the Earth's mantel can be modeled using the finite difference method in mathematics. A simulation model to visualize the resulting temperatures, consisting of millions, perhaps billions, of values, can be developed and displayed on a powerful workstation or supercomputer. Using a program on the workstation, data for each time step is represented as a grid of pixels on the screen, with the pixel colors representing the temperature. These grids can then be displayed in succession to show the evolution of the model over time. In the case of a three-dimensional model, more advanced techniques, such as volumetric rendering, can be used to visualize the data. Once a computer simulation or analysis has been performed, the results must be interpreted by the researcher. Even though looking at the results in a numerical form can provide insights, the quantity of numbers produced by some high-performance computer simulations can make this task impractical. One alternative is to convert the numbers into pictures and animations, because human beings are inherently far better at understanding information in these forms.

Another area requiring visualization is computational fluid dynamics (CFD) modeling. CFD is an excellent tool for improving the design of devices in which flow mechanics plays a crucial role in their operation. CFD can be used to recognize complex flow profiles due to turbulent velocity fields, pressure changes, mass flow rates, and high-temperature zones in and around devices that operate with fluid flow. It basically does this by solving model equations of fluid flow that are often nonlinear partial differential equations over complex geometrical shapes using numerical methods. CFD is also a good research tool for developing better flow models.

The two examples discussed above are some of the numeric-intensive work requiring massive calculations on the computer. Today's computers are very fast and are powerful enough to handle large arrays for updating and displaying high-resolution graphics. With the integration of computers as a component of multimedia and their association with other technologies, such as telecommunications and biotechnology, visualization is now a common medium for describing a problem and its solution.

Visualization is very much associated with simulation. A *simulation* is a study performed to represent a problem and its solution in its real form. A simulation is necessary in cases where the real problem may be too big or too risky to be handled. A complete simulation solves the given problem according to the "simulated" parameters and produces some nice pointers that lead to decision making. Very often, a simulation includes visualization, as this element leads to a good understanding of the whole model.

Figure 2.1 shows the steps in the development of a model for solving a problem. The development consists of seven basic steps. It starts with a comprehensive study on the problem. The study should include its practicality, viability, and cost-effectiveness. A theoretical foundation of the model is then formulated. This step involves literature review and surveys of the existing methods, their relevance, and practicality. One important step in this study is the reduction of the problem into the form of a simulation model. If this is possible, then the rest of the steps in the figure can proceed. Otherwise, a different model needs to be developed to tackle the problem.

It is also important to visualize the problem as an immediate step before developing the simulation model. This visualization models provides a good start in the

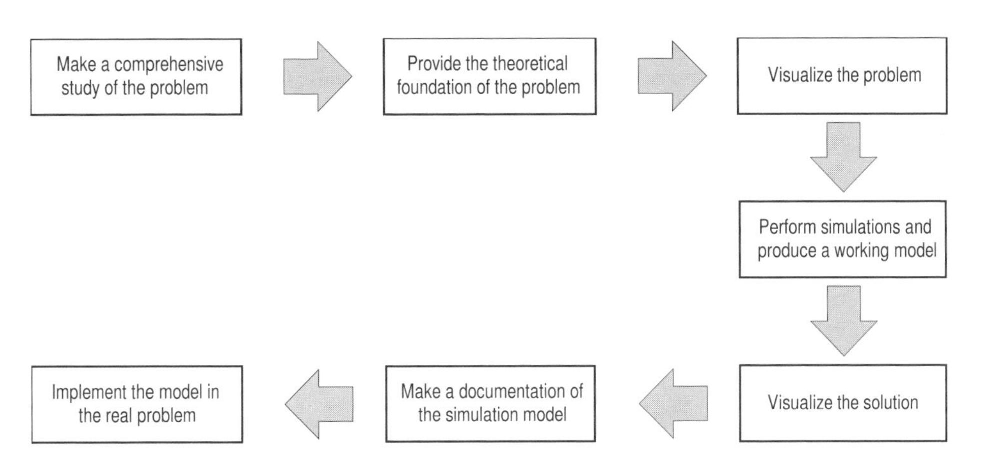

Figure 2.1 Model development involving visualization.

simulation model development as all the inputs and factors that make up the problem can be formulated. The simulation model shows the working mechanism that represents the solution to the problem. As the real problem may be too large to tackle, a simulation model may be sufficient to give a picture of the whole project and show whether it is going to be cost-effective, viable, and practical to be implemented.

The simulation model must also provide a means for visualizing the solution. This is achieved by having the solution represented in the form of text and graphical objects, charts, graphs, and tables. It is also important to include friendly graphical user-interface objects (GUI), such as menus, dialog boxes, buttons, list view windows, and several child windows for displaying the results of the simulation.

The next step is to keep a record of the work by documenting the whole process. Documentation is necessary as a legal binding and to preserve the copyrights. Documentation also stores all the important elements carried out in the simulation model that are useful for future reference. Finally, a simulation model provides good pointers for tackling the real problem. It is through the simulation model that many important decisions are made.

2.2 HANDLING ARRAYS

Many problems in science and engineering involve the use of multidimensional arrays. An array is a data representation in the form of rows and columns, equivalent to a matrix in mathematics. In its simplest form, a one-dimensional array is also called a *vector*. Arrays and vectors are commonly used, especially in applications involving graphics. Arrays and vectors often involve a massive use of memory, which may slow down the computer. Visualization of computational fluid dynamics, for example, requires continuous updating of arrays and vectors. In computer graphics animation, each step of the animation requires copying an image into the memory, displaying it, and refreshing it. A few seconds of animation involve a rapid succession of frames.

Consider a small image of size 100 rows by 100 columns of pixels in an image processing problem. Each pixel holds an integer value that represents the red–green–blue intensity of the pixel. A single array that holds this image has size 100 rows by 100 columns, a total of 10,000 elements. This large array is to be further manipulated and involved in several steps of mathematical calculations. Operations in image processing, such as edge detection, texture segmentation, and object recognition, require several steps of array calculations for finding the eigenvalues, matrix inverse, matrix multiplications, and so on. In real applications, several arrays of equivalent size are involved. Continuous calculations involving these operations definitely take a large portion of the computer memory, which may slow down the computer.

Some tips for getting the maximum performance in scientific computing include effective handling of arrays, good computer memory management, maximizing the use of local variables, and, at the same time, minimizing the use of global variables.

In this chapter, we discuss three common examples of numerical problems. The first problem is about finding the root of a nonlinear equation, whereas the next two are about solving a system of linear equations. Two different models for building the interface for the system of linear equations are discussed.

Dynamic Memory Allocation

One way of reducing the burden of the computer in tackling the memory issue is to allocate the memory to the arrays dynamically. The default setting in the C++ language is the static method of memory allocation. In this strategy, a fixed amount of computer memory is allocated to the array whether the array is fully utilized or not. In the dynamic memory allocation, the computer allocates memory only on variables that are active. Dynamic memory plays an important role in managing the computer memory, especially in cases where the arrays in the program are large.

Dynamic memory allocation is executed in C++ using the command `new`. The variable is declared as a pointer according to the type of variable used. The example in Table 2.1 shows a comparison between the static and dynamic memory allocation methods in a one-dimensional array `x[N+1]`, assuming that `N` is a constant representing the number of rows in the array.

In the dynamic allocation method, the one-dimensional array is declared as a single pointer. In a similar manner, a two-dimensional array is declared as a double pointer according to its data type. For example, a two-dimensional array `a[M+1][N+1]` having `M+1` rows and `N+1` columns is made up of `M+1` one-dimensional arrays, where each row has `N+1` columns, as shown in Table 2.2.

The arrays in use can be destroyed once they are no longer needed. This is necessary so that the memory can be returned to the computer. The one-dimensional array in Table 2.1 is deleted using the command `delete`, as follows:

```
delete x;
```

For the two-dimensional array case, the same command is applied in two stages. The deletion starts with the columns, followed by the rows, as illustrated by the example in Table 2.2:

```
for (int i=0;i<=n;i++)
      delete a[i];
delete a;
```

Table 2.1 Memory allocation in a one-dimensional array

Static allocation method	Dynamic allocation method
`int x[N+1];`	`int *x;` `x = new int [N+1];`

Table 2.2 Memory allocation in a two-dimensional array

Static allocation method	Dynamic allocation method

`int a[M+1][N+1];`	`int **a;`
	`a = new int *[M+1];`
	`for (int i=0;i<=M;i++)`
	`  a[i]=new int [N+1];`

As a general rule, the declaration of a global array is always made in the header file. Normally, memory for an array is allocated dynamically in the constructor, whereas its deletion is in the destructor.

2.3 FINDING THE ROOT OF A NONLINEAR EQUATION

Methods based on iterations are very common in many numerical problems. The iterative method is an important step in numerical methods as it represents a set of repeated steps for the convergence of the problem to its solution, provided the solution exists. In a given problem, iterations represent an improvement and convergence to the solution based on the fact that its solution is bounded within a specified interval according to mathematical rules and properties. One advantage of the iterative methods is the small number of variables (hence, memory) used compared to the noniterative methods. This is well observed as the variables involved in iterations need not store their previous values after the iterations. One disadvantage of the iterative method, on the other hand, is the rate of convergence which could be slow depending on the problem. The rate of convergence is affected by factors such as the suitability of the initial value, size of the increment, and the stopping criteria used.

One well-known problem that involves iterations is the problem of finding the root of a nonlinear equation. We first define a nonlinear equation. A *linear equation* is an equation with n variables that has the following form:

$$a_0x_0 + a_1x_1 + \ldots + a_nx_n = 0 \tag{2.1}$$

where $a_0, a_1, \ldots, a_n$ are constants and $x_0, x_1, \ldots, x_n$ are the variables. It follows that any equation not in the form of Equation (2.1) is called a *nonlinear equation.* For example, $f(x) = 1 - 3x + 4x^3$ and $f(x) = 3 \cos x - 2x$ are nonlinear equations, whereas $2x - 3y = 5$ is a linear equation.

The root of an equation is the point where the continuous function $f(x)$ crosses in the x-axis, or $f(x) = 0$. Several methods have been established for solving this problem. These include the bisection method, false-point position method, Newton–Ralphson's method, secant method, and fixed-point iterative method.

In this section, we illustrate an iterative method using the *bisection method.* Figure 2.2 shows a schematic flowchart for the bisection method. The method searches its solution from a closed interval based on the *intermediate value theorem,* stated as follows:

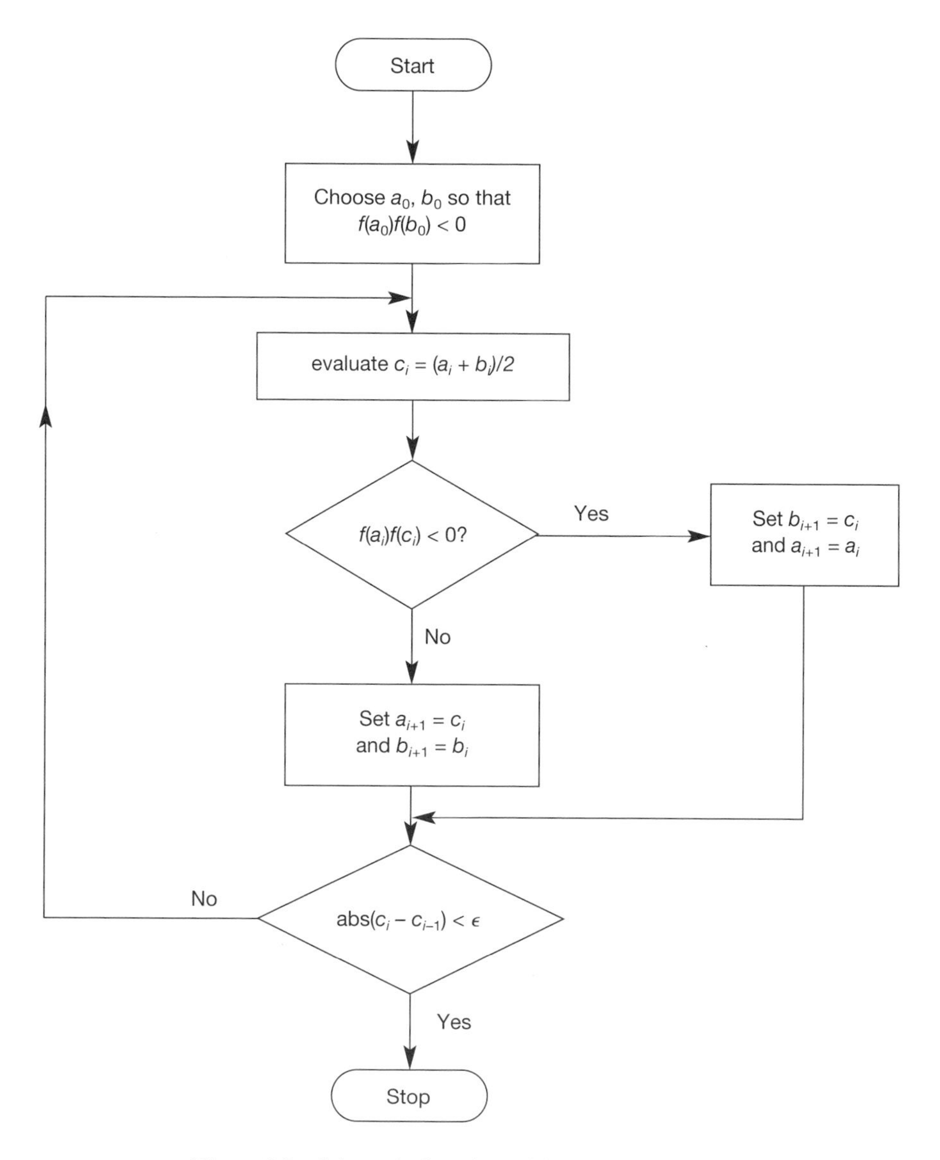

Figure 2.2 Schematic flowchart of the bisection method.

Intermediate Value Theorem: If a function $f(x)$ is continuous in the interval $a_0 < x < b_0$, at least one root exists in the interval if $f(a_0)f(b_0) < 0$.

The relationship $f(a_0)f(b_0) < 0$ means that if $f(a_0) > 0$, then $f(b_0) < 0$, which suggests that the curve is above the x-axis at $x = a_0$ and below the axis at $x = b_0$. Therefore, somewhere along this interval this curve must cross the x-axis since the curve is continuous. Therefore, at least one root exists in this interval. Similarly,

$f(a_0)f(b_0) < 0$ suggests that if $f(a_0) < 0$, then $f(b_0) > 0$, which also works the same way. The closed interval in this case is $a_0 < x < b_0$. Since the interval has been located from this theorem, the task of finding the root is a matter of performing iterations until convergence to its solution is achieved. Convergence to the solution is guaranteed if exactly one root exists in the interval.

The bisection method starts with iteration $i = 0$ by taking two end points a_0 and b_0 of the interval $a_0 \leq x \leq b_0$ as the initial guess points, where $f(a_0)f(b_0) < 0$. Next, we compute the middle point of these two end points, given by

$$c_i = \frac{a_i + b_i}{2} \tag{2.2}$$

where $i = 0$. A test is performed to update the values of a_{i+1} and b_{i+1} according to the following rules:

$$\text{If } f(a_i)f(c_i) < 0 \quad \text{then} \quad b_{i+1} = c_i \quad \text{and} \quad a_{i+1} = a_i \tag{2.3}$$

$$\text{If } f(a_i)f(c_i) > 0 \quad \text{then} \quad a_{i+1} = c_i \quad \text{and} \quad b_{i+1} = b_i \tag{2.4}$$

We obtain the updated values of a_1 and b_1 according Equations (2.3) or (2.4). The iteration is repeated with $i = 1$ and so on until the stopping criteria are met. Both Equations (2.3) and (2.4) imply only one of a_i and b_i will have its value updated at each iteration, whereas the value of the other variable remains unchanged.

The stopping criterion for the iterations is the error $|c_i - c_{i-1}| < \varepsilon$, where ε is a small number close to 0. This criterion is applied at each iteration to determine if the iteration should be continued or stopped. If this criterion is met, then the iterations stop immediately, and the final value of x_{i+1} is then the root of $f(x)$. Otherwise, the iterations continue with the next value of i, applying Equations (2.2), (2.3), and (2.4) for updating the values of a_i and b_i.

Code2A: Bisection Iterative Method

We illustrate the bisection method for finding the root of the function $f(x) = x^3 - x^2 - 2$ using the initial values $a_0 = 1$, $b_0 = 2$, and $|c_i - c_{i-1}| < \varepsilon$ as the stopping criteria, where $\varepsilon = 0.005$. The project is named Code2A. Figure 2.3 shows the output of Code2A with the solution obtained after eight iterations, assuming three decimal places in all calculations.

The project Code2A consists of two main files, Code2A.cpp and Code2A.h. The header file Code2A.h includes the data structure of this application. Only one application class called `CCode2A` is used in this application. Code2A.h includes the preprocessing declarations using `#define` on the maximum number of iterations `N`, the function $f(x)$, and the stopping value `EPSILON`, as follows:

```
#define N 10
#define f(x) (pow(x,3)-pow(x,2)-2)
#define EPSILON 0.005
```

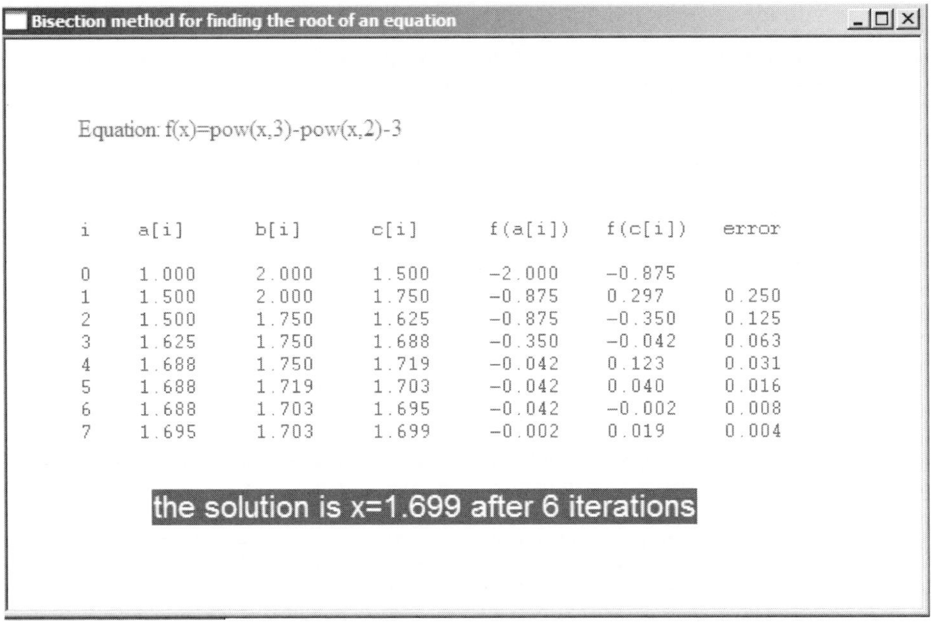

Bisection method for finding the root of an equation

Equation: f(x)=pow(x,3)-pow(x,2)-3

i	a[i]	b[i]	c[i]	f(a[i])	f(c[i])	error
0	1.000	2.000	1.500	−2.000	−0.875	
1	1.500	2.000	1.750	−0.875	0.297	0.250
2	1.500	1.750	1.625	−0.875	−0.350	0.125
3	1.625	1.750	1.688	−0.350	−0.042	0.063
4	1.688	1.750	1.719	−0.042	0.123	0.031
5	1.688	1.719	1.703	−0.042	0.040	0.016
6	1.688	1.703	1.695	−0.042	−0.002	0.008
7	1.695	1.703	1.699	−0.002	0.019	0.004

the solution is x=1.699 after 6 iterations

Figure 2.3 Screen snapshot of Code2A.

The arrays `a[i]`, `b[i]`, and `c[i]`, which represent a_i, b_i, and c_i, respectively, are declared as pointers, as listed in Table 2.3.

The variables are declared as pointers of type double. The memory for these three variables is allocated dynamically in the constructor of Code2A.cpp, as follows:

```
a=new double [N+1];
b=new double [N+1];
c=new double [N+1];
```

The main window for displaying the output is created using `Create()`, which is derived from the class `CFrameWnd`. Besides allocating the memory for the global arrays, the constructor `CCode2A()` also allocates memory for class and includes

Table 2.3 Arrays in Code2A

Variable	Declaration	Description
a_i	`double *a`	Left point of the interval at iteration i
b_i	`double *b`	Right point of the interval at iteration i
c_i	`double *x`	Middle point between a_i and b_i

the initializations of the end-point values of the interval, a[0] and b[0]. The initial values of $a_0 = 1$ and $b_0 = 2$ are suitable since $f(1)f(2) < 0$, according to the mean value theorem. The full code for the constructor is given as follows:

```
CCode2A::CCode2A()
{
    Create(NULL, "Bisection method for finding the root of an equation");
    a=new double [N+1];
    b=new double [N+1];
    c=new double [N+1];
    a[0]=1; b[0]=2;
}
```

Code2A has only one event, namely, output in the main window. The event is detected by WM_PAINT and handled by the function OnPaint(). In order to display text and graphics in the window, a device context object called dc is created from the class CPaintDC as follows:

```
CPaintDC dc(this);
```

Text is displayed by linking dc with TextOut(). For displaying text, the default setting is the black Times New Roman font of size 12. A different font can be designed by first declaring its object derived from the class CFont. This font is created using CreatePointFont() and selected using SelectObject(). Three different fonts, Times New Roman, Courier, and Arial, with sizes of 12, 10, and 20, respectively, are created in this application. The general steps in creating a font are outlined as follows:

```
CFont myFontObject;
myFontObject.CreatePointFont(FontSize,FontName);
dc.SelectObject(myFontObject);
```

In the above steps, *myFontObject* refers to the name of the CFont object. *FontName* refers to the name of a font available in the CFont class, such as Times New Roman, Courier and Arial. *FontSize* is the size of the required font in pixels. The size of the font is determined by multiplying the standard unit by 10, as each unit is represented as 10 pixels. For example, a 10-unit font has *FontSize*=100.

Text can be formatted for display as a CString object according to the identifiers using the function Format(). This function allows text to be formatted according to the correct data type using several identifiers, including those listed in Table 2.4.

An identifier can also include the width, number of decimal places and method of alignment of the data. Table 2.5 shows some examples.

The following code displays a sample output at the Windows coordinates (300, 200), illustrating the above example:

Table 2.4 Some identifiers in `Format()`

Identifier	Variable type
%c	Character
%s	String
%d	Integer
%f	Float
%lf	Double

Table 2.5 Some formatting examples

Identifier	Description
%5d	5 spaces of decimal and right-aligned
%-7s	7 spaces of string and left-aligned
%5.2lf	5 spaces of double with 2 decimal places and right-aligned
%-5.2lf	5 spaces of double with 2 decimal places and left-aligned

```
CPaintDC dc(this);
int i=582;
double x=3.04;
CString s="banana";
s.Format("%5d%7s%-7s%5.2lf%-5.2lf", i, s, s, x, x);
dc.TextOut(300,200,s);
```

The output is shown in Figure 2.4.

The solution provided by the bisection method is based on the flowchart in Figure 2.2, as follows:

```
for (int i=0;i<=N;i++)
{
	c[i]=(a[i]+b[i])/2;
	if (f(a[i])*f(c[i])>0)
	{
		a[i+1]=c[i]; b[i+1]=b[i];
	}
```

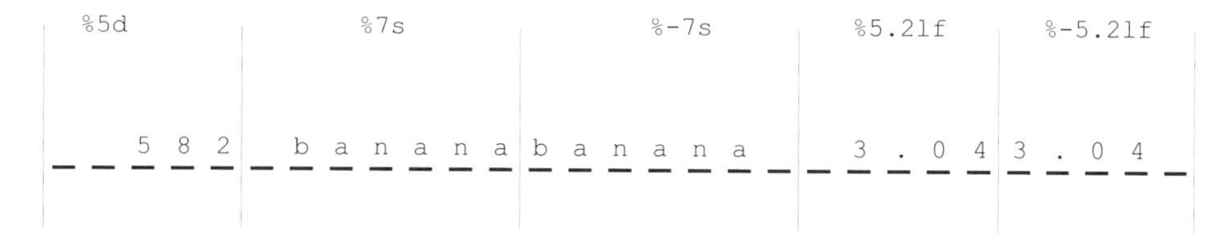

Figure 2.4 Expected output from the example.

```
        else
        {
                b[i+1]=c[i]; a[i+1]=a[i];
        }
        s.Format("%-5d%-10.3lf%-10.3lf%-10.3lf%-10.3lf%-10.3lf",
                i,a[i],b[i],c[i],f(a[i]),f(c[i]));
        dc.TextOut(50,150+15*i,s);
        if (i>0)
        {
                error=fabs(c[i]-c[i-1]);
                s.Format("%-10.3lf",error);
                dc.TextOut(490,150+15*i,s);
                if (error<EPSILON)
                        break;
        }
}
```

The iterations start at `i=0` by setting `a[0]=1` and `b[0]=2`. Equation (2.6) is then applied to determined the value of `c[0]`. With this value, the program evaluates `f(a[0])` and `f(c[0])`, and their product to determine the values of `a[1]` and `b[1]`, according to the update rules in Equations (2.3) and (2.4).

The above step is repeated with `i=1`, which leads to the values of `a[2]` and `b[2]`. A test is then performed to check if the error given by `error=fabs(c[i]-c[i-1])` is less than the constant `EPSILON`. The function `fabs()` is a function declared in the C++ header file `math.h` which returns the absolute value of its argument. If this test is true, then the iterations are stopped, otherwise the process repeats with the next iteration, `i=2`. The steps are repeated until the stopping condition `error<EPSILON` is reached.

Convergence is achieved after six iterations. The loop stops at `i=6` as the stopping criteria has been reached. The final solution is displayed using the Arial font of size 16, as follows:

```
myfont3.CreatePointFont (160,"Arial");
dc.SelectObject (myfont3);
dc.SetBkColor(RGB(100,100,100));
dc.SetTextColor(RGB(255,255,255));
s.Format("the solution is x=%.3lf after %d iterations",c[i],i-1);
dc.TextOut(100,300,s);
```

The program then closes by calling up the destructor `~CCode2B()`, which destroys the class and the arrays `a[]`, `b[]`, and `c[]`, and returns their memory to the computer, as follows:

```
CCode2A::~CCode2A()
{
        delete a,b,c;
}
```

2.4 SOLVING A SYSTEM OF LINEAR EQUATIONS

A system of N linear equations (SLE) is defined as a set of linear equations, given as follows:

$$\begin{aligned} a_{11}x_1 + a_{12}x_2 + \ldots + a_{1N}x_N &= b_1 \\ a_{21}x_1 + a_{22}x_2 + \ldots + a_{2N}x_N &= b_2 \\ &\vdots \\ a_{N1}x_1 + a_{N2}x_2 + \ldots + a_{NN}x_N &= b_N \end{aligned} \tag{2.5}$$

where x_i are variables, and a_{ij} and b_i are constants, for $i, j = 1, 2, \ldots, N$. In matrix form, the above equation can be written as follows:

$$\begin{bmatrix} a_{11} & a_{12} & \ldots & a_{1N} \\ a_{21} & a_{22} & \ldots & a_{2N} \\ \ldots & \ldots & \ldots & \ldots \\ a_{N1} & a_{N2} & \ldots & a_{NN} \end{bmatrix} \begin{bmatrix} x_1 \\ x_2 \\ \ldots \\ x_N \end{bmatrix} = \begin{bmatrix} b_1 \\ b_2 \\ \ldots \\ b_N \end{bmatrix} \tag{2.6}$$

Equation (2.6) is represented in brief form as

$$A\mathbf{x} = \mathbf{b} \tag{2.7}$$

where $A = [a_{ij}]$ is a matrix of size N **x** N, and $\mathbf{b} = (b_1, b_2, \ldots, b_N)^T$ and $\mathbf{x} = (x_1, x_2, \ldots, x_N)^T$ are vectors of size N **x** 1.

The importance of solving the system of linear equation problem can be seen from the fact that many applications in science and engineering are reducible to the form of systems of linear equations before their solutions are obtained. One good example is the heat equation given by $u_t = \alpha^2 u_{xx}$, which is a second-order partial differential equation with boundary conditions. In this problem, $u(x, t)$ is the measure of the heat at position x at time t and α is a constant, for $t > t_0$ and $x_0 < x < x_M$ and M is the number of intervals on the x axis. The numerical solution to this problem consists of a technique called the Crank–Nicholson method. This method involves the finite-difference formula, which implicitly reduces the boundary value problem into a system of linear equations. The solution is then obtained by solving the system of linear equations using a method such as the Gauss elimination method, the LU decomposition method, or the Gauss-Seidel iterative method [1].

In this section, we discuss the Gauss elimination method, which is the most popular technique for solving the system of linear equations problem. The technique involves two major steps. The first is the row operations, which reduces the original coefficient matrix A into an upper triangular matrix U and the vector $\mathbf{b}$ into $\mathbf{v}$. The second step involves the backward substitutions of the triangular matrix to get the solution in the form of $\mathbf{v}'$. These two main steps are summarized in Figure 2.5.

Row operations in the Gauss elimination method involve consecutive reductions of the rows into an upper triangular matrix. For a N **x** N system of linear equations,

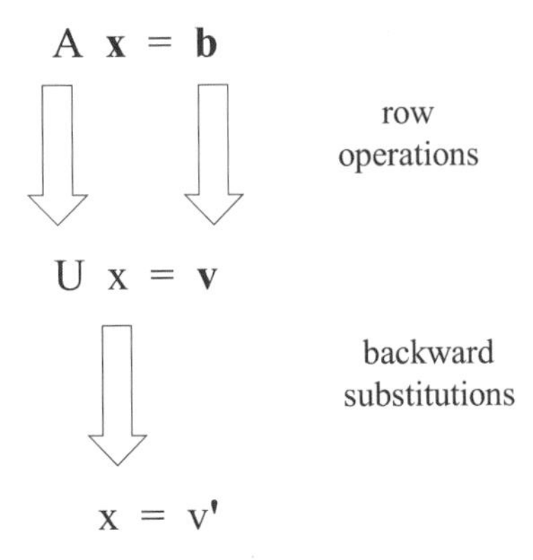

Figure 2.5 Gauss elimination method.

row operations need to be performed consecutively on rows 1, 2, . . . , $N - 1$. For each row k, the diagonal element a_{kk} is called the *pivot element*. This element must have a nonzero value in order for the row operations to continue. If its value is zero or very close to this value, then it is a case called "ill-conditioned" surfaces, which indicates singularity. This problem is normally overcome through a technique of row interchange called *partial pivoting,* which avoids this ill-conditioned effect.

We discuss the Gauss elimination method for solving the problem $A\mathbf{x} = \mathbf{b}$ of Equation (2.6), where $A = [a_{ij}]$ is the coefficient matrix, $\mathbf{x} = (x_1, x_2, \ldots, x_N)^T$ is the unknown vector, and $\mathbf{b} = (b_1, b_2, \ldots, b_N)^T$ is the right-hand-side vector of the equations. Row operations start in ascending order from $k = 1$ to $k = N - 1$ by finding the term $m = a_{ik}/a_{kk}$ at each step. Operations on row k involve updating the values of the elements a_{ij} and b_i for elements at row i and column j according to the following relationships:

$$a_{ij} \leftarrow a_{ij} - m^*a_{kj} \tag{2.8}$$

$$b_i \leftarrow b_i - m^*b_k \tag{2.9}$$

for $k = 1, 2, \ldots, N - 1$; $i = k + 1, k + 2, \ldots, N$; and $j = 1, 2, \ldots, N$. The process has the effect of reducing matrix A to its corresponding upper triangular matrix $U = [u_{ij}]$, and vector $\mathbf{b}$ to its corresponding vector $\mathbf{v} = (v_1, v_2, \ldots, v_N)^T$. Equation (2.6) is reduced to the following form:

$$\begin{bmatrix} u_{11} & u_{12} & \ldots & u_{1N} \\ 0 & u_{22} & \ldots & u_{2N} \\ \ldots & \ldots & \ldots & \ldots \\ 0 & 0 & \ldots & u_{NN} \end{bmatrix} \begin{bmatrix} x_1 \\ x_2 \\ \ldots \\ x_N \end{bmatrix} = \begin{bmatrix} v_1 \\ v_2 \\ \ldots \\ v_N \end{bmatrix} \tag{2.10}$$

In order for the solution to be unique, every diagonal element u_{ii} in U in the above equation must not have a zero value. Otherwise, the solution becomes infinite or does not exist. In C++, the row operations in Equations (2.8) and (2.9) are written as follows:

```
for (k=1;k<=N-1;k++)
     for (i=k+1;i<=N;i++)
     {
          m=a[i][k]/a[k][k];
          for (j=1;j<=N;j++)
               a[i][j]=a[i][j]-m*a[k][j];
          b[i]=b[i]-m*b[k];
     }
```

The next step in the Gauss elimination method is to perform backward substitutions on Equation (2.10). The substitutions start by finding the value of the last element, x_N, as follows:

$$x_N = \frac{v_N}{u_{NN}} \tag{2.11}$$

The rest of the elements are evaluated backward using the following equation:

$$x_i = \frac{v_i - \sum_{j=i+1}^{N} u_{ij} x_j}{u_{ii}} \tag{2.12}$$

for $i = N - 1$ and $N - 2, \ldots, 1$. The C++ code for the backward substitutions in Equations (2.11) and (2.12) is written as follows:

```
for (i=N;i>=1;i--)
{
     Sum=0;
     x[i]=0;
     for (j=i;j<=N;j++)
          Sum += a[i][j]*x[j];
     x[i]=(b[i]-Sum)/a[i][i];
}
```

Code2B: Manual Approach to the SLE Problem

We discuss the development of the C++ solution using the Gauss elimination method on a 3 × 3 system of linear equations. The project is called Code2B and it consists of two files, Code2B.cpp and Code2B.h. The output from this project is shown in Figure 2.6. It consists of a simple window with edit boxes in the first three columns for matrix A, another column of edit boxes for vector **b**, and the static text boxes for the solution vector, **x**. Input is established when the user

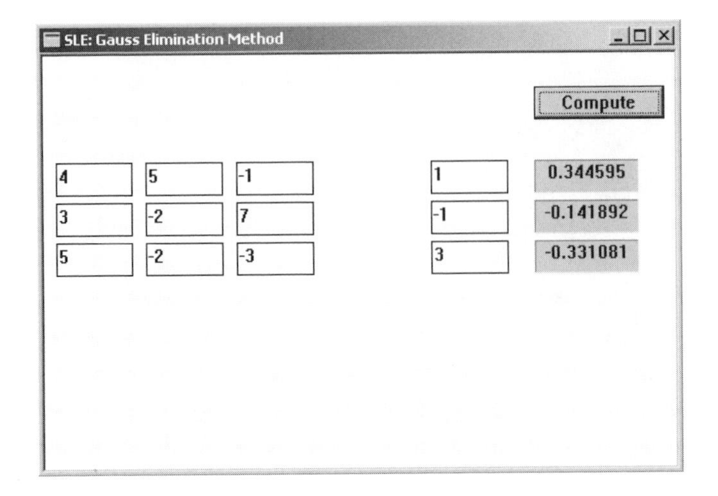

Figure 2.6 Screen snapshot of Code2B.

clicks the mouse on any edit box (white rectangle) and enters a number. This input number is read as a string, and it is converted to the array value. The output is produced in the static boxes (gray rectangle) when a push button called `Compute` is clicked.

The project involves the manual development of the graphical user interface (GUI) facilities, such as edit boxes, static text boxes, and a push button. The approach is termed *manual* as all the child windows involved are created manually using their respective MFC functions. Edit boxes are the dialog boxes that take input from the user. Static text boxes represent the output for displaying the results of the calculations. A push button represents an event handler that responds to the event by calling the appropriate function.

The header file Code2B.h contains the declaration of a class called `CCode2B` and its objects, variables, and member functions. This class is inherited from the MFC class, `CFrameWnd`, which provides the basic window for displaying the results.

Each item shown in Figure 2.6 is an object represented by a name and an id. Table 2.8 lists the variables and objects used in the project Code2B. All the arrays in this project are declared as pointers according to their data types to enable the memory for them to be allocated dynamically. Input for the arrays `a[i][j]` and `b[i]` are represented by the `CEdit` objects `ea[i][j]` and `eb[i]`, respectively. The output `x[i]` is represented by the `CStatic` objects `sx[i]`. Another object in this application is a push button called *Compute,* represented by the object `bCompute`, which is derived from the class `CButton`. It is sufficient to set the scope of all these objects and variables as private since only one class is used in the project.

Each object used in this application has its own control id preceded by the word `idc` or `IDC`. We use `idc_ea`, `idc_eb`, and `idc_sx` to denote the variable ids

Table 2.8 Variables and objects in Code2B

Variable/Object	Type/declaration	Description
a[i][j]	double **	Represents a_{ij} in matrix A
b[i]	double *	Represents b_i in vector **b**
x[i]	double *	Represents x_i in vector **x**
ea[i][j]	CEdit **	Edit box for the input a_{ij} in matrix A
eb[i]	CEdit *	Edit box for the input b_i in vector **b**
sx[i]	CStatic *	Static box for the output x_i in vector **x**
bCompute	CButton	Represents the push button
idc_ea	int	Control id for the edit boxes in matrix A
idc_eb	int	Control id for the edit boxes in vector **b**
idc_sx	int	Control id for the static boxes in vector **x**
IDC_COMPUTE	int	Control id for the push button bCompute

for the objects ea[][], eb[], and sx[], respectively. In comparison, IDC_COMPUTE is a constant id for the object bCompute that is declared using #define. Any integer value may be assigned as the control id. Normally, a value above 300 for the control id is preferred to avoid conflicts with some reserved numbers in MFC.

The variables and objects in Code2B are declared in the header file Code2B.h. Besides the variables and objects, the class CCode2B has a constructor, a destructor, and the Gauss elimination function PGauss() as its member functions.

The constructor CCode2B() allocates sufficient memory for the class to exist in the application. The constructor also allocates memory for all the global arrays and creates all the child windows for the application. The memory for the arrays is allocated dynamically as follows:

```
b=new double [N+1];
eb=new CEdit [N+1];
x=new double [N+1];
sx=new CStatic [N+1];
a=new double *[N+1];
ea=new CEdit *[N+1];
for (int i=1;i<=N;i++)
{
     a[i]=new double [N+1];
     ea[i]=new CEdit [N+1];
}
```

The main window is created in the constructor through Create() as an overlapped rectangle (WS_OVERLAPPEDWINDOW) with the top-left corner at coordinates (0,0) and the bottom-right corner at (500,340) as follows:

```
Create(NULL, "SLE: Gauss Elimination Method",
        WS_OVERLAPPEDWINDOW,CRect(0,0,500,340));
```

Several child windows in the form of edit boxes, static boxes, and a push button, are used in Code2B. The objects are recognized through their ids, which have their initial values defined in the constructor as follows:

```
idc_ea=200; idc_eb=300; idc_sx=400;
```

Child windows are created by linking the objects with the function `Create()`. For example, an edit box is created using the object `ea[i][j]` as follows:

```
ea[i][j].Create(WS_CHILD | WS_VISIBLE | WS_BORDER,
        CRect(CPoint(10+(i-1)*70,80+(j-1)*30),
        CSize(60,25)),this,idc_ea++);
```

The macros `WS_CHILD`, `WS_VISIBLE` and `WS_BORDER` are the options in the window that make it a child window, visible in the parent window, and are displayed with a border. To avoid a conflict, every child window object must have a distinct id. Therefore, it is necessary to increment the value of the control id as `idc_ea++` in the above function to distinguish one object from another. In a similar manner, the edit box `eb[i]` for the object `b[i]` is created as follows:

```
eb[i].Create(WS_CHILD | WS_VISIBLE | WS_BORDER,
        CRect(CPoint(280,80+(i-1)*30),CSize(60,25)),this,idc_eb++);
```

A static box is created from the object `sx[i]`, as follows:

```
sx[i].Create("",WS_CHILD | WS_VISIBLE | SS_SUNKEN | SS_CENTER,
        CRect(CPoint(380,80+(i-1)*30),CSize(80,25)),this,idc_sx++);
```

A static box is displayed in a style slightly different from the edit box. The child window is shown as a sunken object (`SS_SUNKEN`) with its contents centered (`SS_CENTER`). By default, a static box is a gray rectangular box with no border.

A push button called *Compute* with the id `IDC_COMPUTE` is created from the object `bCompute`, as follows:

```
bCompute.Create("Compute",WS_CHILD | WS_VISIBLE | BS_DEFPUSH BUTTON,
    CRect(CPoint(380,25),CSize(100,25)),this, IDC_COMPUTE);
```

The macro `BS_DEFPUSH BUTTON` in the above statement displays a simple animation showing a button being pushed when it is clicked.

The caret can be placed initially at any of the edit boxes by using the function `SetFocus()`. The following code places the caret at the box `ea[1][1]`:

```
ea[1][1].SetFocus();
```

There is only one event in Code2B, namely, the left-button click of the mouse. The event is detected as `BN_CLICKED` and handled by the function `OnCompute()`, as described in the following message map:

```
BEGIN_MESSAGE_MAP(CCode2B, CFrameWnd)
     ON_BN_CLICKED(IDC_COMPUTE, OnCompute)
END_MESSAGE_MAP()
```

The function `OnCompute()` responds to the push-button event on the object `bCompute`. Input from the user in the form of matrix A and vector **b** is read as the objects `ea[i][j]` and `eb[i]`, respectively. `OnCompute()` reads the input data held by the objects `ea[i][j]` and `eb[i]` using the MFC function `GetWindow-Text()` by linking with their objects. This function reads each input datum and stores this value as a string derived from the class `CString`. The string values from the objects are then converted to their corresponding `double` values using the C++ function `atof()`. It is necessary to convert the data to the type `double` since all calculations assume the variables to have this type of data. The following code fragments show how this is done:

```
CString s;
int i,j;
for (i=1;i<=N;i++)
{
       for (j=1;j<=N;j++)
       {
              ea[i][j].GetWindowText(s);
              a[i][j]=atof(s);
       }
       eb[i].GetWindowText(s);
       b[i]=atof(s);
}
```

The function `PGauss()` is called by `OnCompute()` to solve the system of linear equations using the Gauss elimination method:

```
void CCode2B::PGauss()             // compute the SLE
{
     int i,j,k;
     double m,Sum;
     for (k=1;k<=N-1;k++)          // row operations on a
          for (i=k+1;i<=N;i++)
          {
               m=a[i][k]/a[k][k];
               for (j=1;j<=N;j++)
                    a[i][j]=a[i][j]-m*a[k][j];
               b[i]=b[i]-m*b[k];
          }
     for (i=N;i>=1;i–)             // backstitutions on x
```

```
        {
                Sum=0;
                x[i]=0;
                for (j=i;j<=N;j++)
                        Sum +=a[i][j]*x[j];
                x[i]=(b[i]-Sum)/a[i][i];
        }
}
```

The function produces output in the form of the arrays `x[]`. To display these values on the static boxes held by the objects `sx[]`, the values of `x[]` need to be converted to strings. The `CString` object `s` performs the identifier formatting using the function `Format()`. The values of `x[]` are then displayed as strings in the static boxes `sx[]` using the function `SetWindowText()`. The following code fragments show how this is done:

```
PGauss();
for (i=1;i<=N;i++)              // display vector x
{
        s.Format("%lf",x[i]);
        sx[i].SetWindowText(s);
}
```

The last segment of the project is the destructor `~CCode2B()`, which destroys the class and all the arrays used in the program, and returns their memory to the computer. The function is written as follows:

```
CCode2B::~CCode2B()
{
        for (int i=1;i<=N;i++)
                delete a[i],ea[i];
        delete a,ea,b,x,sx;
}
```

Code2C: Resource File Approach for SLE

The interface shown in the last section involves the creation of objects representing items in the main window. Each object is created manually in the header and C++ files. MFC also provides an easier way for creating the dialog interfaces using the resource file (`.rc`) facilities. The facilities include a broad range of control tools for creating child windows, such as edit boxes, static boxes, push buttons, list view windows, combo boxes, radio buttons, menus, and images. Some of these objects are described briefly in Table 2.7. These objects are created and displayed by clicking the object icon, moving the mouse to their location, and clicking at this location. The location and size of the objects are also controlled using the mouse. Each of the

Table 2.7 Some of the common GUI objects for creation of the resource file

Item	Class	Description
Edit box	`CEdit`	Dialog box normally used for data input in the form of a string
Static box	`CStatic`	Output display in the form of a string
Push button	`CButton`	A control push-button event that triggers a command when clicked
Radio button	`CButton`	A list of items; the user can choose only one by clicking on the item
List view window	`CListCtrl`	A tabular display of data showing the fields and records in a horizontal and vertical scrollable manner
Combo box	`CComboBox`	A list of items combined with an edit box

control items is a *container* in the form of a child window that plugs into or is separated from the main window.

This section describes a method for providing a user-friendly interface for the system of linear equations problem using the resource file facilities. The project is called Code2C and the expected output is shown in Figure 2.7. Basically, the project reuses the code from Code2B developed in the previous project. The difference is in the way the input is made and how the output is displayed.

In order to use the dialog resources in MFC, the class `CCode2C` used in this application is inherited from an MFC class called `CDialog`. This base class is an

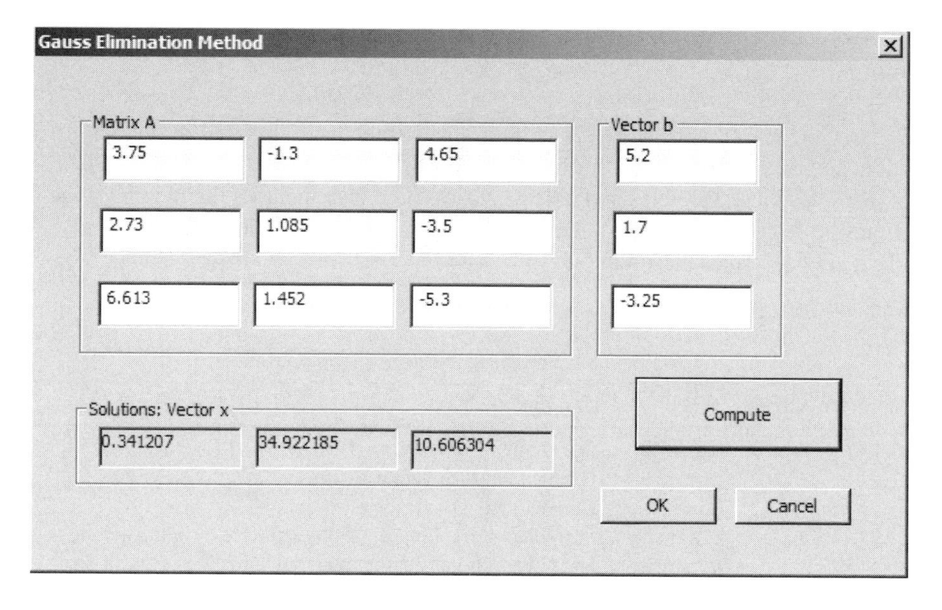

Figure 2.7 Expected output in Code2C.

MFC class for creating two types of dialog windows: modal and modeless. A *modal* window does not allow the user to access to the main window unless the current dialog window is closed. A *modeless* window allows the user to switch back and forth between the dialog window and its main window. Typical examples of modeless dialog windows are the Find and Replace items in the Visual Studio, where the user can search for a word in the program, switch to the editor, and back to searching for another word without exiting from either window.

The project Code2C is about the creation of a modal window. Besides the files Code2C.cpp and Code2C.h, the project also includes a resource file called Code2C.rc. This resource file builds the dialog window with id `IDD_GAUSSDIALOG`. This dialog window also hosts `CEdit`, `CStatic`, and `CButton` objects for allowing interaction using edit boxes, static boxes, and a push button, respectively. It is obvious that the interface in this project differs from the one in the last section as the host is a dialog window derived from the class `CDialog`, instead of `CFrameWnd`.

The file Code2C.h declares the class `CCode2C` as the inherited class and `CDialog` as its base class. In order to use the resources, the MFC header file called Afxdisp.h must be included. This file has the prototypes of several member functions, variables, and objects used for creating the resources. Code2C.h also refers to the resources that are declared in the file Resource.h. The resource file is created automatically by Visual C++ when the resource file Code2C.rc is created. Resource.h is a read-only file that contains things like the ids of the objects created in Code2C.rc. Therefore, this file must be included through `#include` in the header file.

Since Code2C does not have a parent window, the constructor is declared in Code2C.h as `CCode2C(CWnd* pParent = NULL)`. The dialog window with the id `IDD_GAUSSDIALOG` now represents the main window, as indicated in the statement

```
enum { IDD = IDD_GAUSSDIALOG };
```

In Code2C.h, several global variables and objects are declared in the application, and they are described briefly in Table 2.8.

Besides these variables and objects, Code2C.h also includes the member functions `DoDataExchange()`, `OnCompute()`, and the solution file `PGauss()`. Table 2.9 describes these functions.

Table 2.8 Variables/objects in Code2C.rc

Variables/Object	Type	Description
`a[i][j]`	`double`	The element a_{ij} in matrix A
`b[i]`	`double`	The element b_i in vector **b**
`x[i]`	`double`	The element x_i in vector **x**
`sa[i][j]`	`CString`	The object for getting the input value for a_{ij} in the edit box
`sb[i]`	`CString`	The object for getting the input value for b_i in the edit box
`sx[i]`	`CString`	The object for putting the output value for x_i in the static box

Table 2.9 The member functions in `CCode2C`

Function	Description
`CCode2C()`	The constructor
`~CCode2C()`	The destructor
`DoDataExchange()`	Exchanges data between its pointer argument and the `CString` objects in the child window
`PGauss()`	Solves the system of linear equations
`OnCompute()`	Event handler that responds to the push-button event

We will now discuss the creation of Code2C.rc. Several steps are involved, including the use of the resource editor for creating and displaying the objects in the dialog window.

Step 1: In the workspace area, right-click *Resource Files,* followed by *Add* and *Add Files.* The window shown in Figure 2.8 appears. Highlight *Resource File* from the available selections and name the file Code2C.rc. Click `Open` to confirm the selection. This step creates the file Code2C.rc.

Step 2: From the workspace area, right-click *Resource Files* and choose *Open With* from the menu. Next, click *Resource Editor* to open the resource editor.

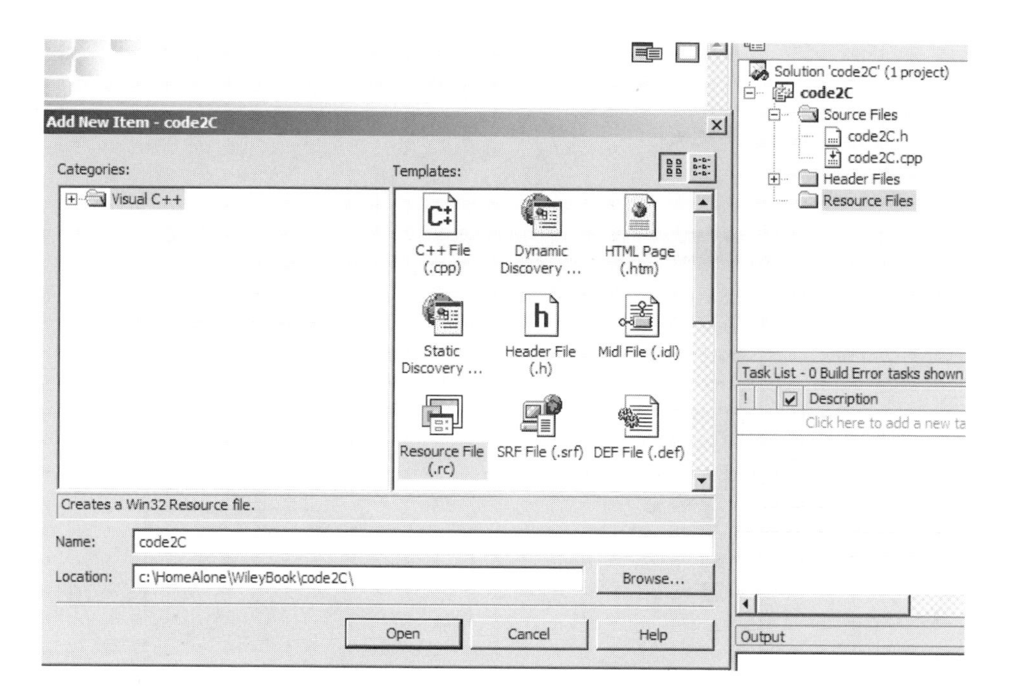

Figure 2.8 Creating the file Code2C.rc.

The dialog window resources appear, as shown in Figure 2.9. Double click *Dialog* to create the dialog window.

Step 3: Increase the size of the dialog window by dragging its right-hand corner boundary, as shown in Figure 2.10. The figure also shows all the available icons for creating objects in the box on the left window. Right-click and choose *Properties* to set the properties of this window. The *Properties* window appears on the right side, as shown in Figure 2.10. From the *Properties,* rename the window to *Gauss Elimination Mtd* by entering this name in the *Caption*. Change the id of the dialog window to `IDD_GAUSSDIALOG` by entering this name in *ID*.

Step 4: We are now ready to use the resource editor to create and display the visual objects. Start by moving the buttons `Ok` and `Cancel` to a new location, as shown in Figure 2.11. Click the *Button* icon, then draw and name this button `Compute`. Assign the id `IDC_COMPUTE` to this new object.

Step 5: Click the *Edit Control* button and draw an edit box, as shown in Figure 2.12. Assign the id `IDC_A11` to this object to denote it is an input box for the array `a[1][1]`. The input in this box is recognized as a text string represented by the `CString` object `sa[1][1]`. Repeat this step with the rest of the elements in matrix A and vector **b** to produce the full input dialogs, as shown in Figure 2.13.

Step 6: Create a static box by clicking the *Static Text* button, as shown in Figure 2.13. Set the caption to nothing, the *Client Edge* to *true,* and the id of the object as `IDC_x1`. This box displays the array `x[1]` through the `CString` object

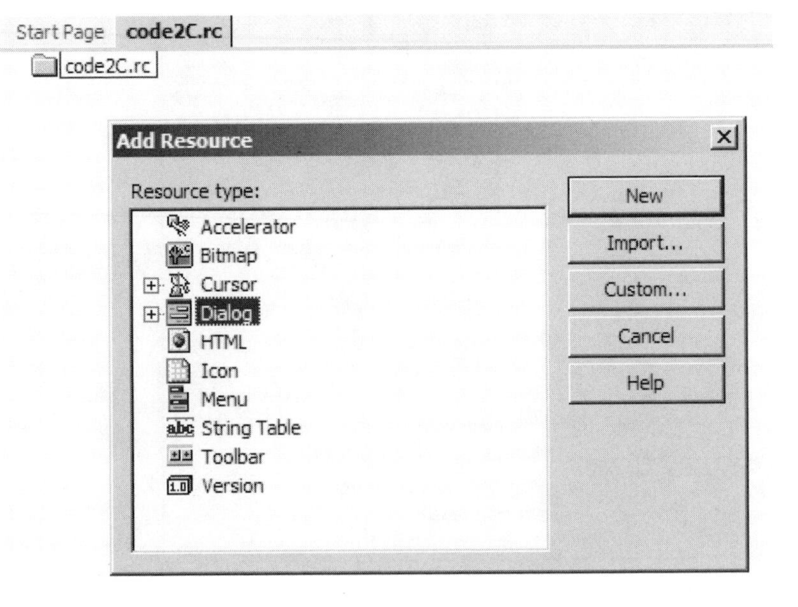

Figure 2.9 Resources available in the resource editor.

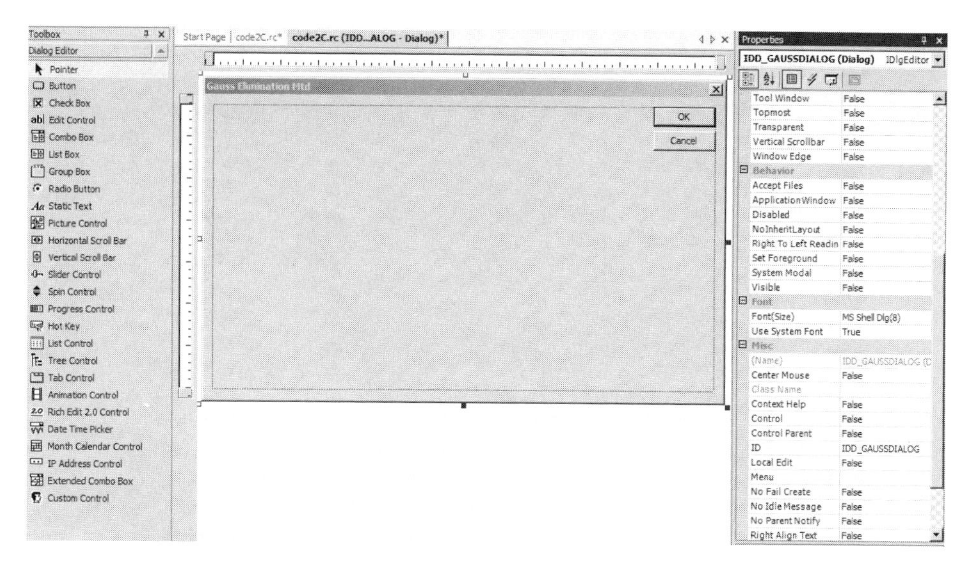

Figure 2.10 Dialog window with id IDD_GAUSSDIALOG.

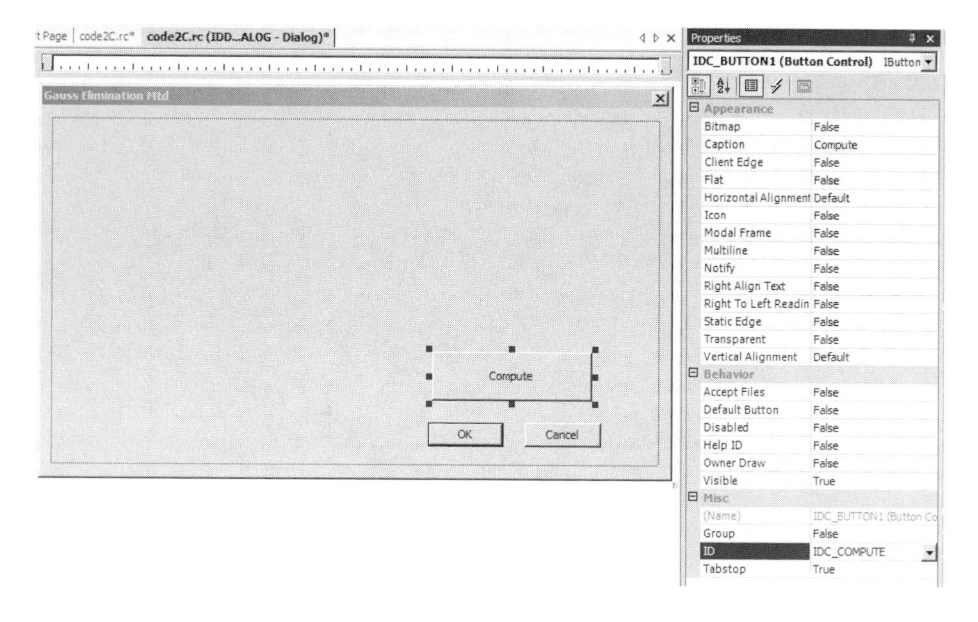

Figure 2.11 Creating the push button `Compute`.

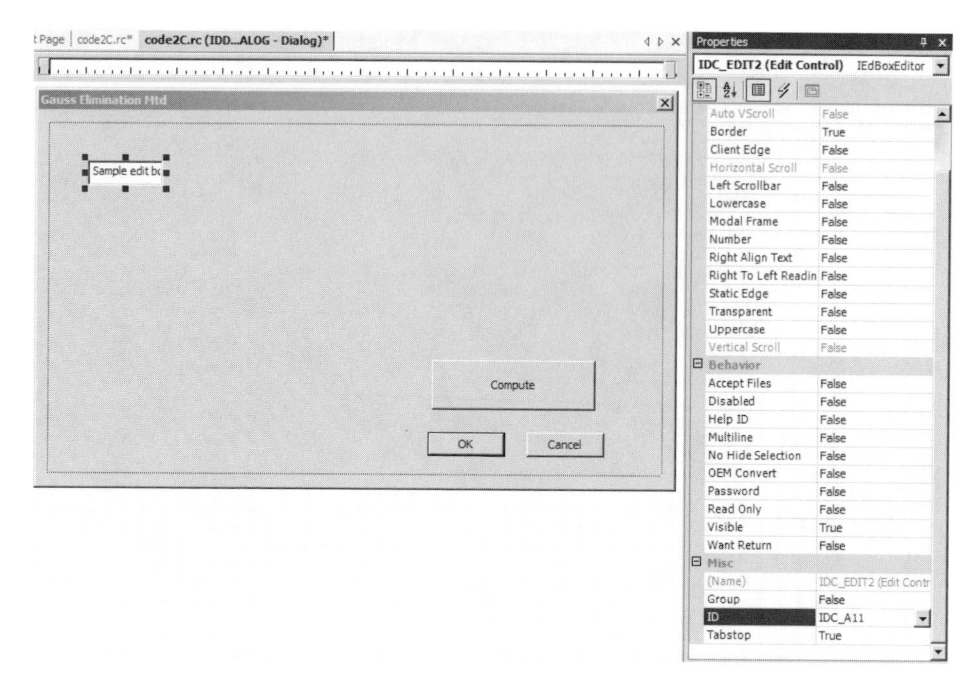

Figure 2.12 Creating an edit box for the array `a[1][1]`.

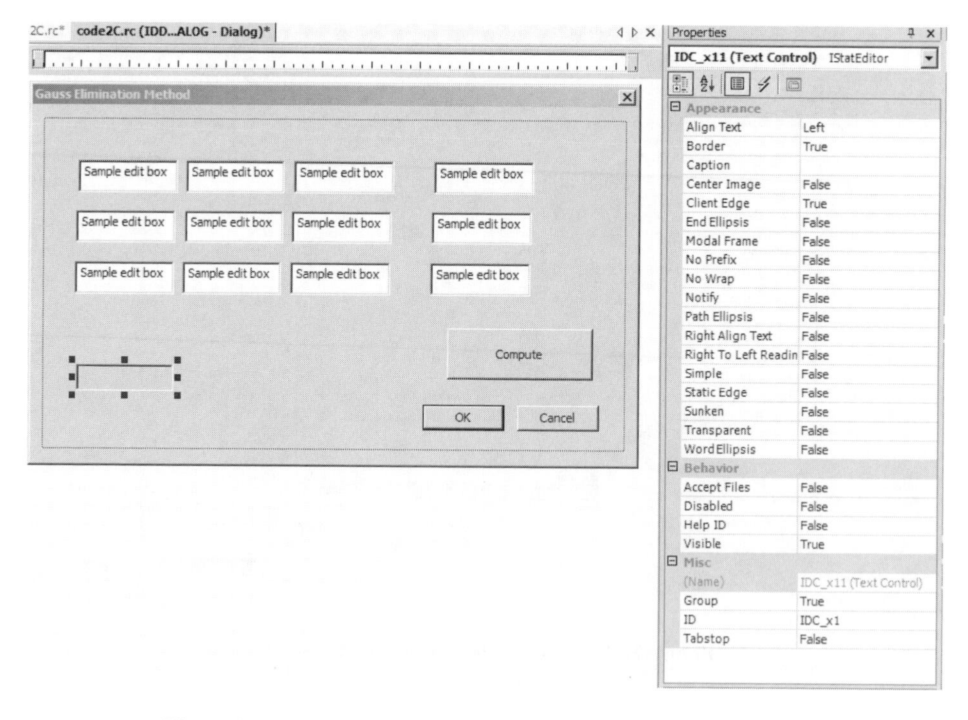

Figure 2.13 Creating the static box for displaying the array `x[1]`.

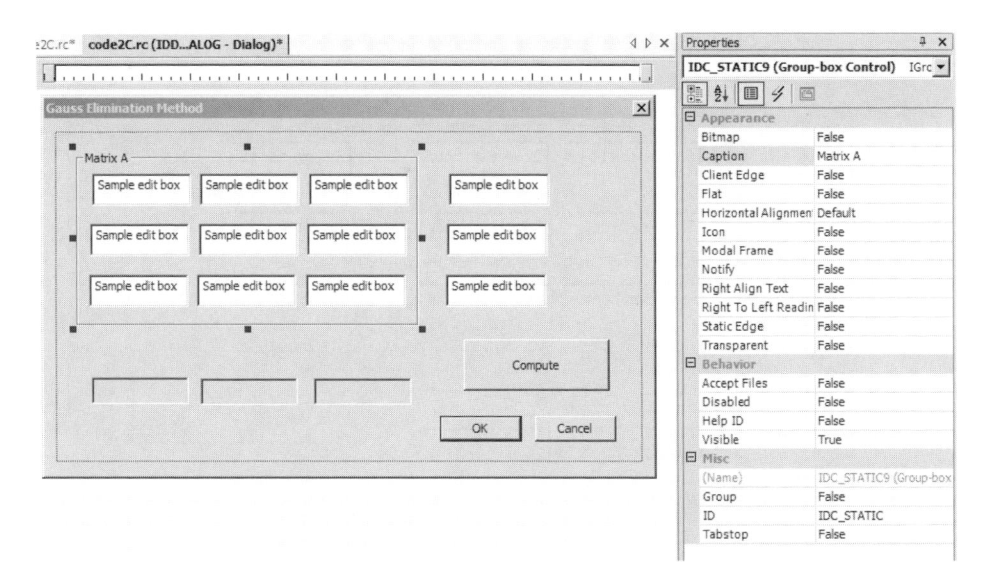

Figure 2.14 Labeling the group as *Matrix A*.

sx[1]. Continue with the other two boxes in vector x[] by repeating this step, to produce an interface as shown in Figure 2.14.

Step 7: The next step is to label the objects according to their group in the dialog window, which is an optional step in the creation of the window. To label the first three rows and columns of the edit boxes as *Matrix A,* click the *Group Box* icon and draw the area. Name the group by choosing *Caption.* Repeat the step for the other vectors to produce an interface as shown in Figure 2.15.

The resource file Code2C.rc is referred as a blank form by Code2C.cpp. As mentioned earlier, the constructor creates the class CCode2C and this class is inherited from CDialog. Besides this, the constructor also allocates memory for the arrays a[][], b[], x[], sa[][], sb[], and sx[], as follows:

```
CCode2C::CCode2C(CWnd* pParent): CDialog(CCode2C::IDD, pParent)
{
       b=new double [N+1];
       sb=new CString [N+1];
       x=new double [N+1];
       sx=new CString [N+1];
       a=new double *[N+1];
       sa=new CString *[N+1];
       for (int i=1;i<=N;i++)
       {
              a[i]=new double [N+1];
```

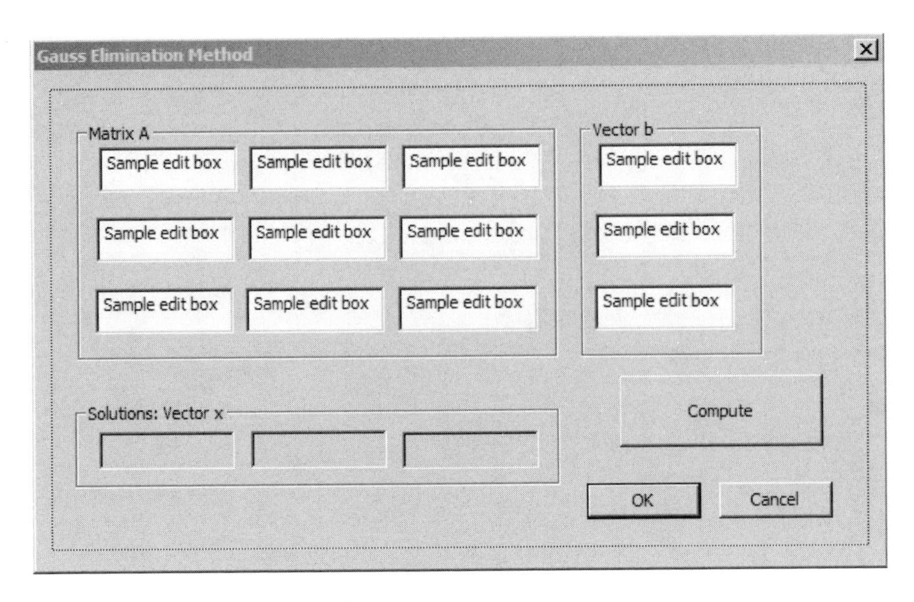

Figure 2.15 The complete dialog window.

```
                sa[i]=new CString [N+1];
        }
}
```

The parent window in this application is the modal dialog window, as shown in Figure 2.15. This window is created in the application class `CWinApp`, as follows:

```
BOOL CMyWinApp::InitInstance()
{
        AfxEnableControlContainer();
        Enable3dControlsStatic();
        CCode2C dlg;
        m_pMainWnd = &dlg;
        int nResponse = dlg.DoModal();
        return FALSE;
}
```

The above code fragments represent the creation of a modal window by calling up two MFC functions, `AfxEnableControlContainer()` and `Enable3d-ControlsStatic()`. A modal window is created through the function `Do-Modal()` from the class `CDialog`. In creating the window, an object called `dlg`, derived from `CCode2C`, is linked to `DoModal()`. The class `CCode2C`, in turn, is inherited from `CDialog`, and this makes it possible for the dialog window to be created in the application.

The input for matrix A and vector **b** in the application is achieved through the function `DoDataExchange()`, as follows:

```
void CCode2C::DoDataExchange(CDataExchange* pDX)
{
      CDialog::DoDataExchange(pDX);
      DDX_Text(pDX, IDC_A11, sa[1][1]);
      DDX_Text(pDX, IDC_A12, sa[1][2]);
      DDX_Text(pDX, IDC_A13, sa[1][3]);
      DDX_Text(pDX, IDC_A21, sa[2][1]);
      DDX_Text(pDX, IDC_A22, sa[2][2]);
      DDX_Text(pDX, IDC_A23, sa[2][3]);
      DDX_Text(pDX, IDC_A31, sa[3][1]);
      DDX_Text(pDX, IDC_A32, sa[3][2]);
      DDX_Text(pDX, IDC_A33, sa[3][3]);
      DDX_Text(pDX, IDC_b1, sb[1]);
      DDX_Text(pDX, IDC_b2, sb[2]);
      DDX_Text(pDX, IDC_b3, sb[3]);
      DDX_Text(pDX, IDC_x1, sx[1]);
      DDX_Text(pDX, IDC_x2, sx[2]);
      DDX_Text(pDX, IDC_x3, sx[3]);
}
```

This function has an argument in the form of an object called `pDX`, derived from the MFC class `CDataExchange`. This object is a pointer for reading the input from the user at the corresponding edit box. Data from the user is read as a string by the function `DDX_Text()` at the edit box identified through its id. For example, the following statement reads the data entered at the edit box with id `IDC_A23` and assigns it to the string `sa[2][3]`:

```
DDX_Text(pDX, IDC_A23, sa[2][3]);
```

The push-button event at the `Compute` button invokes the message handler `ON_BN_CLICKED`, which then calls the function `OnCompute()`, as follows:

```
void CCode2C::OnCompute()
{
      int i,j;
      CString s;
      UpdateData(TRUE);
      for (i=1;i<=N;i++)
      {
           b[i]=atof(sb[i]);
           for (j=1;j<=N;j++)
                a[i][j]=atof(sa[i][j]);
      }
```

```
        PGauss();
        for (i=1;i<=N;i++)
                sx[i].Format("%lf", x[i]);
        UpdateData(FALSE);
}
```

In `OnCompute()`, data from the edit boxes are first read through the MFC function `UpdateData()`. This function reads data from the objects in the dialog window or writes data to these objects. This function has a Boolean argument in the form of 1 (TRUE) or 0 (FALSE). A value of 1 means this function calls `DoDataExchange()` to get the latest data from the edit boxes entered by the user. On the other hand, a value of 0 does not read data from the edit boxes. Instead, it passes the latest data to these objects for display on the static boxes.

Data in the form of text strings is read from the edit boxes and converted to the arrays `a[][]` and `b[]` using the C++ function `atof()`. `OnCompute()` then calls the function `PGauss()` to solve the linear equations and produce the results as `x[]`. The formatted values of `x[]` are then converted to the strings `sx[]` and displayed in the static boxes.

The program ends through the destructor `~CCode2C()`, which deallocates the memories assigned to the arrays and returns the memory to the computer. The destructor also destroys the class `CCode2C`.

2.5 SUMMARY AND CONCLUSION

We discussed three models for solving linear and nonlinear problems in this chapter. Both linear and nonlinear equations are considered fundamental problems in numerical computing. These problems exist in many applications in science and engineering. One way to get people to appreciate the importance of numerical problems is to present the problems in a friendly and acceptable manner. This encompasses the need for a tool that allows the visual representation of both the problem and its solution.

The first problem in this chapter is the bisection method, which is an iterative method for finding the root of an equation. The steps may look easy but they are very tedious to solve manually as they involve a lot of repetition. The programming approach solves this problem by a more systematic and convenient method. A friendly numerical interface that displays the solution visually on the computer screen makes the problem accessible to the general audience.

Problems involving iterations are very common in real life. One such problem is optimization. A typical solution to a problem in optimization requires finding the minimum energy or cost function. This energy function has to be designed in such a way that it models the problem significantly. The solution to this problem is obtained through massive iterations of the energy function until convergence to the global minimum is achieved. One potential problem in an iterative method is the possibility of getting trapped in a local minimum. A suitable algorithm is needed to take care of this issue. This is another challenge that implies that an iterative method alone does not guarantee the ultimate solution.

The other two topics discussed are the different styles for providing the interfaces in solving a system of linear equations. The first method involves a single window that hosts several child windows in the form of edit boxes, static boxes, and a push button. The second approach creates a dialog window for providing an interaction between the user and the program through the use of a resource file. We will continue the discussion of a third method for the same problem through the use of a tool called the Wizard in the next chapter.

BIBLIOGRAPHY

1. R. L. Burden and J. D. Faires, *Numerical Analysis,* Brooks Cole, 2000.
2. S. C. Chapra and R. Canale, *Numerical Methods for Engineers,* McGraw-Hill Science, 2001.
3. K. W. Morton and D. F. Mayers, *Numerical Solution of Partial Differential Equations,* Cambridge University Press, 1994.
4. W. H. Press and S. A. Teukolsky (Eds), *Numerical Recipes in C++: The Art of Scientific Computing,* 2nd ed., Cambridge University Press, 2002.

CODE LISTINGS

Code2A: Bisection Method

```
// Code2A.h
#include <afxwin.h>
#include <math.h>
#define N 10                        // maximum number of iterations
#define f(x) (pow(x,3)-pow(x,2)-2)
#define EPSILON 0.005

class CCode2A : public CFrameWnd
{
private:
      double *a,*b,*c;
public:
      CCode2A();
      ~CCode2A();
      afx_msg void OnPaint();
      DECLARE_MESSAGE_MAP();
};

class CMyWinApp : public CWinApp
{
public:
  virtual BOOL InitInstance();
};

#include "code2A.h"
```

```
BEGIN_MESSAGE_MAP(CCode2A, CFrameWnd)
      ON_WM_PAINT()
END_MESSAGE_MAP()

CMyWinApp  MyApplication;

BOOL CMyWinApp::InitInstance()
{
      CCode2A* pFrame = new CCode2A;
      m_pMainWnd = pFrame;
      pFrame->ShowWindow(SW_SHOW);
      pFrame->UpdateWindow();
      return TRUE;
}

CCode2A::CCode2A()
{
      Create(NULL, "Bisection method for finding the root of an equation");
      a=new double [N+1];
      b=new double [N+1];
      c=new double [N+1];
      a[0]=1; b[0]=2;
}

CCode2A::~CCode2A()
{
      delete a,b,c;
}

void CCode2A::OnPaint()
{
      CPaintDC dc(this);
      CString s;
      double error;

      CFont myfont1,myfont2,myfont3;

      myfont1.CreatePointFont (120,"Times New Roman");
      dc.SelectObject (myfont1);
      dc.SetTextColor(RGB(100,100,100));
      dc.TextOut(50,50,"Equation: f(x)=pow(x,3)-pow(x,2)-3");

      myfont2.CreatePointFont (100,"Courier");
      dc.SelectObject (myfont2);
      s.Format("%-5s%-10s%-10s%-10s%-10s%-10s%-10s",
               "i","a[i]","b[i]","c[i]","f(a[i])","f(c[i])","error");
      dc.TextOut(50,120,s);

      for (int i=0;i<=N;i++)
      {
            c[i]=(a[i]+b[i])/2;
            if (f(a[i])*f(c[i])>0)
            {
                  a[i+1]=c[i]; b[i+1]=b[i];
            }
```

```
            else
            {
                   b[i+1]=c[i]; a[i+1]=a[i];
            }
            s.Format("%-5d%-10.3lf%-10.3lf%-10.3lf%-10.3lf%-10.3lf",
                   i,a[i],b[i],c[i],f(a[i]),f(c[i]));
            dc.TextOut(50,150+15*i,s);
            if (i>0)
            {
                   error=fabs(c[i]-c[i-1]);
                   s.Format("%-10.3lf",error);
                   dc.TextOut(490,150+15*i,s);
                   if (error<EPSILON)
                         break;
            }
      }
      myfont3.CreatePointFont (160,"Arial");
      dc.SelectObject (myfont3);
      dc.SetBkColor(RGB(100,100,100));
      dc.SetTextColor(RGB(255,255,255));
      s.Format("the solution is x=%.3lf after %d iterations",
c[i],i-1);
      dc.TextOut(100,300,s);
}
```

Code2B: Solving a System of Linear Equations

```
// code2C.h
#include <afxwin.h>
#define IDC_COMPUTE   500
#define N 3

class CCode2B : public CFrameWnd
{
private:
      int idc_ea, idc_eb, idc_sx;
      double **a,*x,*b;
      CEdit **ea,*eb;
      CStatic *sx;
      CButton bCompute;
public:
      CCode2B();
      ~CCode2B();
      void PGauss();
      afx_msg void OnCompute();
      DECLARE_MESSAGE_MAP()
};

class CMyWinApp : public CWinApp
{
public:
      virtual BOOL InitInstance();
};

// code2B.cpp
```

```
#include "code2B.h"
CmyWinApp  MyApplication;

BOOL CMyWinApp::InitInstance()
{
        CCode2B* pFrame = new CCode2B;
        m_pMainWnd = pFrame;
        pFrame->ShowWindow(SW_SHOW);
        pFrame->UpdateWindow();
        return TRUE;
}

BEGIN_MESSAGE_MAP(CCode2B, CFrameWnd)
    ON_BN_CLICKED (IDC_COMPUTE,OnCompute)
END_MESSAGE_MAP()

CCode2B::CCode2B()
{
        b=new double [N+1];
        eb=new CEdit [N+1];
        x=new double [N+1];
        sx=new CStatic [N+1];
        a=new double *[N+1];
        ea=new CEdit *[N+1];
        for (int i=1;i<=N;i++)
        {
              a[i]=new double [N+1];
              ea[i]=new CEdit [N+1];
        }
        idc_ea=200; idc_eb=300; idc_sx=400;
        Create(NULL, "SLE: Gauss Elimination Method",
                WS_OVERLAPPEDWINDOW,CRect(0,0,500,340));
        bCompute.Create("Compute",WS_CHILD | WS_VISIBLE | BS_DEFPUSH BUTTON,
                CRect(CPoint(380,25),CSize(100,25)),this, IDC_COMPUTE);
        for (int i=1;i<=N;i++)
        {
              for (int j=1;j<=N;j++)
                    ea[i][j].Create(WS_CHILD | WS_VISIBLE | WS_BORDER,

                          CRect(CPoint(10+(i-1)*70,80+(j-1)*30),
                          CSize(60,25)),this,idc_ea++);
              eb[i].Create(WS_CHILD | WS_VISIBLE | WS_BORDER,
                          CRect(CPoint(280,80+(i-1)*30),CSize(60,25)),this,idc_eb++);
              sx[i].Create("",WS_CHILD | WS_VISIBLE | SS_SUNKEN | SS_CENTER,
                          CRect(CPoint(380,80+(i-1)*30),CSize(80,25)),this,idc_sx++);
        }
        ea[1][1].SetFocus();    // caret starts blinking here
}

CCode2B::~CCode2B()
{
        for (int i=1;i<=N;i++)
              delete a[i],ea[i];
        delete a,ea,b,x,sx;
}
```

```
void CCode2B::OnCompute()
{
	CString s;
	int i,j;
	for (i=1;i<=N;i++)
	{
		for (j=1;j<=N;j++)
		{
			ea[i][j].GetWindowText(s);
			a[i][j]=atof(s);
		}
		eb[i].GetWindowText(s);
		b[i]=atof(s);
	}
	PGauss();
	for (i=1;i<=N;i++)           // display vector x
	{
		s.Format("%lf",x[i]);
		sx[i].SetWindowText(s);
	}
}

void CCode2B::PGauss()           // compute the SLE
{
	int i,j,k;
	double m,Sum;
     for (k=1;k<=N-1;k++)        // row operations on a
	  for (i=k+1;i<=N;i++)
		{
			m=a[i][k]/a[k][k];
			for (j=1;j<=N;j++)
				a[i][j]=a[i][j]-m*a[k][j];
			b[i]=b[i]-m*b[k];
		}
     for (i=N;i>=1;i–)           // backward substitutions on x
     {
	  Sum=0;
		x[i]=0;
		for (j=i;j<=N;j++)
		Sum +=a[i][j]*x[j];
		x[i]=(b[i]-Sum)/a[i][i];
     }
}
```

Code2C: Resource File Approach to the SLE Problem

```
// code2C.h
#include <afxwin.h>
#include <afxdisp.h>
#include "resource.h"
#define N 3

class CCode2C : public CDialog
```

```
{
protected:
	double **a,*b,*x;
	CString **sa,*sb,*sx;
public:
	CCode2C(CWnd* pParent = NULL);
	~CCode2C();
	enum { IDD = IDD_GAUSSDIALOG };
	virtual void DoDataExchange(CDataExchange* pDX);
	afx_msg void OnCompute();
	DECLARE_MESSAGE_MAP()
	void PGauss();
};

class CMyWinApp : public CWinApp
{
public:
	virtual BOOL InitInstance();
};

#include "code2C.h"

CMyWinApp MyApplication;

BOOL CMyWinApp::InitInstance()
{
	AfxEnableControlContainer();
	Enable3dControlsStatic();
	CCode2C dlg;
	m_pMainWnd = &dlg;
	int nResponse = dlg.DoModal();
	return FALSE;
}

BEGIN_MESSAGE_MAP(CCode2C,CDialog)
	ON_BN_CLICKED(IDC_COMPUTE, OnCompute)
END_MESSAGE_MAP()


CCode2C::CCode2C(CWnd* pParent): CDialog(CCode2C::IDD, pParent)
{
	b=new double [N+1];
	sb=new CString [N+1];
	x=new double [N+1];
	sx=new CString [N+1];
	a=new double *[N+1];
	sa=new CString *[N+1];
	for (int i=1;i<=N;i++)
	{
		a[i]=new double [N+1];
		sa[i]=new CString [N+1];
	}
}
```

```
CCode2C::~CCode2C()
{
      for (int i=1;i<=N;i++)
            delete a[i],sa[i];
      delete a,sa,b,x,sb,sx;
}

void CCode2C::DoDataExchange(CDataExchange* pDX)
{
      CDialog::DoDataExchange(pDX);
      DDX_Text(pDX, IDC_A11, sa[1][1]);
      DDX_Text(pDX, IDC_A12, sa[1][2]);
      DDX_Text(pDX, IDC_A13, sa[1][3]);
      DDX_Text(pDX, IDC_A21, sa[2][1]);
      DDX_Text(pDX, IDC_A22, sa[2][2]);
      DDX_Text(pDX, IDC_A23, sa[2][3]);
      DDX_Text(pDX, IDC_A31, sa[3][1]);
      DDX_Text(pDX, IDC_A32, sa[3][2]);
      DDX_Text(pDX, IDC_A33, sa[3][3]);
      DDX_Text(pDX, IDC_b1, sb[1]);
      DDX_Text(pDX, IDC_b2, sb[2]);
      DDX_Text(pDX, IDC_b3, sb[3]);
      DDX_Text(pDX, IDC_x1, sx[1]);
      DDX_Text(pDX, IDC_x2, sx[2]);
      DDX_Text(pDX, IDC_x3, sx[3]);
}

void CCode2C::OnCompute()
{
      int i,j;
      CString s;
      UpdateData(TRUE);
      for (i=1;i<=N;i++)
      {
            b[i]=atof(sb[i]);
            for (j=1;j<=N;j++)
                  a[i][j]=atof(sa[i][j]);
      }
      PGauss();
      for (i=1;i<=N;i++)
            sx[i].Format("%lf", x[i]);
      UpdateData(FALSE);
}

void CCode2C::PGauss()
{
      int i,j,k;
      double m,Sum;

      // Perform row operations
      for (k=1;k<=N-1;k++)
            for (i=k+1;i<=N;i++)
            {
                  m=a[i][k]/a[k][k];
```

```
                for (j=1;j<=N;j++)
                        a[i][j]=a[i][j]-m*a[k][j];
                b[i]=b[i]-m*b[k];
        }

    // Perform back substitutions
    for (i=N;i>=1;i—)
    {
        Sum=0;
        x[i]=0;
        for (j=i;j<=N;j++)
                Sum +=a[i][j]*x[j];
        x[i]=(b[i]-Sum)/a[i][i];
    }
}
```

CHAPTER 3

MATRIX OPERATIONS USING WIZARD

3.1 DOCUMENT/VIEW ARCHITECTURE USING WIZARD

The previous chapter discusses the non-Wizard solution to creating three Windows applications. This approach is good as it provides a solution with very few lines of code. However, the approach may be difficult for a beginning programmer who is burdened with the task of understanding Windows deeply before developing an application. It is also not easy for the programmer to produce a professional-quality program if extensive use of Windows resources is required.

A professional-looking application requires the document/view application. The *document/view architecture* is a program development approach that integrates documents that hold the data and present this data as an output using the view facility in MFC. The approach involves the use of several classes in MFC and each one of them has access to several functions for utilizing the Windows resources. A program using the Windows interface can be written and developed faster using a tool known as Wizard. With Wizard, an application can be written to produce a professional quality presentation, ready for deployment in the market. This tool is available in Visual Studio to help a programmer concentrate on the application code, leaving the details about communication with Windows to the compiler. Wizard helps a programmer by automatically generating the relevant code using a series of menus, buttons, and other resources in Windows. This facility frees the programmer from the burden of having to know the details about Windows before coding. An analogy can be drawn here: a driver need not know the details about his car engine in order to drive the car. The system in the car has been designed in such a way that a driver can get used to it easily without the need to study its technical details.

Wizard does not provide everything that a programmer may expect. A programmer still needs to know the language, its program development, and, most important of all, the requirements of the program with respect to Windows as the programming environment. The programmer still has to learn and understand some basic components and the mechanism of the working of Windows. There is a steep learning curve in using Wizard as well. The programmer needs to get used to some terminology and jargon used in Windows in order to understand the role of wizard. There are also many lines of code involved, although many of them may not be relevant to the application. Wizard also takes a huge amount of disk space as many standby files are generated during the compiling and linking processes.

This chapter discusses two models involving the use of Wizard. The first is a model of matrix arithmetic that includes the multiplication and the inverse of a matrix. This model illustrates the maximum use of local variables in the form of arrays, and their transfer from one function to another. The second is the rewriting of the earlier examples of a system of linear equations in Chapter two, this time using MFC Wizard.

3.2 MATRIX ALGEBRA

Several solutions to difficult problems in mathematics and their applications need a substantial reduction to the form of their vectors and matrices. This reduction is often achieved through a successful implementation of several algorithms. The overall solution is obtained after some vigorous mathematical operations on these vectors and matrices. Therefore, matrix operations are some of the core elements in numerical computations. Mathematical operations involving matrices can be very tedious and difficult, especially when their size is large. This complexity is shown by the powerful computers often used in performing calculations involving large matrices. For example, in modeling the wind flow that forms a flux in weather forecasting, powerful machines such as the Cray supercomputers are used. A powerful computer alone is not sufficient to produce a fast result. More important than that is the software part. A program needs to be written in such a way as to optimize the use of resources, and this greatly contributes to high-performance results.

A *matrix* is a two-dimensional table that allows a set of data to be arranged in rows and columns. Each element in a matrix has a row and column connection that relates to this element. A rectangular matrix is very commonly represented as a form of data entry as it maps physically and logically to the real problem. For example, a matrix may collectively represent the x and y geographical coordinates of an object. This representation allows a direct mapping from the ground location of the object onto the mapping board, useful for purposes like tracking the path of this object. In another example, a matrix may represent an image, with its pixels as the elements in the matrix. The quality of the image can be improved by performing some operations on the matrix.

In a programming language, an array is a broad term that includes matrix. An array can be created into several dimensions for supporting several different requirements of a problem. An array can also be linked as a variable or object to other data members in a class. A one-dimensional form of an array is called a vector. In fact, a matrix is made up of columns of vectors. Besides having magnitude and length, a vector has direction, and this feature differentiates a vector from a scalar.

Matrix arithmetic involves all the four basic tools in algebra: addition, subtraction, multiplication, and division. Operations involving addition and subtraction are quite trivial. Therefore, they are not discussed here as the user can easily design these items from the idea discussed in this current application. In this section, we discuss operations involving matrix multiplication and the problem of finding the inverse of a matrix. We further illustrate the concepts of dynamic memory allocation discussed in Chapter 2 and data passing between functions.

We discuss a project having an expression involving matrix multiplication and finding the inverse of a matrix using the following example:

$$Z = A^2 B^{-1} A^{-1} \tag{3.1}$$

In Equation (3.1), A, B, and Z are square matrices of size $N \times N$. In this example, we set $N = 3$, but the solution is scalable upward and downward to support any other reasonable value of N. An easy way of solving the problem in Equation (3.1) is to break down the problem into several small components, or tokens, according to the following sequence, then solve each component one by one:

Step 1: Let $P = A^2$ and compute P through the multiplication $P = A \cdot A$.
Step 2: Let $Q = B^{-1}$ and compute this matrix inverse.
Step 3: Let $R = A^{-1}$ and compute this matrix inverse.
Step 4: Let $Y = QR$ and compute Y through the multiplication $Y = Q \cdot R$.
Step 5: Let $Z = PY$ and compute Z through the multiplication $Z = P \cdot Y$.

By performing the above steps one by one in their order, we obtain the final solution, Z. The motivation for writing the solution according to the above steps is to make full use of the two matrix operations, namely, multiplication and inverse. Two functions, one each for the matrix multiplication and matrix inverse, are to be created in the project files. These functions are called repeatedly for solving each component in the above steps, which will then lead to the ultimate solution.

Data Passing Between Functions

One important strategy in numerical programming for achieving high performance is to maximize the use of local variables instead of global variables. The advantage of this strategy can be seen from the fact that local variables are easier to manage as they are confined inside the given functions only. These functions are not used all the time during the program runtime and, therefore, the use of local variables will optimize the overall memory usage inside the computer. A global variable may also cause some problems in some areas of the program. A global variable is visible everywhere in the class and even in another class. Some slight assignment on its value in a function may alter the value of another variable in another function.

One problem with having local variables in several functions is the way a function can use the local variables declared in another function. The problem is solved through a proper data-passing mechanism between functions. Data in the form of arrays can be passed from one function to another through the arguments in the two functions. The general rule for passing the data is that when one function sends data through an argument, the receiving function must accept this data in another argument. The two arguments must have the same data type, array dimension, and size.

Matrix Multiplication

Two matrices in the form of two-dimensional arrays are multiplied using the same rule that governs their operation in mathematics. We discuss an example with two matrices $A = [a_{ij}]$ and $B = [b_{ij}]$, of sizes 3 × 4 and 4 × 2, respectively. Their multiplication produces matrix $C = [c_{ij}]$ of size 3 × 2, as follows:

$$C = \begin{bmatrix} c_{11} & c_{12} \\ c_{21} & c_{22} \\ c_{31} & c_{32} \end{bmatrix} = \begin{bmatrix} c_{11} & a_{12} & a_{13} & a_{14} \\ a_{21} & a_{22} & a_{23} & a_{24} \\ a_{31} & a_{32} & a_{33} & a_{34} \end{bmatrix} \begin{bmatrix} b_{11} & b_{12} \\ b_{21} & b_{22} \\ b_{31} & b_{32} \\ b_{41} & b_{42} \end{bmatrix}$$
$$= \begin{bmatrix} a_{11}b_{11} + a_{12}b_{21} + a_{13}b_{31} + a_{14}b_{41} & a_{11}b_{12} + a_{12}b_{22} + a_{13}b_{32} + a_{14}b_{42} \\ a_{21}b_{11} + a_{22}b_{21} + a_{23}b_{31} + a_{24}b_{41} & a_{21}b_{12} + a_{22}b_{22} + a_{23}b_{32} + a_{24}b_{42} \\ a_{31}b_{11} + a_{32}b_{21} + a_{33}b_{31} + a_{34}b_{41} & a_{31}b_{12} + a_{32}b_{22} + a_{33}b_{32} + a_{34}b_{42} \end{bmatrix} \quad (3.2)$$

The solution in a compact form representing the manual calculations above is shown, as follows:

$$[c_{ij}] = \begin{bmatrix} \sum_{k=1}^{4} a_{1k}b_{k1} & \sum_{k=1}^{4} a_{1k}b_{k2} \\ \sum_{k=1}^{4} a_{2k}b_{k1} & \sum_{k=1}^{4} a_{2k}b_{k2} \\ \sum_{k=1}^{4} a_{3k}b_{k1} & \sum_{k=1}^{4} a_{3k}b_{k2} \end{bmatrix} \quad (3.3)$$

for rows $i = 1, 2, 3$ and columns $j = 1, 2$, which further reduces to the following:

$$[c_{ij}] = \left[\sum_{k=1}^{4} a_{ik}b_{kj} \right] \quad (3.4)$$

In C++, Equation (3.4) is written simply as

```
for (i=1;i<=3;i++)
      for (j=1;j<=2;j++)
      {
            c[i][j]=0;
            for (k=1;k<=4;k++)
                  c[i][j]+=a[i][k]*b[k][j];
      }
```

Finding the Inverse of a Matrix

A matrix A of size $N \times N$ is said to have an inverse given by $X = A^{-1}$ if their product is an identity matrix, as follows:

$$AX = I \tag{3.5}$$

In the above equation I is an identity matrix. The rule of mathematics governs the steps in finding the inverse of a matrix: the inverse of a square matrix exists only if the matrix is not singular. A *singular* matrix is a matrix whose determinant is zero.

The inverse of a matrix is computed using the Gaussian elimination method, similar to the method used in solving a system of linear equations. This is observed by replacing the vectors **x** and **b** in Equation (2.7) with the matrices X and I, respectively. The method starts by assigning a new matrix B with the identity matrix I in Equation (3.5) as its initial value, as follows:

$$AX = B \tag{3.6}$$

Steps similar to Equation (2.8) in Chapter 2 are applied to reduce matrix A into its upper triangular matrix form U through a series of row operations. Backward substitutions then follow using steps similar to Equation (2.9) to generate the solution, X. These steps are shown in Figure 3.1.

For the case of A with size 3×3, the steps in Figure 3.1 are illustrated as follows:

$$AX = I: \begin{bmatrix} a_{11} & a_{12} & a_{13} \\ a_{21} & a_{22} & a_{23} \\ a_{31} & a_{32} & a_{33} \end{bmatrix} \begin{bmatrix} x_{11} & x_{12} & x_{13} \\ x_{21} & x_{22} & x_{23} \\ x_{31} & x_{32} & x_{33} \end{bmatrix} = \begin{bmatrix} 1 & 0 & 0 \\ 0 & 1 & 0 \\ 0 & 0 & 1 \end{bmatrix}$$

$\downarrow$ Row operations

$$\begin{bmatrix} a'_{11} & a'_{12} & a'_{13} \\ 0 & a'_{22} & a'_{23} \\ 0 & 0 & a'_{33} \end{bmatrix} \begin{bmatrix} x_{11} & x_{12} & x_{13} \\ x_{21} & x_{22} & x_{23} \\ x_{31} & x_{32} & x_{33} \end{bmatrix} = \begin{bmatrix} v_{11} & v_{12} & v_{13} \\ v_{21} & v_{22} & v_{23} \\ v_{31} & v_{32} & v_{33} \end{bmatrix}$$

$\downarrow$ Backward substitutions

$$\begin{bmatrix} x_{11} & x_{12} & x_{13} \\ x_{21} & x_{22} & x_{23} \\ x_{31} & x_{32} & x_{33} \end{bmatrix} = \begin{bmatrix} v'_{11} & v'_{12} & v'_{13} \\ v'_{21} & v'_{22} & v'_{23} \\ v'_{31} & v'_{32} & v'_{33} \end{bmatrix}$$

We now discuss the method. Suppose $A = [a_{ij}]$ is a square matrix of size $N \times N$ and $B = [b_{ij}]$ is an identity matrix having elements with the initial values given by

$$b_{ij} = \begin{cases} 1 & \text{if } i = j \\ 0 & \text{if } i \neq j \end{cases}$$

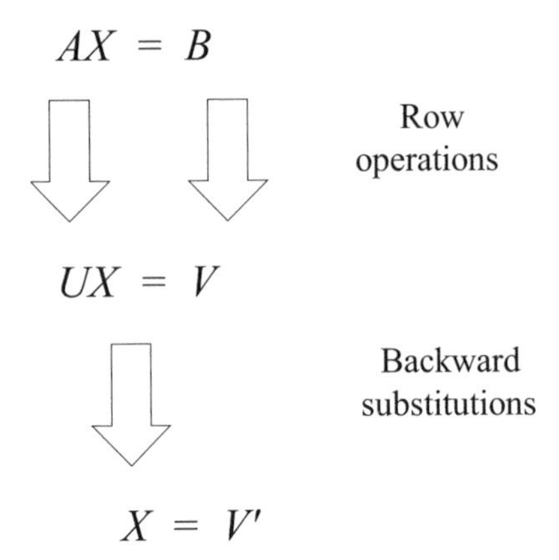

Figure 3.1 Gaussian elimination method for finding the inverse of a matrix.

The ith row operations on row k are performed by first finding the value of m, which is the ratio of the ith element in the row to its pivot element, given as follows:

$$m = a_{ik}/a_{kk} \tag{3.7}$$

Row operations then follow on a_{ij} and b_{ij} according to the following steps:

$$a_{ij} \leftarrow a_{ij} - m \cdot a_{kj} \tag{3.8}$$

$$b_{ij} \leftarrow b_{ij} - m \cdot b_{kj} \tag{3.9}$$

for $k = 1, 2, \ldots, N - 1$, followed by $i = k + 1, k + 2, \ldots, N$, and $j = 1, 2, \ldots, N$. Equations (3.7), (3.8), and (3.9) collectively reduce matrix A to its upper triangular form U while B becomes V.

The second step in finding the inverse of a matrix is to apply backward substitutions on the upper triangular matrix U to get the solution V', as follows:

$$x_{Ni} = \frac{v_{Ni}}{u_{NN}} \tag{3.10}$$

$$x_{ij} = \frac{v_{ij} - \sum_{k=i+1}^{N} u_{ij} x_{kj}}{u_{ii}} \tag{3.11}$$

for $i = N - 1, N - 2, \ldots, 1$, and $j = 1, 2, \ldots, N$.

Code3A: Matrix Operations

We illustrate matrix operations involving multiplication and matrix inverse in solving the problem $Z = A^2B^{-1}A^{-1}$ through a project called Code3A. The application assumes all the matrices have size 3×3. The output of Code3A is shown in Figure 3.2. The display consists of a single window showing the values of the input matrices A and B, the components P, Q, R, and Y, and the solution Z. The project demonstrates the use of dynamic memory allocation and how data in the form of arrays is passed effectively from one function to another. There are no global variables used in this application.

We illustrate the development of the application with the help of Wizard. MFC Wizard consists of a guided-development approach to building an application. The user is presented with a series of questions and must choose the options suitable for the application. The process begins by first creating a blank application using the document/view architecture. It then proceeds with inserting the code for the application.

Our main function in this application is `OnDraw()`, which declares all the required arrays in the problem. This function gets the input data from the function `InputData()` and performs matrix operations by passing the data to the functions `MatMultiply()` and `MatInverse()`.

Step 1: The process begins by creating a new project using Wizard for a blank

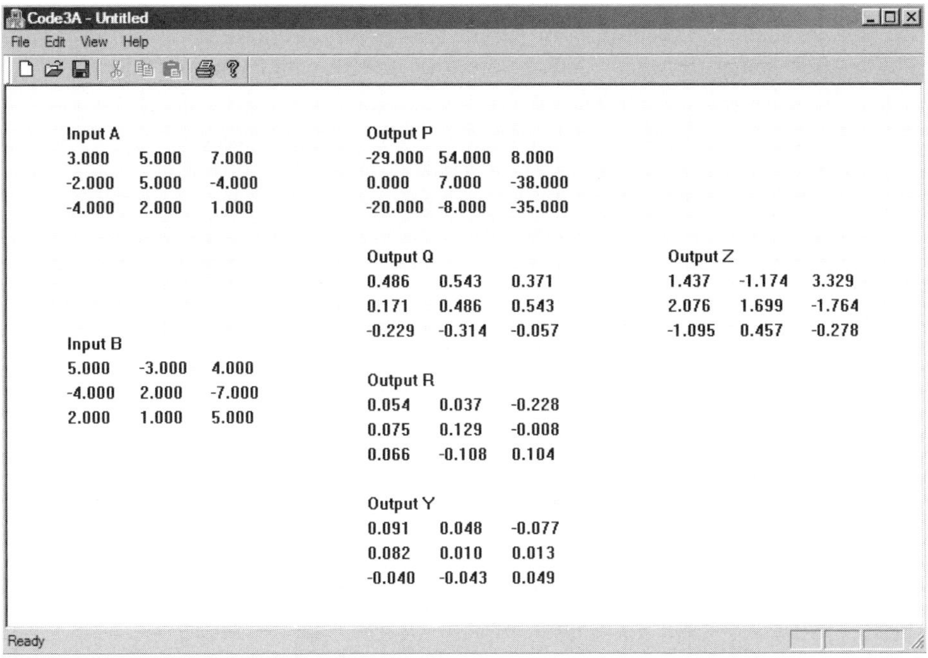

Code3A - Untitled
File Edit View Help

Input A
3.000 5.000 7.000
-2.000 5.000 -4.000
-4.000 2.000 1.000

Input B
5.000 -3.000 4.000
-4.000 2.000 -7.000
2.000 1.000 5.000

Output P
-29.000 54.000 8.000
0.000 7.000 -38.000
-20.000 -8.000 -35.000

Output Q
0.486 0.543 0.371
0.171 0.486 0.543
-0.229 -0.314 -0.057

Output R
0.054 0.037 -0.228
0.075 0.129 -0.008
0.066 -0.108 0.104

Output Y
0.091 0.048 -0.077
0.082 0.010 0.013
-0.040 -0.043 0.049

Output Z
1.437 -1.174 3.329
2.076 1.699 -1.764
-1.095 0.457 -0.278

Ready

Figure 3.2 Output of Code3A.

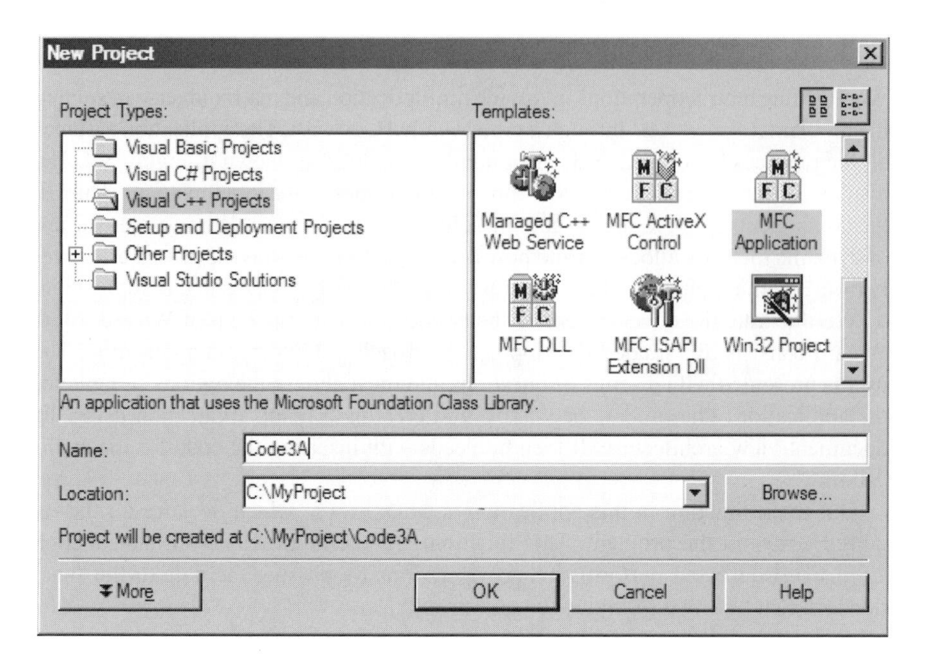

Figure 3.3 Starting menu for creating the application using Wizard.

application. The display in Figure 3.3 appears. Choose *MFC Application* from the choices and name the project Code3A with the proper folder location. Press the *Ok* button to continue.

Step 2: A menu with six items and several other subitems appears, as shown in Figure 3.4. For our application, it is only necessary to choose the items *Application Type* and *Advanced Features* as they are relevant here. Choose *Application Type*. Several radio buttons appear as options with the defaults shown for the application. Our application requires a single document only since a single window is sufficient for displaying the output. Hence, the choice should be *Single Document*. The application is also producing a single EXE file. Therefore, choose *Use MFC in a Static Library* for this option.

Step 3: An interface, shown in Figure 3.5, appears. Choose *Advanced Features* and deselect *ActiveX controls*.

Step 4: The final step in the document/view blank application is to confirm the selection by viewing the main classes generated from the choices: `CCode3AView`, `CCode3AApp`, `CCode3ADoc`, and `CMainFrame`. The classes are shown in Figure 3.6. By default, all the generated classes in MFC start with the letter `C`. At this stage, the application can be compiled, linked, and run to produce a blank window that is not displaying anything.

Step 5: The files generated from the choices made in the blank application using the Solution Explorer are shown in Figure 3.7. From the listing, each of the generated classes produces a C++ and header files. In our application, only

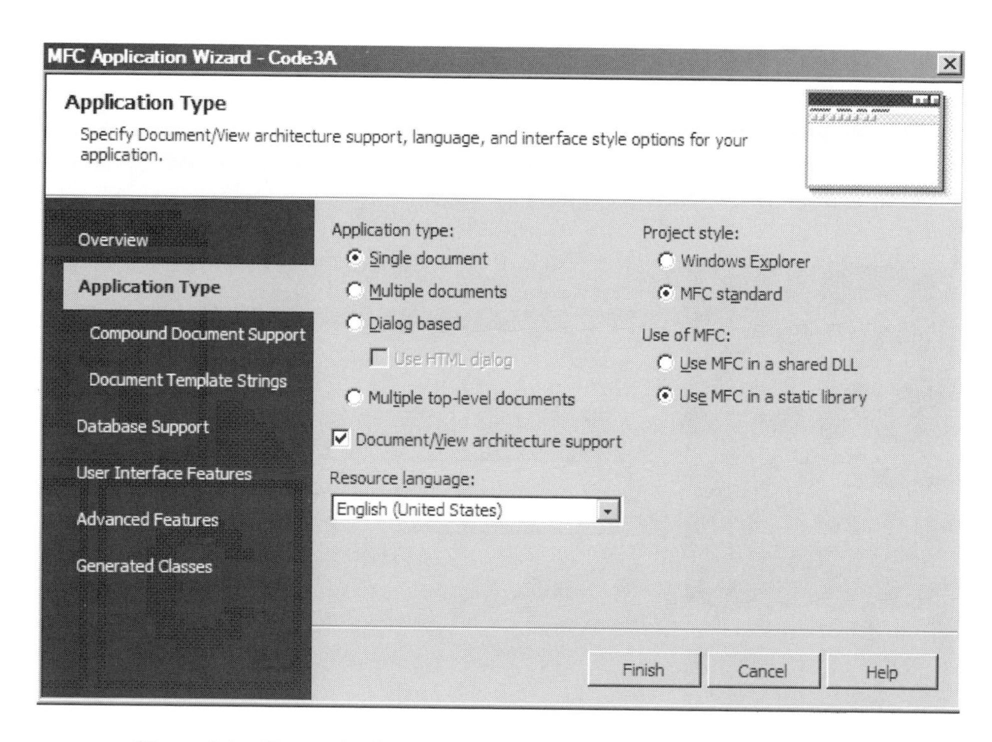

Figure 3.4 Choose *Application Type*, and enter the choices as shown above.

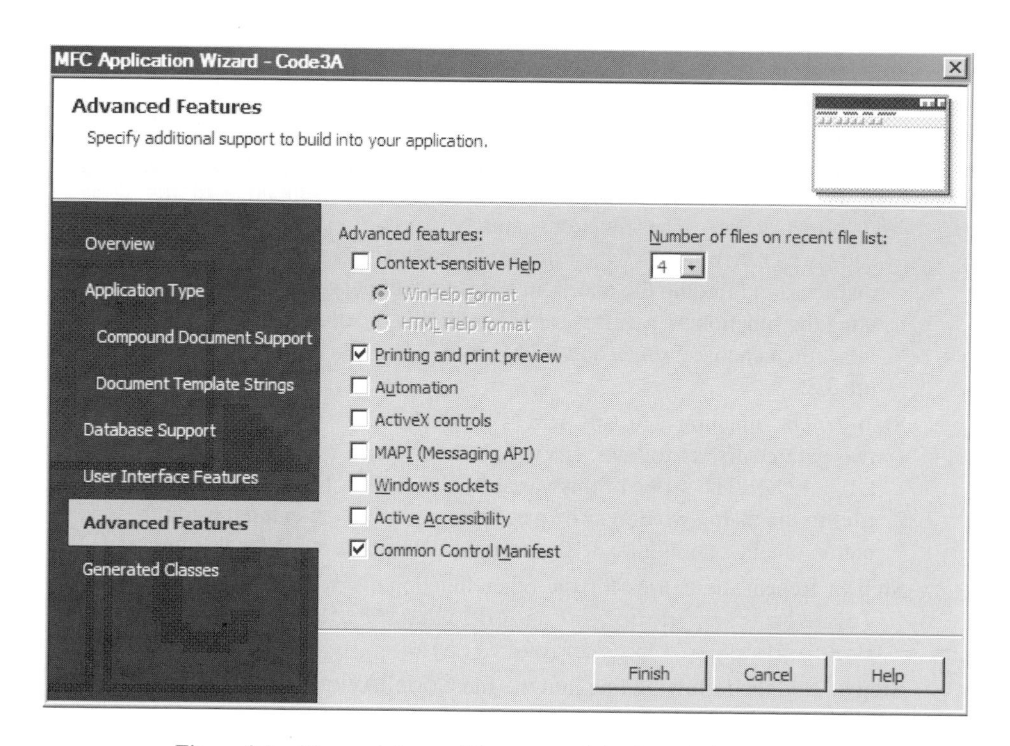

Figure 3.5 Choose Advanced Features and deselect ActiveX Control.

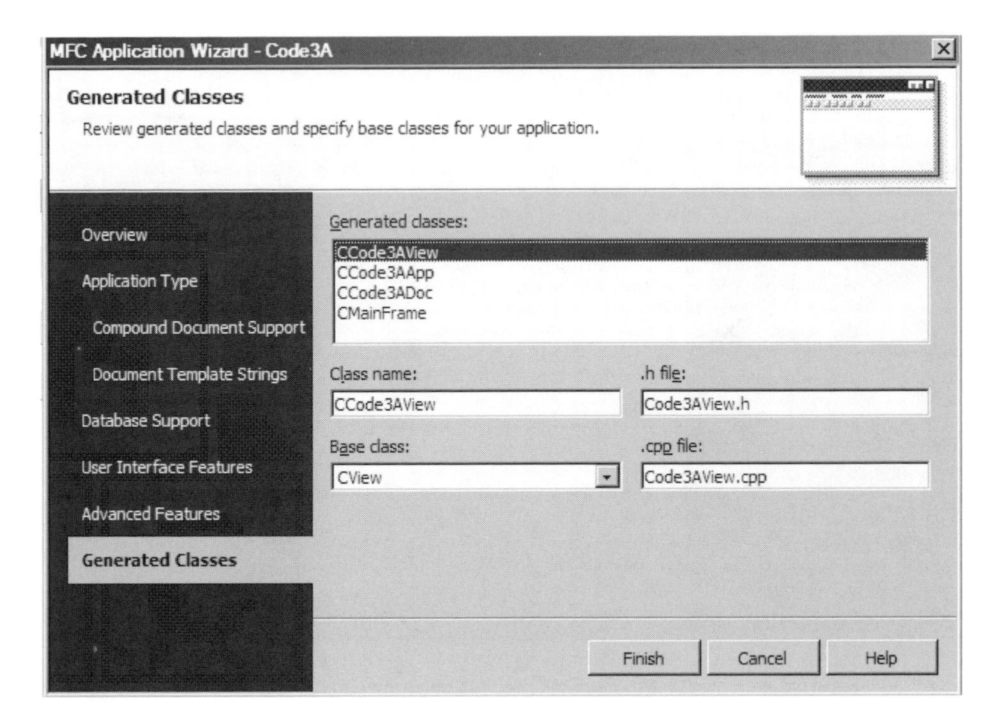

Figure 3.6 Choose Generated Classes to view the main classes generated from the choices.

the view files, `CCode3AView.h` and `CCode3AView.cpp`, are relevant as we are only concerned with displaying the output on the main window.

Step 6: The next step is to add the user member functions to the class `CCode3AView`. The functions are `InputData()`, `MatMultiply()`, and `MatInverse()`, which are functions for data input, multiplication of matrices, and finding the matrix inverse, respectively. The step begins by creating the function `InputData()`. Highlight `CCode3AView` from the class view, then choose *Project* and *Add Function* from the menu, as shown in Figure 3.8.

Step 7: The function `InputData()` is declared as type double and it has two parameters, as follows: `double InputData(double **a,double **b)`. The scope of this variable is `public`. Figure 3.9 shows the entries in the dialog window. The parameters `a` and `b` are entered using the *Add* button one by one. Figure 3.10 shows the complete entries for this function.

Step 8: Repeat the entries for the other functions: `double MatMultiply (double **c, double **a, double **b)` and `double MatInverse(double **x, double **a)`.

Step 9: Add the following line into the file Code3AView.h to define the size of all the matrices:

```
#define N 3
```

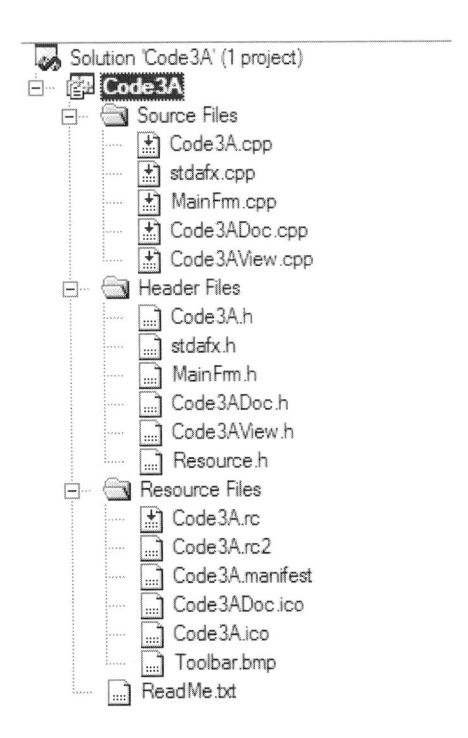

Figure 3.7 The files associated with the classes in the project.

Step 10: Write the application code into the function `OnDraw()`. Begin by creating the two-dimensional arrays `A`, `B`, `P`, `Q`, `R`, `Y`, and `Z`, and dynamically allocating their memory, as follows:

```
double **A,**B,**P,**Q,**R,**Y,**Z;
A=new double *[N+1];
B=new double *[N+1];
P=new double *[N+1];
Q=new double *[N+1];
```

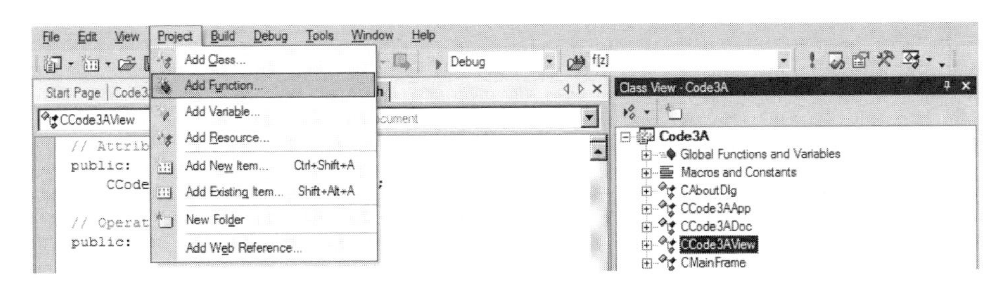

Figure 3.8 Adding a member function to the class `CCode3AView`.

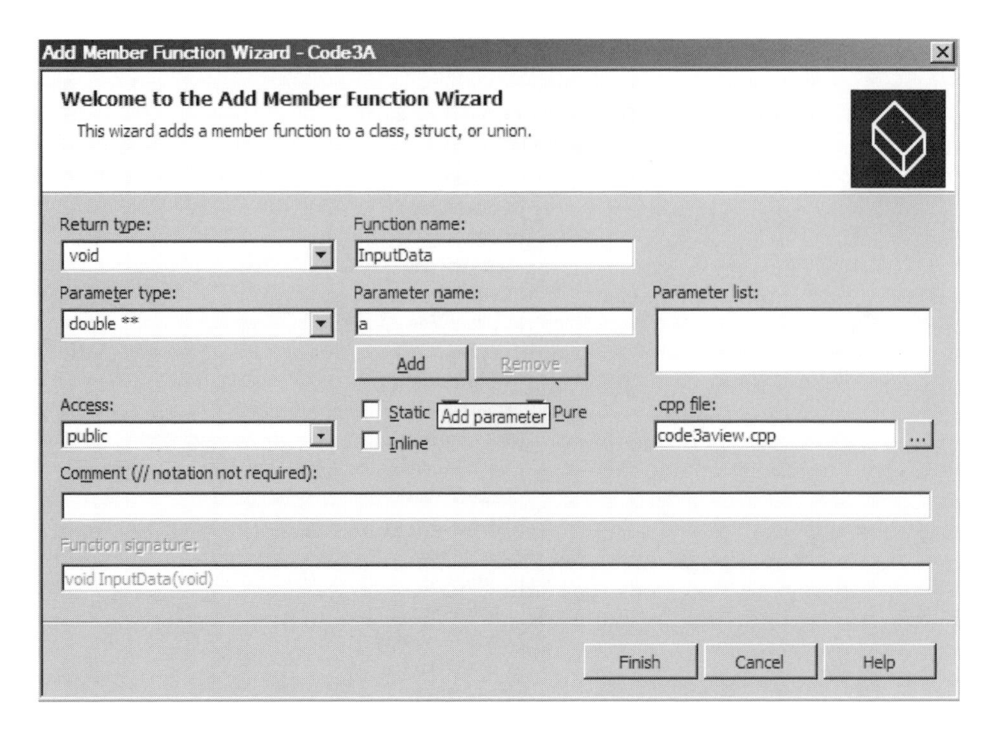

Figure 3.9 Adding a member function to the class.

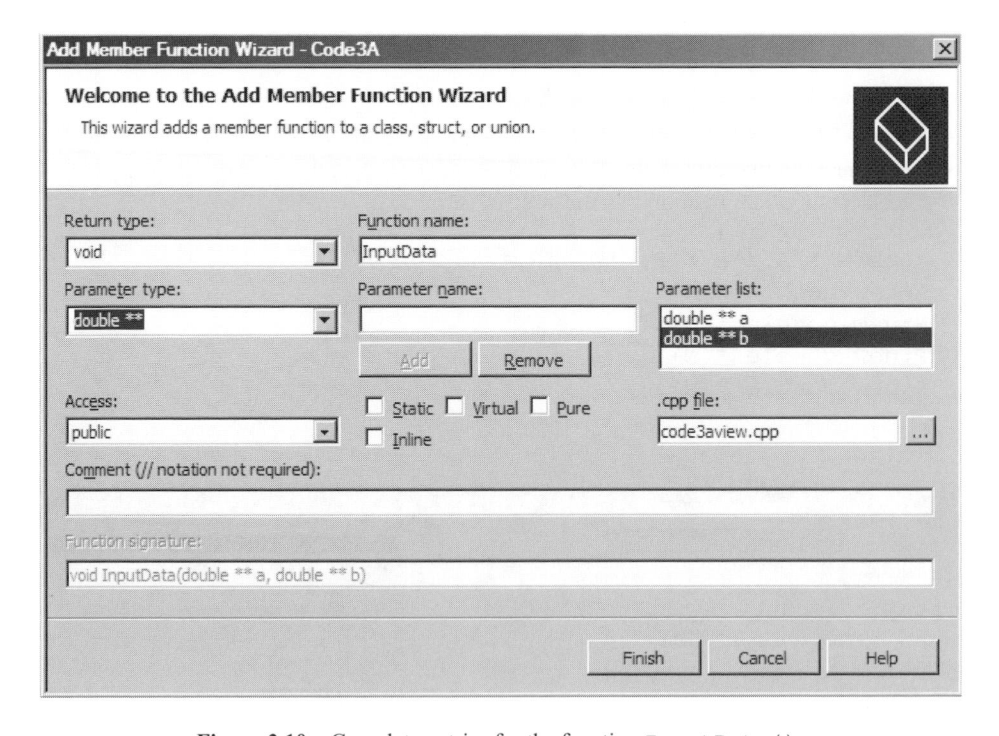

Figure 3.10 Complete entries for the function `InputData()`.

```
R=new double *[N+1];
Y=new double *[N+1];
Z=new double *[N+1];
for (i=1;i<=N;i++)
{
      A[i]=new double [N+1];
      B[i]=new double [N+1];
      P[i]=new double [N+1];
      Q[i]=new double [N+1];
      R[i]=new double [N+1];
      Y[i]=new double [N+1];
      Z[i]=new double [N+1];
}
```

The next step is to get the input data as `A` and `B` from the function `InputData()`:

```
InputData(A,B);
```

This is followed by the code to display the `A` and `B` matrices, as follows:

```
pDC->TextOut(100,30,"Input A");
pDC->TextOut(400,30,"Input B");
for (i=1;i<=N;i++)
      for (j=1;j<=N;j++)
      {
            s.Format("%.3lf",A[i][j]);
            pDC->TextOut(50+(j-1)*60,50+(i-1)*20,s);
            s.Format("%.3lf",B[i][j]);
            pDC->TextOut(350+(j-1)*60,50+(i-1)*20,s);
      }
```

The real algebraic operations follow by passing data to the related functions and getting back the results from these functions. This is done as follows:

```
MatMultiply(P,A,A);
MatInverse(Q,B);
MatInverse(R,A);
MatMultiply(Y,Q,R);
MatMultiply(Z,P,Y);
```

Next, we write the code for displaying the results and destroying the arrays at the end of the function:

```
pDC->TextOut(50,130,"Output P");
```

```
pDC->TextOut(50,230,"Output Q");
pDC->TextOut(50,330,"Output R");
pDC->TextOut(50,430,"Output Y");
pDC->TextOut(400,430,"Output Z");
for (i=1;i<=N;i++)
      for (j=1;j<=N;j++)
      {
            s.Format("%.3lf",P[i][j]);
            pDC->TextOut(50+(j-1)*60,150+(i-1)*20,s);
            s.Format("%.3lf",Q[i][j]);
            pDC->TextOut(50+(j-1)*60,250+(i-1)*20,s);
            s.Format("%.3lf",R[i][j]);
            pDC->TextOut(50+(j-1)*60,350+(i-1)*20,s);
            s.Format("%.3lf",Y[i][j]);
            pDC->TextOut(50+(j-1)*60,450+(i-1)*20,s);
            s.Format("%.3lf",Z[i][j]);
            pDC->TextOut(350+(j-1)*60,450+(i-1)*20,s);
      }
for (i=1;i<=N;i++)
      delete A[i],B[i],P[i],Q[i],R[i],Y[i],Z[i];
delete A,B,P,Q,R,Y,Z;
```

The complete code for the function `OnDraw()` is shown below with the code written by the user in the shaded area:

```
void CCode3AView::OnDraw(CDC* pDC)
{
      CCode3ADoc* pDoc = GetDocument();
      ASSERT_VALID(pDoc);

      // TODO: add draw code for native data here
      CString s;
      int i,j;
      double **A,**B,**P,**Q,**R,**Y,**Z;

      A=new double *[N+1];
      B=new double *[N+1];
      P=new double *[N+1];
      Q=new double *[N+1];
      R=new double *[N+1];
      Y=new double *[N+1];
      Z=new double *[N+1];
      for (i=1;i<=N;i++)
      {
            A[i]=new double [N+1];
            B[i]=new double [N+1];
            P[i]=new double [N+1];
```

```
        Q[i]=new double [N+1];
        R[i]=new double [N+1];
        Y[i]=new double [N+1];
        Z[i]=new double [N+1];
    }
    InputData(A,B);
    pDC->TextOut(100,30,"Input A");
    pDC->TextOut(400,30,"Input B");
    for (i=1;i<=N;i++)
        for (j=1;j<=N;j++)
        {
            s.Format("%.3lf",A[i][j]);
            pDC->TextOut(50+(j-1)*60,50+(i-1)*20,s);
            s.Format("%.3lf",B[i][j]);
            pDC->TextOut(350+(j-1)*60,50+(i-1)*20,s);
        }
    MatMultiply(P,A,A);
    MatInverse(Q,B);
    MatInverse(R,A);
    MatMultiply(Y,Q,R);
    MatMultiply(Z,P,Y);
    pDC->TextOut(50,130,"Output P");
    pDC->TextOut(50,230,"Output Q");
    pDC->TextOut(50,330,"Output R");
    pDC->TextOut(50,430,"Output Y");
    pDC->TextOut(400,430,"Output Z");
    for (i=1;i<=N;i++)
        for (j=1;j<=N;j++)
        {
            s.Format("%.3lf",P[i][j]);
            pDC->TextOut(50+(j-1)*60,150+(i-1)*20,s);
            s.Format("%.3lf",Q[i][j]);
            pDC->TextOut(50+(j-1)*60,250+(i-1)*20,s);
            s.Format("%.3lf",R[i][j]);
            pDC->TextOut(50+(j-1)*60,350+(i-1)*20,s);
            s.Format("%.3lf",Y[i][j]);
            pDC->TextOut(50+(j-1)*60,450+(i-1)*20,s);
            s.Format("%.3lf",Z[i][j]);
            pDC->TextOut(350+(j-1)*60,450+(i-1)*20,s);
        }
    for (i=1;i<=N;i++)
        delete A[i],B[i],P[i],Q[i],R[i],Y[i],Z[i];
    delete A,B,P,Q,R,Y,Z;
}
```

Step 11: The next step is to write the code for the function `InputData()`. This function defines the input values for the arrays `a[][]` and `b[]`. The values for the two arrays are defined using a simple assignment as follows:

```
void CCode3AView::InputData(double **a,double **b)
{
     a[1][1]=3; a[1][2]=5; a[1][3]=7;
     a[2][1]=-2; a[2][2]=5; a[2][3]=-4;
     a[3][1]=-4; a[3][2]=2; a[3][3]=1;
     b[1][1]=5; b[1][2]=-3; b[1][3]=4;
     b[2][1]=-4; b[2][2]=2; b[2][3]=-7;
     b[3][1]=2; b[3][2]=1; b[3][3]=5;
}
```

Step 12: The next step is to write the code for the function `MatMultiply()`. This function receives input as the arrays `a[][]` and `b[]`, performs the multiplication using Equation (3.3), and returns the results as the array `c[][]`. `MatMultiply()` is a client function to be called up from `OnDraw()` specifically for computing the multiplication. For example, `MatMultiply(P,A,A)` causes matrix `A[][]` from `OnDraw()` to be passed as `a[][]` and `b[]` in `MatMultiply()`. These values are executed in `MatMultiply()` and returned as `c[][]` which is received by `OnDraw()` as `P[][]`. The code in the function `MatMultiply()` is written as follows:

```
void CCode3AView::MatMultiply(double **c,double **a,double **b)
{
     int i,j,k;
     for (i=1;i<=N;i++)
          for (j=1;j<=N;j++)
          {
               c[i][j]=0;
               for (k=1;k<=N;k++)
                    c[i][j]+=a[i][k]*b[k][j];
          }
}
```

Step 13: Write the code for the function `MatInverse()`. This function computes the inverse of a matrix. The function takes the input array as `a[][]` and produces the results as `x[][]`, according to Equations (3.10) and (3.11). The function is written as follows:

```
void CCode3AView::MatInverse(double **x,double **a)
{
     int i,j,k;
     double Sum,m;
     double **b;

     b=new double *[N+1];
     for (i=1;i<=N;i++)
          b[i]=new double [N+1];
```

```
        for (i=1;i<=N;i++)      // form an identity matrix B
        for (j=1;j<=N;j++)
                {
                        b[i][j]=0;
                        if (i==j)
                                b[i][j]=1;
                }
        for (k=1;k<=N-1;k++)   // perform row operations on A
        for (i=k+1;i<=N;i++)
                {
                        m=a[i][k]/a[k][k];
                        for (j=1;j<=N;j++)
                        {
                                a[i][j]-=m*a[k][j];
                                b[i][j]-=m*b[k][j];
                        }
                }
        for (i=N;i>=1;i–)       // perform backstitutions on X
                for (j=1;j<=N;j++)
                {
                        Sum=0;
                        x[i][j]=0;
                        for (k=i+1;k<=N;k++)
                                Sum += a[i][k]*x[k][j];
                        x[i][j]=(b[i][j]-Sum)/a[i][i];
                }
}
```

Code3A is now complete and ready to be run to produce the desired results. It is easy to extend the method for solving any matrix problem involving multiplication and the inverse of a matrix from Code3A. For example, a problem such as $Z = A^2B^{-1} - B^4A^{-1}B^2 + A^2B$ is just a simple extension by adding a new function for adding two matrices (which also applies to substraction). The overall solution to this problem is obtained by breaking down the problem into several components, then solving each component one by one by calling up its respective function, similar to the steps shown in Code3A. The details are left to the reader as an exercise.

3.3 SYSTEM OF LINEAR EQUATIONS PROBLEM REVISITED

In Chapter 2, two interfaces were presented for creating a dialog window with regard to solving the problem involving a system of linear equations. The two methods are based on the non-Wizard approach. We look at this problem again, but this time we apply the Wizard approach with the document/view architecture. The work continues from the matrix operations problem discussed in the last section.

Code3B: Solving the SLE Problem Using Wizard

We extend the work in Code3A by adding an interface for solving the problem of a system of linear equations discussed in Chapter 2. This time we apply the document/view architecture in developing this application using Wizard. The project is named Code3B and the expected output is shown in Figure 3.11. It consists of the main window and a dialog window. The main window is generated from the blank application using the same steps as in Code3A. A menu with item *Our Work* is added in this application to activate the dialog window. The dialog window appears when the user selects item *Our Work* and subitem *Code3B* from the menu. This prompts the user to key in the input values in the edit boxes of the dialog window. Once the input is completed, the results can be obtained by clicking the *Compute* push button.

The dialog window shown in Figure 3.11 is created from a new class called `CGaussDlg`. For relevancy, this new class has been designed in such a way that it retains most of the code in project Code2C. The dialog window has been assigned the id `IDD_GAUSSDLG`. This window has been designed using the visual tools in the resource file, as discussed in Code2C.

We discuss the Wizard approach in developing the application in Code3B. The steps are more complicated here than in the previous application as the present application involves a few additional classes and the use of several resources.

Step 1: Start the project by selecting *New Project* and *MFC Class*, just as in the previous application. Name the project Code3B. Repeat steps 1–12 in Code3A. This generates the main classes, as shown in Figure 3.12.

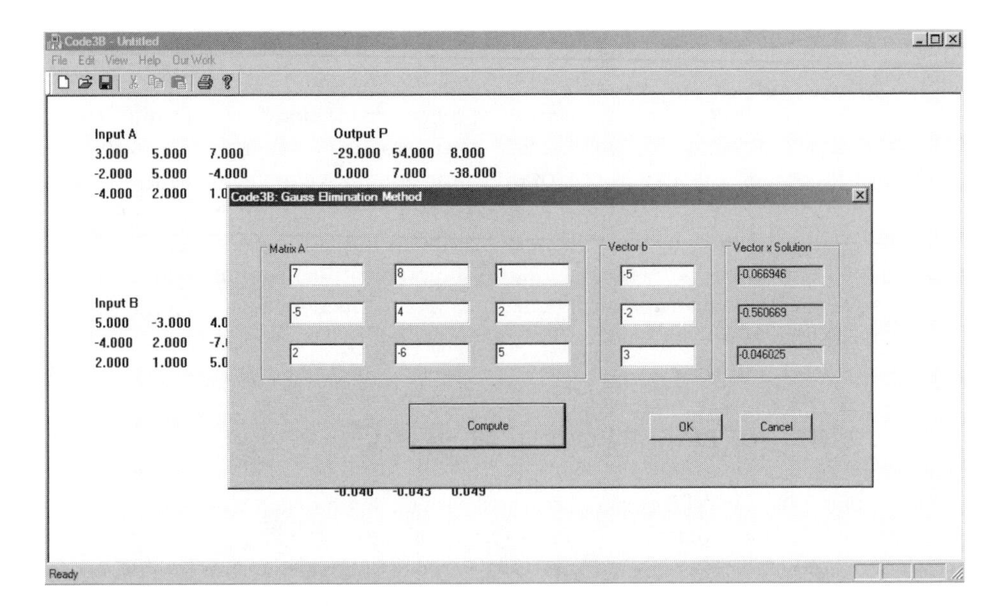

Figure 3.11 Output from Code3B.

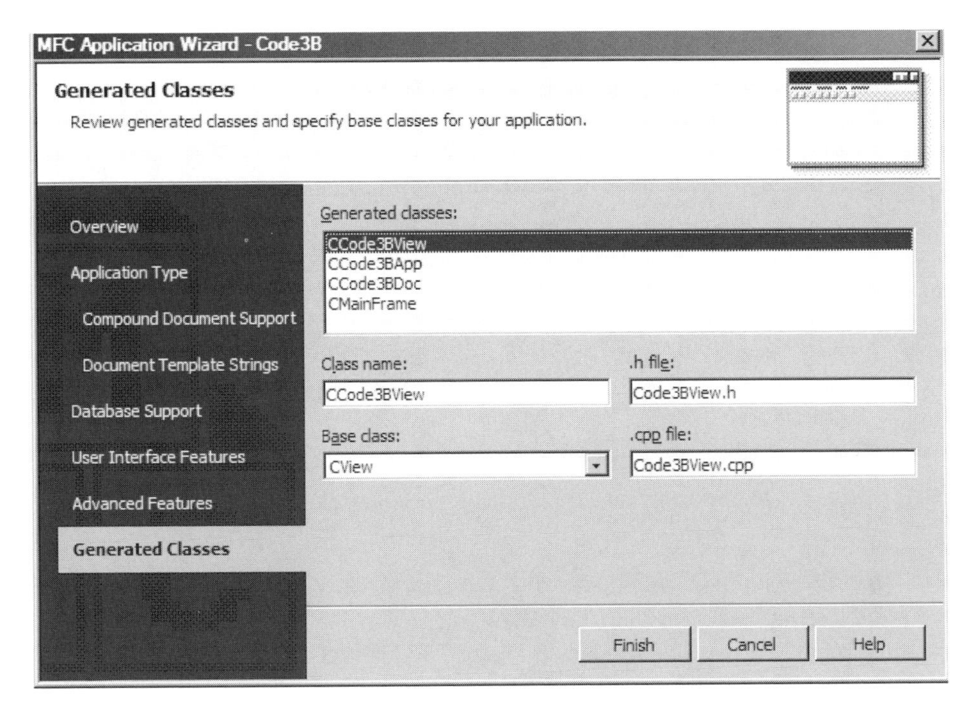

Figure 3.12 Main classes generated in Code3B.

Step 2: Add a dialog window into the resource file Code3B.rc and assign to this window the id `IDD_GAUSSDLG`. This is shown in Figure 3.13.

Step 3: Add the edit boxes, static boxes, and the *Compute* push button into the dialog window using the same procedures as in Code2C discussed earlier. We should be getting a nice interface as shown in Figure 3.14.

Figure 3.13 Resource View showing the new dialog window `IDD_GAUSSDLG`.

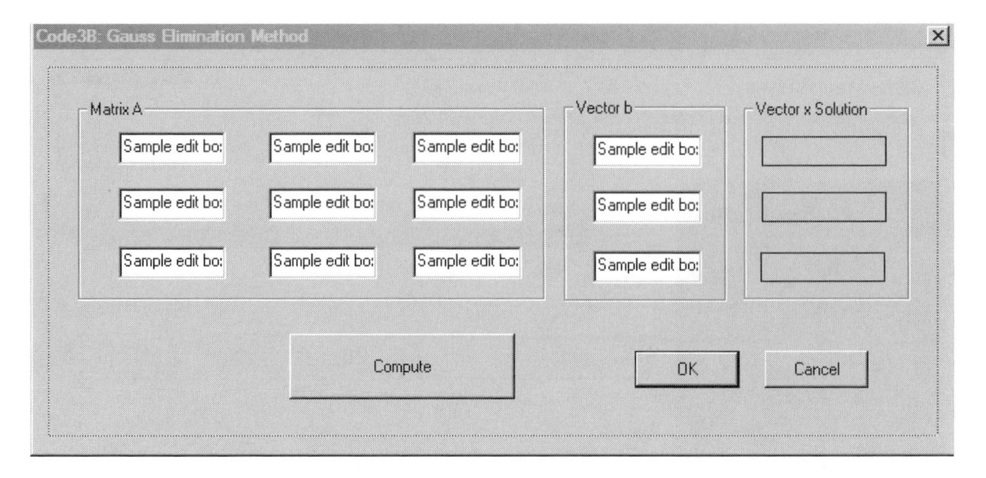

Figure 3.14 Adding the resources into the dialog window.

Step 4: Add the new class `CGaussDlg` by first highlighting the class `CCode3Bview` in the Class View, then choosing *Project* and *Add Class* from the menu. The window in Figure 3.15 appears. Enter the class name as `CGaussDlg` and the base class as `CDialog`. The dialog id `IDD_GAUSS-DLG` is automatically assigned as it is the only dialog window that has been created so far. The files `GaussDlg.h` and `GaussDlg.cpp` are also created automatically. The project Code3B now has a list of classes and files as shown in Figure 3.16.

Step 6: The next step is to add new items to the present menu and link these items with the dialog window. The menu has a default id `IDR_MAINFRAME`. From the Resource View choose Menu and double-click `IDR_MAINFRAME` (see Figure 3.17).

Step 7: Add a new item in the menu called *Our Work* and a subitem called *Code3B*, as shown in Figure 3.18. Right-click on the subitem *Code3B* and select *Add Event Handler* from the menu to add an event handler to *Code3B*. The event handler provides a link from the subitem to the class `CCode3Bview` (see Figure 3.19).

Step 8: The event-handler dialog window appears with most of its entries automatically assigned, as shown in Figure 3.20. The subitem menu has the id `ID_OURWORK_CODE3B` and it will call up the function `OnOurwork-Code3B()` when invoked.

Step 9: An event handler is to be added to the *Compute* push button so that it responds to the event by calling the appropriate function. This is achieved by opening the dialog window and right-clicking on the *Compute* push button, as shown in Figure 3.21. Choose *Event Handler* from the menu.

Step 10: The dialog window for this event appears, as shown in Figure 3.22. The *Compute* push button has been automatically assigned with the id `IDC_BUT-`

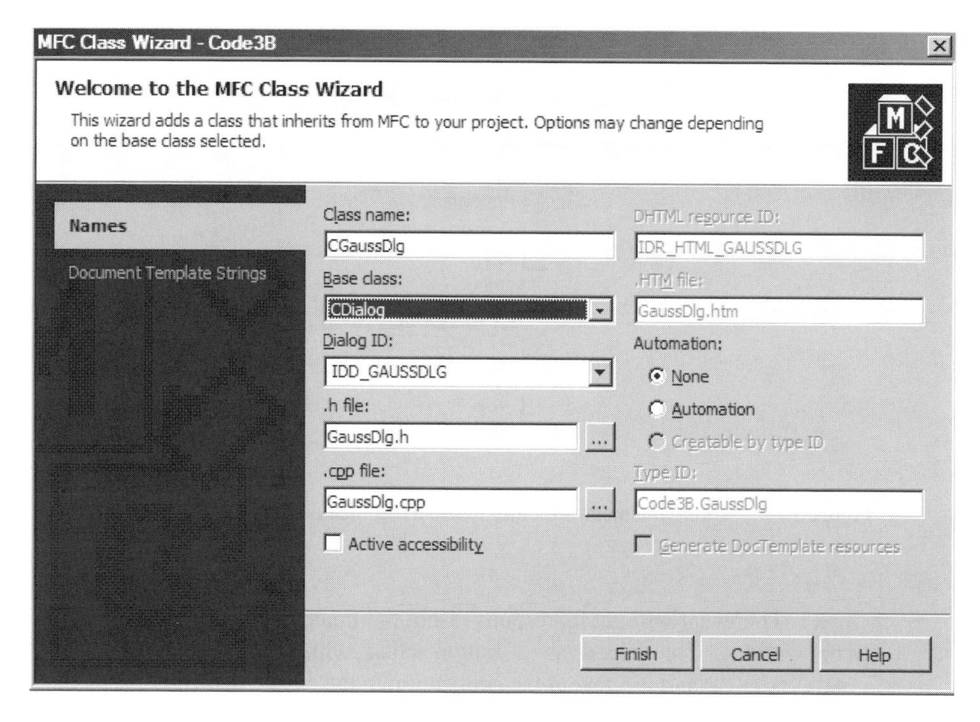

Figure 3.15 Class creation dialog window.

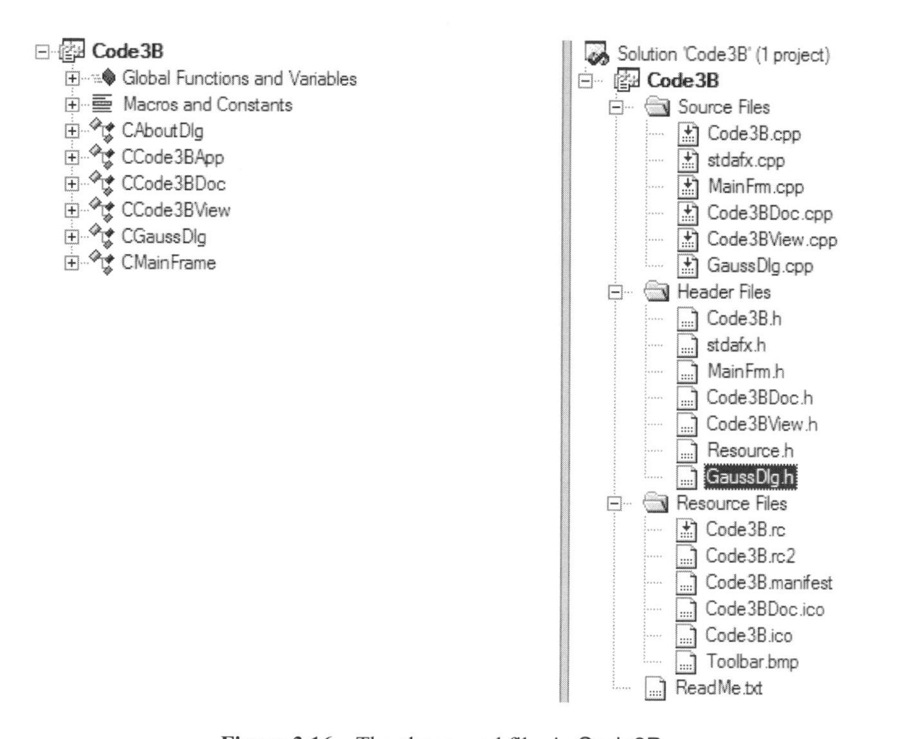

Figure 3.16 The classes and files in Code3B.

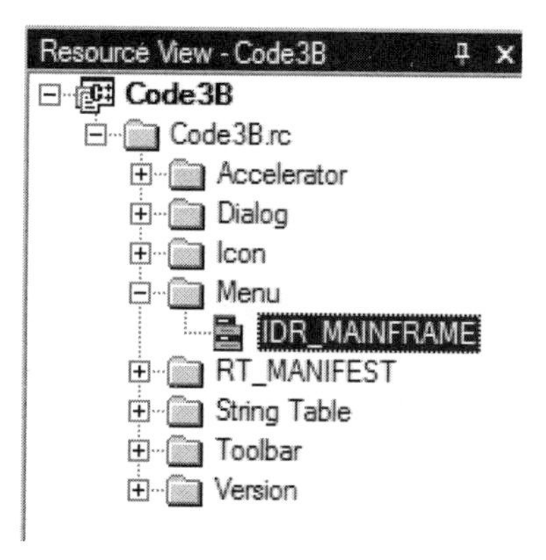

Figure 3.17 Editing the menu structure to add a resource on `IDR_MAINFRAME`.

`TON1`. The event will get its response from the function `OnBnCompute()`. The event is recognized as a button click with the message handler `BN_CLICKED` and it refers to the application in the class `CGaussDlg`.

Step 11: All the events for the application have been added so far. We now create the function `PGauss()`, which will solve the system of linear equations problem. In the Class View, highlight the class `CGaussDlg`. Choose *Project* and *Add Function* from the menu. Enter the information as shown in Figure 3.23 and click Finish to complete.

Step 12: The next few steps involve code entry into some files, which will be performed manually. We start with the file CGaussDlg.h. Open this file and enter the lines marked in the shaded area, as follows:

```
#pragma once
#define N 3

// CGaussDlg dialog

class CGaussDlg : public CDialog
```

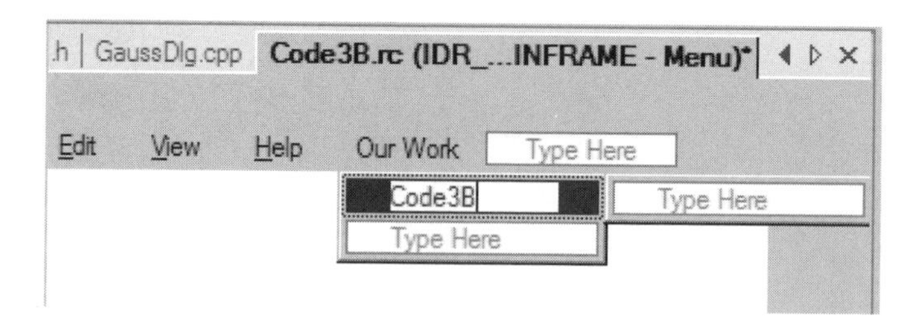

Figure 3.18 Creating a new sub-item in the menu called *Code3B*.

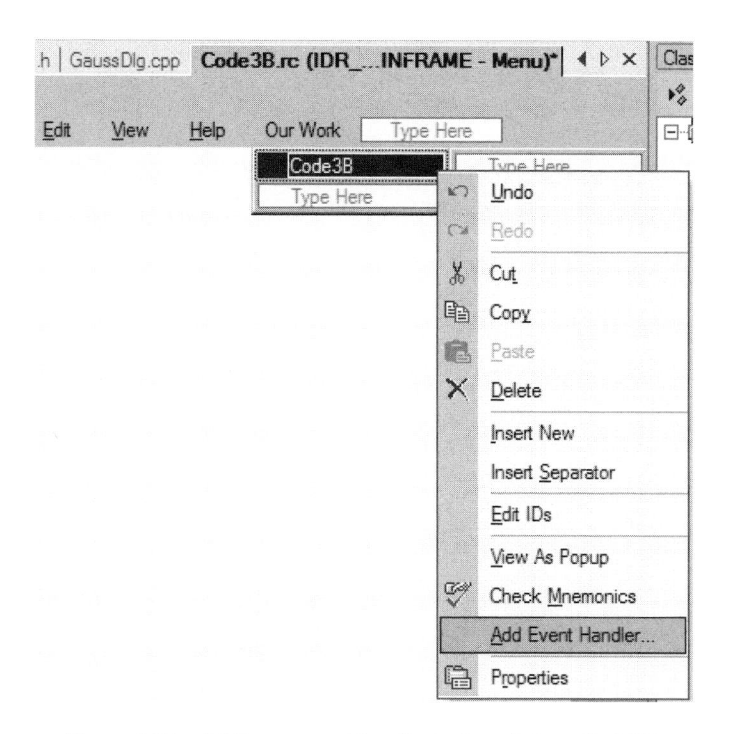

Figure 3.19 Adding an event handler to the sub-item *Code3B*.

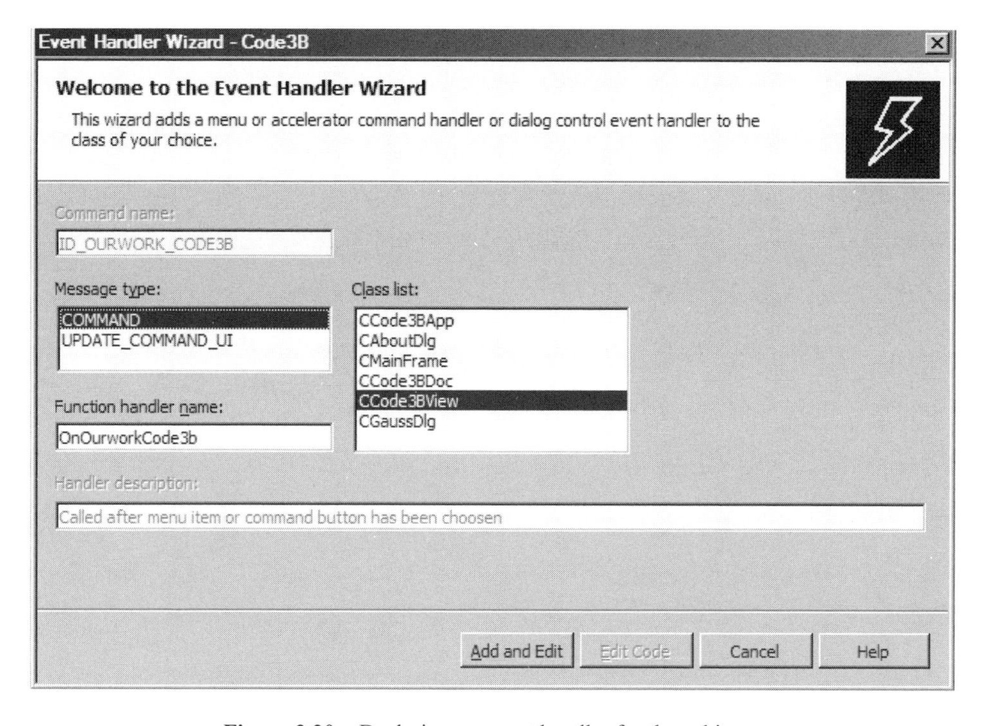

Figure 3.20 Declaring an event handler for the subitem.

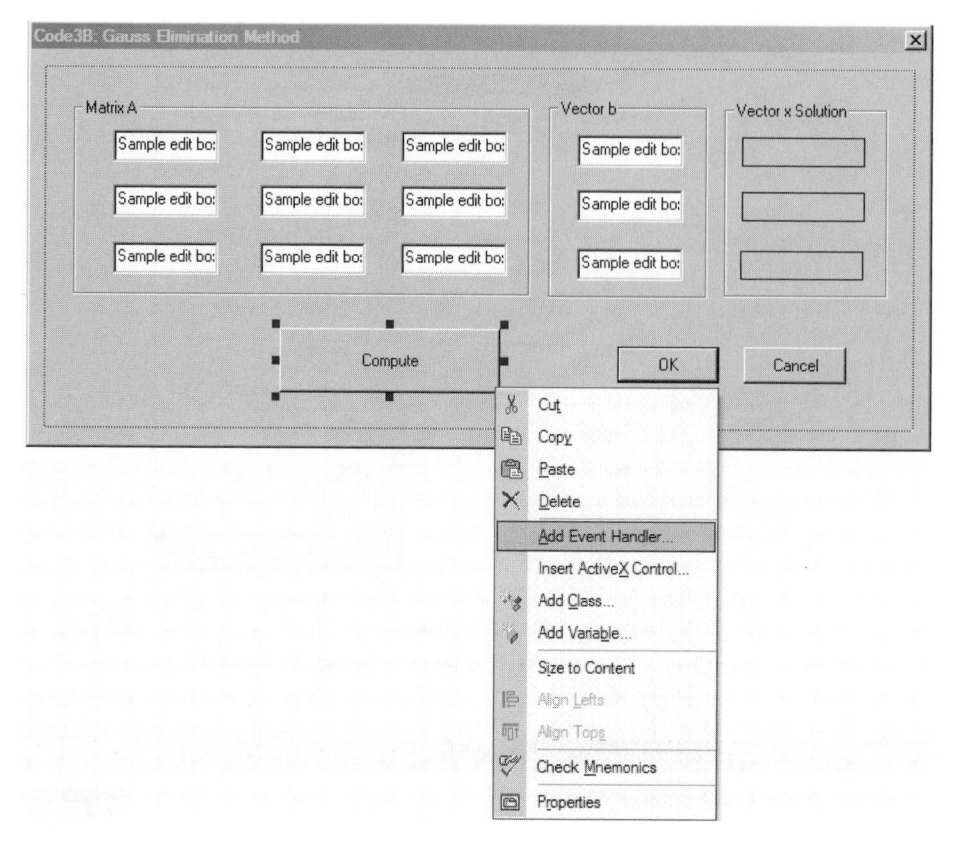

Figure 3.21 To activate the *Compute* button.

```
{
       DECLARE_DYNAMIC(CGaussDlg)

public:
       CGaussDlg(CWnd* pParent = NULL);  // standard constructor
       virtual ~CGaussDlg();

// Dialog Data
       enum { IDD = IDD_GAUSSDLG };

protected:
       virtual void DoDataExchange(CdataExchange* pDX); // DDX/DDV support

       DECLARE_MESSAGE_MAP()
public:
       afx_msg void OnBnCompute();
       void PGauss(void);
       double **a,*b,*x;
       CString **sa,*sb,*sx;
};
```

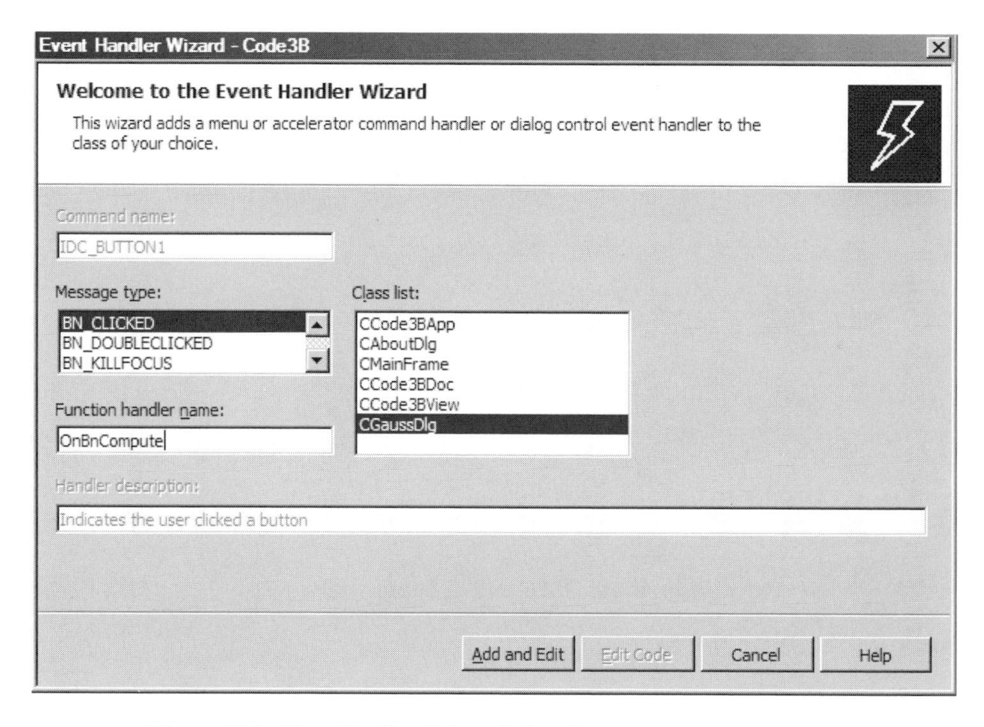

Figure 3.22 Event handler dialog window for the *Compute* push button.

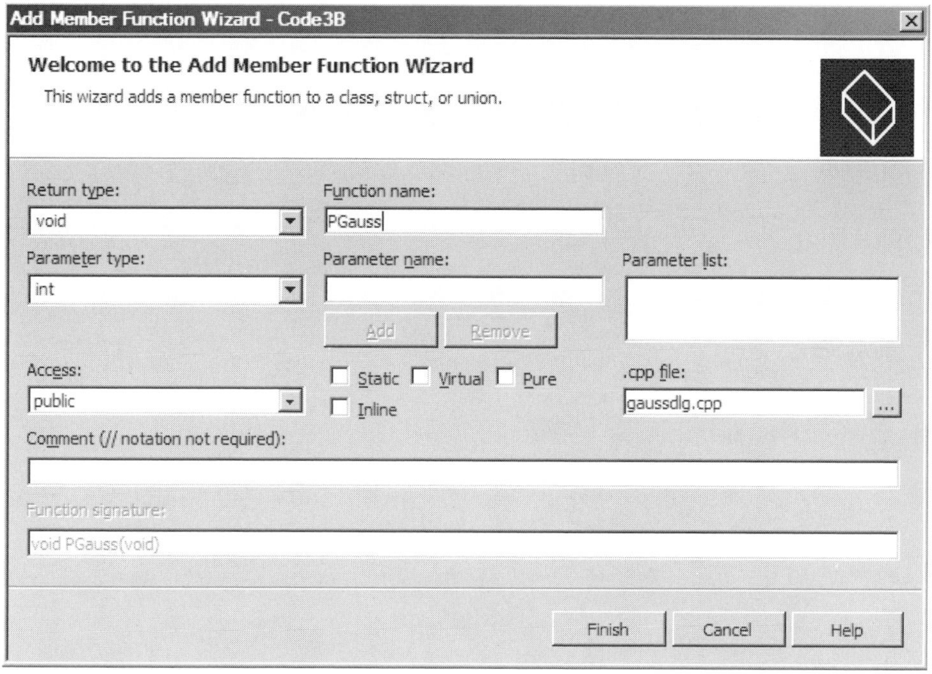

Figure 3.23 Add the function `PGauss()`.

Step 13: Now open the file GaussDlg.cpp and add the lines shown in the shaded area below into the constructor and the destructor of the class `CGaussDlg`:

```
CGaussDlg::CGaussDlg(CWnd* pParent /*=NULL*/)
      : CDialog(CGaussDlg::IDD, pParent)
{
      a=new double *[N+1];
      b=new double [N+1];
      x=new double [N+1];
      sa=new CString *[N+1];
      sb=new CString [N+1];
      sx=new CString [N+1];
      for (int j=1;j<=N;j++)
      {
            a[j]=new double [N+1];
            sa[j]=new CString [N+1];
      }
}

CGaussDlg::~CGaussDlg()
{
      for (int j=1;j<=N;j++)
            delete a[j],sa[j];
      delete a,sa,b,sb,x,sx;
}
```

Enter the lines of code for the function `PGauss()` as marked in the shaded area:

```
void CGaussDlg::PGauss()
{
      int i,j,k;
      double m,Sum;

      // Perform row operations
      for (k=1;k<=N-1;k++)
      for (i=k+1;i<=N;i++)
            {
                  m=a[i][k]/a[k][k];
                  for (j=1;j<=N;j++)
                        a[i][j]=a[i][j]-m*a[k][j];
                  b[i]=b[i]-m*b[k];
            }

      // Perform back substitutions
      for (i=N;i>=1;i—)
```

```
        {
        Sum=0;
                x[i]=0;
                for (j=i;j<=N;j++)
                        Sum +=a[i][j]*x[j];
                x[i]=(b[i]-Sum)/a[i][i];
        }
}
```

And, finally, add the code for creating the modal window into the function `OnOurworkCode3B()`. This code responds to the *Compute* push button event by activating the dialog window for the application.

```
void CCode3Bview::OnOurworkCode3B()
{
        CGaussDlg Dlg(this);
        int nResponse=Dlg.DoModal();
}
```

The coding for the application is now complete. It can now be compiled, linked, and run to produce the desired results.

Code3B: Discussion

The Wizard approach is a convenient way of developing an application especially for big projects. Wizard provides a full guided approach to developing an application and this reduces the burden of programming. Not only that, Wizard also provides ways to explore the resources available in MFC through the friendly dialog windows and menus. A professional-looking application can be generated through the use of the MFC Wizard, which offers a better presentation, suitable for commercial marketing.

However, there is a steep learning curve in using Wizard. It may not look easy to a beginner, who must know exactly the program flow and the steps necessary before using the facilities in Wizard. The path to the successful implementation of Wizard may be disrupted if the steps are not implemented well.

Code3B produces an interesting interface for creating an application. The project illustrates the use of dialog windows and menus for solving a numerical problem. We discussed a system of linear equations that is very small in size. In real applications, a system of linear equations may be very large so that the input will not be in the form of edit boxes. The document/view architecture provides a way for the input to be made in the form of one or more files. The document part of the architecture reads the input files and makes this data available for further processing and viewing through a process known as serialization. We will not be discussing this topic in this book as our main priority is to discuss ways for solving several numerical simulation topics using MFC. Interested readers can refer to several books specializing in Visual C++ such as Shepherds [3] to explore this topic further.

3.4 SUMMARY AND CONCLUSION

Matrix algebra is one of the most important elements in scientific computing. Basically, this topic involves the normal tools of algebra: addition, substraction, multiplication, and division. But what makes matrix algebra different from other problems is the presence of arrays and vectors. As the size of the arrays and vectors become large, the task of solving a problem involving matrix algebra becomes very difficult. In this case, a computer becomes an indispensable tool for solving these problems.

A powerful computer alone is not enough to provide the solution to several difficult matrix-related problems. More important than that is the way the computer is used to solve these problems. It is the software that provides the ultimate solution to the problems. Good software is capable of handling matrices, managing the resources in the computer, and providing a systematic way for solving the problems.

We discussed two problems in matrix algebra involving the use of Wizard in this chapter. The first problem describes matrix operations involving the use of arrays and data passing between the functions. We presented a linear approach in which a given problem was broken down into tokens, then solved each one of them based on the fundamental rules of algebra. We illustrated the data-passing mechanism between functions, which has the advantage of maximizing the use of local variables. The use of local variables, rather than global variables, makes the program more modular so that the role of each function can be further enhanced.

We also revisited the problem of solving a system of linear equations. Several improvements were made to the interface presented earlier in Chapter 2 through the use of Wizard. Wizard may look difficult to beginners, but this just involves another learning process. Many benefits are obtained through Wizard, which makes it an indispensable tool for a programmer.

BIBLIOGRAPHY

1. R. L. Burden and J. D. Faires, *Numerical Analysis,* Brooks Cole, 2000.
2. S. C. Chapra and R. Canale, *Numerical Methods for Engineers,* McGraw-Hill, 2001.
3. G. Shepherd, *Programming with Microsoft Visual C++.Net,* Microsoft Press, 2003.

CODE LISTINGS

Code3AView

```
// Code3AView.h : interface of the CCode3AView class
//
#define N 3
```

```
#pragma once


class CCode3AView : public CView
{
protected: // create from serialization only
     CCode3AView();
     DECLARE_DYNCREATE(CCode3AView)

// Attributes
public:
     CCode3ADoc* GetDocument() const;

// Operations
public:

// Overrides
     public:
     virtual void OnDraw(CDC* pDC);  // overridden to draw this view
virtual BOOL PreCreateWindow(CREATESTRUCT& cs);
protected:
     virtual BOOL OnPreparePrinting(CPrintInfo* pInfo);
     virtual void OnBeginPrinting(CDC* pDC, CPrintInfo* pInfo);
     virtual void OnEndPrinting(CDC* pDC, CPrintInfo* pInfo);

// Implementation
public:
     virtual ~CCode3AView();
#ifdef _DEBUG
     virtual void AssertValid() const;
     virtual void Dump(CDumpContext& dc) const;
#endif

protected:

// Generated message map functions
protected:
     DECLARE_MESSAGE_MAP()
public:
     void InputData(double ** a, double ** b);
     void MatMultiply(double ** c, double ** a, double ** b);
     void MatInverse(double ** x, double ** a);
};

#ifndef _DEBUG  // debug version in Code3AView.cpp
inline CCode3ADoc* CCode3AView::GetDocument() const
   { return reinterpret_cast<CCode3ADoc*>(m_pDocument); }
#endif

// Code3AView.cpp : implementation of the CCode3AView class
//

#include "stdafx.h"
#include "Code3A.h"
```

```
#include "Code3ADoc.h"
#include "Code3AView.h"

#ifdef _DEBUG
#define new DEBUG_NEW
#endif


// CCode3AView

IMPLEMENT_DYNCREATE(CCode3AView, CView)

BEGIN_MESSAGE_MAP(CCode3AView, CView)
     // Standard printing commands
     ON_COMMAND(ID_FILE_PRINT, CView::OnFilePrint)
     ON_COMMAND(ID_FILE_PRINT_DIRECT, CView::OnFilePrint)
     ON_COMMAND(ID_FILE_PRINT_PREVIEW, CView::OnFilePrintPreview)
END_MESSAGE_MAP()

// CCode3AView construction/destruction

CCode3AView::CCode3AView()
{
     // TODO: add construction code here

}

CCode3AView::~CCode3AView()
{
}

BOOL CCode3AView::PreCreateWindow(CREATESTRUCT& cs)
{
     // TODO: Modify the Window class or styles here by modifying
     //  the CREATESTRUCT cs

     return CView::PreCreateWindow(cs);
}

// CCode3AView drawing

void CCode3AView::OnDraw(CDC* pDC)
{
     CCode3ADoc* pDoc = GetDocument();
     ASSERT_VALID(pDoc);

     // TODO: add draw code for native data here
     CString s;
     int i,j;
     double **A,**B,**P,**Q,**R,**Y,**Z;

     A=new double *[N+1];
     B=new double *[N+1];
```

```
    P=new double *[N+1];
    Q=new double *[N+1];
    R=new double *[N+1];
    Y=new double *[N+1];
    Z=new double *[N+1];
    for (i=1;i<=N;i++)
    {
        A[i]=new double [N+1];
        B[i]=new double [N+1];
        P[i]=new double [N+1];
        Q[i]=new double [N+1];
        R[i]=new double [N+1];
        Y[i]=new double [N+1];
        Z[i]=new double [N+1];
    }
    InputData(A,B);
    pDC->TextOut(50,30,"Input A");
    pDC->TextOut(50,200,"Input B");
    for (i=1;i<=N;i++)
        for (j=1;j<=N;j++)
        {
            s.Format("%.3lf",A[i][j]);
            pDC->TextOut(50+(j-1)*60,50+(i-1)*20,s);
            s.Format("%.3lf",B[i][j]);
            pDC->TextOut(50+(j-1)*60,220+(i-1)*20,s);
        }
    MatMultiply(P,A,A);
    MatInverse(Q,B);
    MatInverse(R,A);
    MatMultiply(Y,Q,R);
    MatMultiply(Z,P,Y);
    pDC->TextOut(300,30,"Output P");
    pDC->TextOut(300,130,"Output Q");
    pDC->TextOut(300,230,"Output R");
    pDC->TextOut(300,330,"Output Y");
    pDC->TextOut(550,130,"Output Z");
    for (i=1;i<=N;i++)
        for (j=1;j<=N;j++)
        {
            s.Format("%.3lf",P[i][j]);
            pDC->TextOut(300+(j-1)*60,50+(i-1)*20,s);
            s.Format("%.3lf",Q[i][j]);
            pDC->TextOut(300+(j-1)*60,150+(i-1)*20,s);
            s.Format("%.3lf",R[i][j]);
            pDC->TextOut(300+(j-1)*60,250+(i-1)*20,s);
            s.Format("%.3lf",Y[i][j]);
            pDC->TextOut(300+(j-1)*60,350+(i-1)*20,s);
            s.Format("%.3lf",Z[i][j]);
            pDC->TextOut(550+(j-1)*60,150+(i-1)*20,s);
        }
    for (i=1;i<=N;i++)
        delete A[i],B[i],P[i],Q[i],R[i],Y[i],Z[i];
    delete A,B,P,Q,R,Y,Z;
}
```

```
// CCode3AView printing

BOOL CCode3AView::OnPreparePrinting(CPrintInfo* pInfo)
{
     // default preparation
     return DoPreparePrinting(pInfo);
}

void CCode3AView::OnBeginPrinting(CDC* /*pDC*/, CPrintInfo* /*pInfo*/)
{
     // TODO: add extra initialization before printing
}

void CCode3AView::OnEndPrinting(CDC* /*pDC*/, CPrintInfo* /*pInfo*/)
{
     // TODO: add cleanup after printing
}


// CCode3AView diagnostics

#ifdef _DEBUG
void CCode3AView::AssertValid() const
{
     CView::AssertValid();
}

void CCode3AView::Dump(CDumpContext& dc) const
{
     CView::Dump(dc);
}

CCode3ADoc* CCode3AView::GetDocument() const // non-debug version is inline
{
     ASSERT(m_pDocument->IsKindOf(RUNTIME_CLASS(CCode3ADoc)));
     return (CCode3ADoc*)m_pDocument;
}
#endif //_DEBUG


// CCode3AView message handlers

void CCode3AView::InputData(double ** a, double ** b)
{
     a[1][1]=3; a[1][2]=5; a[1][3]=7;
     a[2][1]=-2; a[2][2]=5; a[2][3]=-4;
     a[3][1]=-4; a[3][2]=2; a[3][3]=1;
     b[1][1]=5; b[1][2]=-3; b[1][3]=4;
     b[2][1]=-4; b[2][2]=2; b[2][3]=-7;
     b[3][1]=2; b[3][2]=1; b[3][3]=5;
}

void CCode3AView::MatMultiply(double ** c, double ** a, double ** b)
```

```
{
    int i,j,k;
    for (i=1;i<=N;i++)
        for (j=1;j<=N;j++)
        {
            c[i][j]=0;
            for (k=1;k<=N;k++)
            c[i][j]+=a[i][k]*b[k][j];
        }
}

void CCode3AView::MatInverse(double ** x, double ** a)
{
    int i,j,k;
    double Sum,m;
     double **b;

     b=new double *[N+1];
     for (i=1;i<=N;i++)
          b[i]=new double [N+1];
    for (i=1;i<=N;i++)         // form an identity matrix B
        for (j=1;j<=N;j++)
          {
              b[i][j]=0;
              if (i==j)
                  b[i][j]=1;
          }
    for (k=1;k<=N-1;k++)       // perform row operations on A
        for (i=k+1;i<=N;i++)
          {
              m=a[i][k]/a[k][k];
              for (j=1;j<=N;j++)
              {
                  a[i][j]-=m*a[k][j];
                  b[i][j]-=m*b[k][j];
              }
          }
    for (i=N;i>=1;i–)          // perform backstitutions on X
        for (j=1;j<=N;j++)
          {
              Sum=0;
              x[i][j]=0;
              for (k=i+1;k<=N;k++)
                  Sum += a[i][k]*x[k][j];
              x[i][j]=(b[i][j]-Sum)/a[i][i];
          }
}
```

Code3BView

```
// Code3BView.h : interface of the CCode3BView class
//

#pragma once
```

```
class CCode3BView : public CView
{
protected: // create from serialization only
      CCode3BView();
      DECLARE_DYNCREATE(CCode3BView)

// Attributes
public:
      CCode3BDoc* GetDocument() const;

// Operations
public:

// Overrides
      public:
      virtual void OnDraw(CDC* pDC);  // overridden to draw this view
virtual BOOL PreCreateWindow(CREATESTRUCT& cs);
protected:
      virtual BOOL OnPreparePrinting(CPrintInfo* pInfo);
      virtual void OnBeginPrinting(CDC* pDC, CPrintInfo* pInfo);
      virtual void OnEndPrinting(CDC* pDC, CPrintInfo* pInfo);

// Implementation
public:
      virtual ~CCode3BView();
#ifdef _DEBUG
      virtual void AssertValid() const;
      virtual void Dump(CDumpContext& dc) const;
#endif

protected:

// Generated message map functions
protected:
      DECLARE_MESSAGE_MAP()
public:
      afx_msg void OnOurworkCode3b();
};

#ifndef _DEBUG  // debug version in Code3BView.cpp
inline CCode3BDoc* CCode3BView::GetDocument() const
    { return reinterpret_cast<CCode3BDoc*>(m_pDocument); }
#endif

// Code3BView.cpp : implementation of the CCode3BView class
//

#include "stdafx.h"
#include "Code3B.h"

#include "Code3BDoc.h"
#include "Code3BView.h"
#include "GaussDlg.h"
```

```
#ifdef _DEBUG
#define new DEBUG_NEW
#endif


// CCode3BView

IMPLEMENT_DYNCREATE(CCode3BView, CView)

BEGIN_MESSAGE_MAP(CCode3BView, CView)
	// Standard printing commands
	ON_COMMAND(ID_FILE_PRINT, CView::OnFilePrint)
	ON_COMMAND(ID_FILE_PRINT_DIRECT, CView::OnFilePrint)
	ON_COMMAND(ID_FILE_PRINT_PREVIEW, CView::OnFilePrintPreview)
	ON_COMMAND(ID_OURWORK_CODE3B, OnOurworkCode3b)
END_MESSAGE_MAP()

// CCode3BView construction/destruction

CCode3BView::CCode3BView()
{
	// TODO: add construction code here

}

CCode3BView::~CCode3BView()
{
}

BOOL CCode3BView::PreCreateWindow(CREATESTRUCT& cs)
{
	// TODO: Modify the Window class or styles here by modifying
	//  the CREATESTRUCT cs

	return CView::PreCreateWindow(cs);
}

// CCode3BView drawing

void CCode3BView::OnDraw(CDC* pDC)
{
	CCode3BDoc* pDoc = GetDocument();
	ASSERT_VALID(pDoc);

	// TODO: add draw code for native data here
}


// CCode3BView printing

BOOL CCode3BView::OnPreparePrinting(CPrintInfo* pInfo)
{
	// default preparation
	return DoPreparePrinting(pInfo);
}
```

```
void CCode3BView::OnBeginPrinting(CDC* /*pDC*/, CPrintInfo* /*pInfo*/)
{
      // TODO: add extra initialization before printing
}

void CCode3BView::OnEndPrinting(CDC* /*pDC*/, CPrintInfo* /*pInfo*/)
{
      // TODO: add cleanup after printing
}


// CCode3BView diagnostics

#ifdef _DEBUG
void CCode3BView::AssertValid() const
{
      CView::AssertValid();
}

void CCode3BView::Dump(CDumpContext& dc) const
{
      CView::Dump(dc);
}

CCode3BDoc* CCode3BView::GetDocument() const // non-debug version is inline
{
      ASSERT(m_pDocument->IsKindOf(RUNTIME_CLASS(CCode3BDoc)));
      return (CCode3BDoc*)m_pDocument;
}
#endif //_DEBUG


// CCode3BView message handlers

void CCode3BView::OnOurworkCode3b()
{
      CGaussDlg Dlg(this);
      int nResponse=Dlg.DoModal();
}
```

CHAPTER 4

DIFFERENTIAL EQUATIONS PROBLEMS

4.1 DIFFERENTIAL EQUATIONS

Differential equations describe the behavior of a system that arises from its static and dynamic properties. For example, the flow of fluid, the dissipation of heat in an engine, and the vibrations in a moving aircraft, are described by one or more differential equations. The differential equations that describe this behavior are autonomously formulated as a mathematical model. A good mathematical model consists of a complete set of equations that correctly match the given problem.

There are two approaches to solving a differential equation problem: analytical and numerical. The analytical approach to a differential equation is based on the fundamental properties of the equation that normally lead to accurate solutions. Numerical methods complement the analytical approach by providing approximate solutions that strongly contribute to their implementation on the computer. Numerical methods are very useful, especially in cases where the analytical solutions are difficult or impossible to find.

Differential equations are classified as either ordinary differential equations (ODE) or partial differential equations (PDE). An ordinary differential equation describes a function of a single independent variable that contains only the variable and the ordinary derivatives of the function with respect to the variable. In comparison, a partial differential equation is an equation that has two or more independent variables and derivatives that refer to one variable in that function.

In this chapter, we explore a few problems commonly discussed in the undergraduate syllabus, one involving ordinary differential equations and another one involving partial differential equations. We solve these problems and display the solutions visually using the tools available in the MFC libraries. The solutions shown on these selected topics should provide a good foundation and understanding for the rest of the work involving problems in differential equations.

4.2 ORDINARY DIFFERENTIAL EQUATIONS

In its general form, an ordinary differential equation of order n with one dependence x is expressed as follows:

$$f\left(x, y(x), \frac{dy}{dx}, \frac{d^2y}{dx^2}, \ldots, \frac{d^{(n)}}{dx^n}\right) = 0 \tag{4.1}$$

In Equation (4.1), dy/dx is also written as y', d^2y/dx^2 as y'', and so on. A first-order ordinary differential equation problem has the following general form:

$$f(x, y, y') = 0 \tag{4.2}$$

A first-order ordinary differential equation of the form of Equation (4.2) is said to be linear if it can be written as follows:

$$a(x)y' + b(x)y = c(x) \tag{4.3}$$

where $a(x)$, $b(x)$, and $c(x)$ are functions of x. The corresponding second-order ordinary differential equation is stated as follows:

$$f(x, y, y', y'') = 0 \tag{4.4}$$

A linear second-order differential equation has the following form:

$$a(x)y'' + b(x)y' + c(x)y = d(x) \tag{4.5}$$

where $d(x)$ is another function of x.

The first-order ordinary differential equation problem requires only one initial value for its solution. Hence, the problem is called an *initial value problem,* stated as follows: Given a function $f(x, y)$ continuous and differentiable in $x_0 < x < x_N$ and $y_0 < y < y_N$, the initial value problem in a first-order ordinary differential problem consists of finding the values of y in $x_0 < x < x_N$ from the equation $dy/dx = f(x, y)$ with an initial value condition given by $y(x_0) = y_0$.

An important foundation in many mathematical problems is the Taylor series, stated as follows:

$$y_{i+1} \approx y_i + \frac{h}{1!} y_i' + \frac{h^2}{2!} y_i'' + \ldots + \frac{h^n}{n!} y_i^{(n)} \tag{4.6}$$

Equation (4.6) is an expansion of order n of the Taylor series with an incremental value of h. The Taylor series is used primarily to approximate a function into the form of a series. Many methods for solving the first-order ordinary differential equation problems have their root in the Taylor series. They include the Taylor series method, Euler, Heun, Runge–Kutta, and Adams–Bashforth multistep methods.

In this section, we discuss the development of a user-friendly interface solution using the fourth-order Runge–Kutta method. The idea from this method can easily be incorporated into solving a similar problem using the other methods.

Fourth-Order Runge–Kutta Method (RK4)

The fourth-order Runge–Kutta method is one of the most commonly used techniques for solving the first-order initial value problem. The Runge–Kutta method is derived from the Taylor series (refer to Burden and Faires [1] for details) and it has many variations in the form of the nth-order expansion of the series.

We discuss the fourth-order Runge–Kutta method. As the order suggests, RK4 is based on the fourth-order Taylor series expansion with the terms k_1, k_2, k_3, and k_4 given as follows:

$$k_1 = hf(x_i, y_i) \tag{4.7a}$$

$$k_2 = hf(x_i + h/2, y_i + k_1/2) \tag{4.7b}$$

$$k_3 = hf(x_i + h/2, y_i + k_2/2) \tag{4.7c}$$

$$k_4 = hf(x_i + h, y_i + k_3) \tag{4.7d}$$

The solution is provided according to the following formula:

$$y_{i+1} = y_i + \tfrac{1}{6}(k_1 + 2k_2 + 2k_3 + k_4) \tag{4.8}$$

RK4 is sequential in nature, where the value of k_2 is dependent on k_1, k_3 on k_2, and so on. The method is summarized in the following algorithm:

Input: $f(x, y)$, x_0, y_0, h, and N
Output: $y_1, y_2, \ldots, y_N$
Steps:

for $i = 1$ to N *// i=iteration number, N=number of intervals*
Compute $k_1 = hf(x_i, y_i)$;
Compute $k_2 = hf(x_i + h/2, y_i + k_1/2)$;
Compute $k_3 = hf(x_i + h/2, y_i + k_2/2)$;
Compute $k_4 = hf(x_i + h, y_i + k_3)$;
Compute $y_{i+1} = y_i + \tfrac{1}{6}(k_1 + 2k_2 + 2k_3 + k_4)$;
Update $x_{i+1} = x_i + h$;

■ Example 1

We discuss one example initial value problem involving the fourth-order Runge–Kutta method as follows:

Solve $\dfrac{dy}{dx} = x^2 \cos y + 1$ for $0 < x \leq 0.2$, given $y(0) = -1$ and the increment $h = \Delta x = 0.1$

Solution

From the given initial values, the x increment is $h = \Delta x = 0.1$, and the discrete form of $f(x, y)$ is $f(x_i, y_i) = x_i^2 \cos y_i + 1$. It follows that:

$$
\begin{aligned}
k_1 &= hf(x_i, y_i) = h(x_i^2 \cos y_i + 1) \\
k_2 &= hf(x_i + h/2, y_i + k_1/2) = h[(x_i + h/2)^2 \cos(y_i + k_1/2) + 1] \\
k_3 &= hf(x_i + h/2, y_i + k_2/2) = h[(x_i + h/2)^2 \cos(y_i + k_2/2) + 1] \\
k_4 &= hf(x_i + h, y_i + k_3) = h[(x_i + h)^2 \cos(y_i + k_3) + 1]
\end{aligned}
$$

Start the iteration at $i = 0$ and the initial values, $x_0 = 0$ and $y_0 = -1$. We get

$$
\begin{aligned}
k_1 &= 0.1(0 + 1) = 0.1 \\
k_2 &= 0.1[0.05^2 \cos(-0.95) + 1] = 0.100145 \\
k_3 &= 0.1[0.05^2 \cos(-0.949927) + 1] = 0.100145 \\
k_4 &= 0.1[0.1^2 \cos(-0.899855) + 1] = 0.100622 \\
\therefore y(0.1) &= y_1 = y_0 + \tfrac{1}{6}[k_1 + 2k_2 + 2k_3 + k_4] = -0.899799
\end{aligned}
$$

Continue with $i = 1$ and the computed values, $x_1 = 0.2$ and $y_1 = -0.899799$:

$$
\begin{aligned}
k_1 &= 0.1[0.1^2 \cos(-0.899799) + 1] = 0.100622 \\
k_2 &= 0.1[0.15^2 \cos(-0.849489) + 1] = 0.101486 \\
k_3 &= 0.1[0.15^2 \cos(-0.849057) + 1] = 0.101487 \\
k_4 &= 0.1[0.2^2 \cos(-0.798313) + 1] = 0.102792 \\
\therefore y(0.2) &= y_2 = y_1 + \tfrac{1}{6}[k_1 + 2k_2 + 2k_3 + k_4] = -0.798240
\end{aligned}
$$

■

Code4A: Small Window for Displaying Large Amounts of Data

We illustrate the fourth-order Runge–Kutta method for solving the first-order initial value problem in the project Code4A. The numerical solution may involve many iterations and this requires a large display area in order to view them. The display comes in the form of a table that clearly shows all the parameters involved as well as the results from the calculations. A user-friendly interface called the list view window is used to display the solution in this application. A *list view window* is a scrollable child window that displays data in the form of a table. A *table* is defined as a set of organized data arranged in columns and rows classified according to the fields and records of the data, respectively. A *field* is a set of data classified according to its category, whereas a *record* is one row of a complete set of data belonging to the item. The list view window has both the horizontal and vertical scroll buttons to allow a large table to be displayed. A set of tables is called a *database,* which forms a component of the database management system. A table provides a two-dimensional visualization of the whole set of data, useful for presenting data and information in applications such as in statistics, numerical analysis, business forecasting, and engineering data analysis.

We discuss an example of a problem requiring a large display area. The project Code4A consists of the solution to the problem in Example 1 using the fourth-order Runge–Kutta method. We expand the domain of the problem in this example as follows:

Solve $\frac{dy}{dx} = x^2 \cos y + 1$ for $0 < x \le 5$, given $y(0) = -1$ and the increment $h = \Delta x = 0.1$

The output of the project is shown in Figure 4.1. Data in the form of a table is displayed in the list view window. The output from the list view window consists of the fields i, x_i, y_i, k_1, k_2, k_3, and k_4, displayed under the titles `i`, `x`, `y`, `k1`, `k2`, `k3`, and `k4`, respectively. The number of intervals N is evaluated from the relationship $N = (x_N - x_0)/h = 50$, so this requires a display of each of the items i, x_i, y_i, k_1, k_2, k_3, and k_4, for $i = 0, 1, 2, \ldots, 50$ in the list view window.

Code4A consists of the files Code4A.h and Code4A.cpp. We employ the method of creating the list view window on the parent window, without the use of a resource file. This requires the main window `CCode4A` to be derived from the MFC class `CFrameWnd`.

The member variables and objects in the class are declared in the header file Code4A.h, as shown in Table 4.1. The MFC header file afxcmn.h is included in

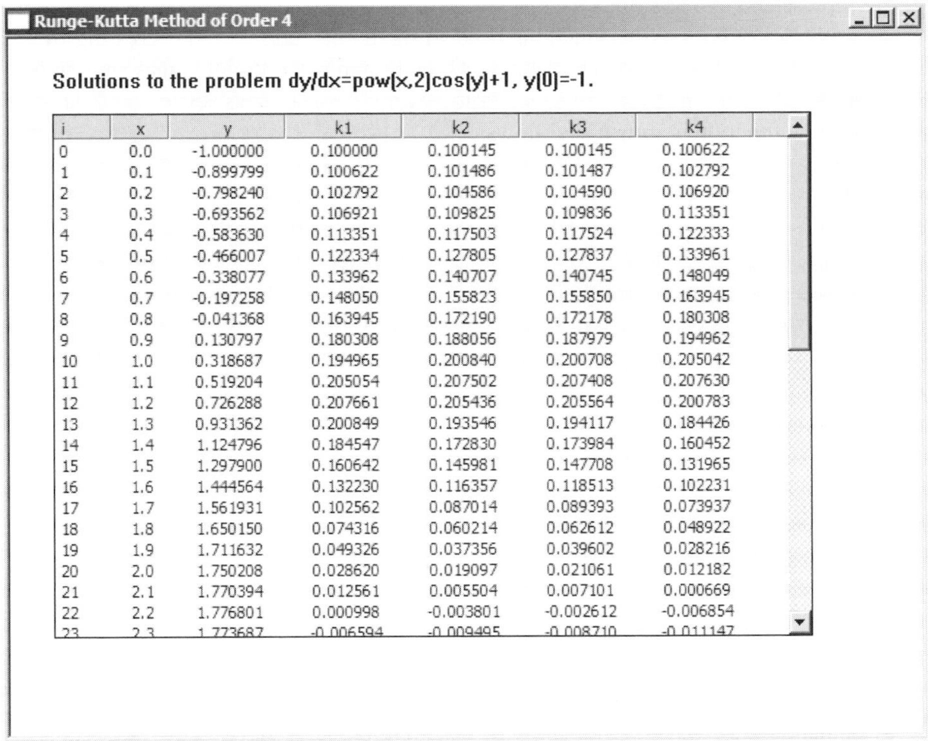

Runge-Kutta Method of Order 4

Solutions to the problem dy/dx=pow(x,2)cos(y)+1, y(0)=-1.

i	x	y	k1	k2	k3	k4
0	0.0	-1.000000	0.100000	0.100145	0.100145	0.100622
1	0.1	-0.899799	0.100622	0.101486	0.101487	0.102792
2	0.2	-0.798240	0.102792	0.104586	0.104590	0.106920
3	0.3	-0.693562	0.106921	0.109825	0.109836	0.113351
4	0.4	-0.583630	0.113351	0.117503	0.117524	0.122333
5	0.5	-0.466007	0.122334	0.127805	0.127837	0.133961
6	0.6	-0.338077	0.133962	0.140707	0.140745	0.148049
7	0.7	-0.197258	0.148050	0.155823	0.155850	0.163945
8	0.8	-0.041368	0.163945	0.172190	0.172178	0.180308
9	0.9	0.130797	0.180308	0.188056	0.187979	0.194962
10	1.0	0.318687	0.194965	0.200840	0.200708	0.205042
11	1.1	0.519204	0.205054	0.207502	0.207408	0.207630
12	1.2	0.726288	0.207661	0.205436	0.205564	0.200783
13	1.3	0.931362	0.200849	0.193546	0.194117	0.184426
14	1.4	1.124796	0.184547	0.172830	0.173984	0.160452
15	1.5	1.297900	0.160642	0.145981	0.147708	0.131965
16	1.6	1.444564	0.132230	0.116357	0.118513	0.102231
17	1.7	1.561931	0.102562	0.087014	0.089393	0.073937
18	1.8	1.650150	0.074316	0.060214	0.062612	0.048922
19	1.9	1.711632	0.049326	0.037356	0.039602	0.028216
20	2.0	1.750208	0.028620	0.019097	0.021061	0.012182
21	2.1	1.770394	0.012561	0.005504	0.007101	0.000669
22	2.2	1.776801	0.000998	-0.003801	-0.002612	-0.006854
23	2.3	1.773687	-0.006594	-0.009495	-0.008710	-0.011147

Figure 4.1 List view window interface for the Runge–Kutta method.

Table 4.1 Variables/Objects used in Code4A

Variable/object	Class/type	Description
x[i]	double	The variable x_i
y[i]	double	The variable y_i
h	double	The variable h
sx[i]	CString	The string for holding the value of x_i
sy[i]	CString	The string for holding the value of y_i
sk1[i]	CString	The string for holding the value of k_1
sk2[i]	CString	The string for holding the value of k_2
sk3[i]	CString	The string for holding the value of k_3
sk4[i]	CString	The string for holding the value of k_4

Code4A.h as it has the prototypes for several variables and member functions associated with the creation of the list view window.

In the list view table, data is displayed in the form of reports as strings using objects derived from the class CString. Therefore, to display the values of the arrays x[] and y[], the CString array objects sx[] and sy[], respectively, are used. The same argument applies to the local variables k1, k2, k3, and k4, which are represented by the strings sk1[], sk2[], sk3[], and sk4, respectively.

Table 4.2 describes the functions used in the application. Besides creating the class CCode4A, the constructor CCode4A() provides the initial values for some variables, allocates memory for several global arrays, and creates several child windows. The function is shown as follows:

```
CCode4A::CCode4A()
{
     x=new double [N+1];
     y=new double [N+1];
     sx=new CString [N+1];
     sy=new CString [N+1];
     sk1=new CString [N+1];
     sk2=new CString [N+1];
     sk3=new CString [N+1];
     sk4=new CString [N+1];
     x[0]=0; y[0]=-1; h=0.1;
     Create(NULL,"Runge-Kutta Method of Order 4");
     TableView.Create(WS_VISIBLE | WS_CHILD | WS_BORDER | LVS_REPORT
       | LVS_NOSORTHEADER, CRect(30,50, 550, 400), this, IDC_TABLEVIEW);
     CreateListColumns();
     RK4();
     AddListItems();
}
```

In the above function, the list view window is a child window created by linking the function Create() with the CListCtrl object TableView (see Table 4.3). The parameters LVS_REPORT and LVS_NOSORTHEADER in the function are the optional list view control styles in the list view window. LVS_NOSORTHEAD-

Table 4.2 Member functions in Code4A

Function	Description
`CCode4A()`	The constructor
`~CCode4A()`	The destructor
`CreateListColumns()`	Creates the frame of the list view window in the form of columns, and sets their widths, styles, and titles
`RK4()`	Runs the Runge–Kutta method and produces the results for display in `AddListColumns()`
`AddListColumns()`	Adds the results from `RK4()` for display according to their classification in the columns of the list view window
`OnPaint()`	Displays messages in the parent window

ER disables the column header buttons in the report, whereas LVS_REPORT creates the list in the report view.

The list view window has seven columns, as defined by the constant nFIELDS. The columns in the list view window are created as a frame using the function CreateListColumns():

```
void CCode4A::CreateListColumns()
{
      char* column[nFIELDS+1]={"i","x","y","k1","k2","k3","k4"};
      int columnWidth[nFIELDS+1]={40,40,80,80,80,80,80};
      LV_COLUMN lvColumn;
      lvColumn.mask = LVCF_WIDTH | LVCF_TEXT | LVCF_FMT | LVCF_SUBITEM;
      lvColumn.fmt = LVCFMT_CENTER;
      lvColumn.cx = 85;
      for (int i=0;i<=nFIELDS;i++)
      {
            lvColumn.iSubItem = 0;
            lvColumn.pszText = column[i];
            TableView.InsertColumn(i, &lvColumn);
            TableView.SetColumnWidth(i,columnWidth[i]);
      }
}
```

This function assigns the width, style and title of each column using the MFC structure LV_COLUMN. The object lvColumn is derived from this structure and this

Table 4.3 Some common member functions of the class CListCtrl

Member function	Description
`Create()`	Creates the list view child window and display on the main window
`InsertColumn()`	Inserts a new column
`InsertItem()`	Inserts items into the list view table
`SetColumnWidth()`	Sets the width of the column

provides a linkage to the members of the structure. The items in the structure are described briefly in Table 4.4.

Runge–Kutta calculations are performed in the function `RK4()`. The work is very straightforward, involving `N` iterations. The results from the calculations are converted to the string objects `sx[]`, `sy[]`, `sk1[]`, `sk2[]`, `sk3[]`, and `sk4[]` using the MFC function `Format()`:

```
void CCode4A::RK4()
{
	int i;
	double k1,k2,k3,k4;
	for (i=0;i<=N;i++)
	{
		k1=h*f(x[i],y[i]);
		k2=h*f(x[i]+h/2,y[i]+k1/2);
		k3=h*f(x[i]+h/2,y[i]+k2/2);
		k4=h*f(x[i]+h,y[i]+k3);
		y[i+1]=y[i]+(k1+2*k2+2*k3+k4)/6;
		x[i+1]=x[0]+(double)(i+1)*h;   // increment for x
		// convert the values to strings
		sx[i].Format("%.1lf",x[i]);
		sy[i].Format("%lf",y[i]);
		sk1[i].Format("%lf",k1);
		sk2[i].Format("%lf",k2);
		sk3[i].Format("%lf",k3);
		sk4[i].Format("%lf",k4);
	}
}
```

The function `AddListItems()` displays the results from the function `RK4()` on the list view table. This is performed through the `CListCtrl` function `SetItemText()` which adds the string objects `sx[]`, `sy[]`, `sk1[]`, `sk2[]`, `sk3[]`, and `sk4[]` onto their respective columns. The items are recognized as the object `lvItem` derived from the MFC structure `LV_ITEM`. Some members of `LV_ITEM` are described briefly in Table 4.5.

Table 4.4 Members of the MFC structure `LV_COLUMN`

Structure member	Description
`mask`	A flag for specifying which member in the structure provides the information
`fmt`	A flag for the style of the text alignment
`cx`	Default width of the columns
`iSubItem`	Index of the item
`pszText`	A pointer to the string for the column's heading

Table 4.5 Some members of the MFC structure `LV_ITEM`

Structure member	Description
`mask`	A flag for specifying which member in the structure provides the information
`state`	A flag for the style of the text alignment
`stateMask`	A flag to specify which member has the information
`iItem`	Index of the main item
`iSubItem`	Index of the subitem
`pszText`	A pointer to the item's text

```
void CCode4A::AddListItems()
{
      CString iteration;
      LV_ITEM lvItem;
      lvItem.mask = LVIF_TEXT | LVIF_STATE;
      lvItem.state = 0;
      lvItem.stateMask = 0;
      for (int i=0;i<=N;i++)
      {
           lvItem.iItem=i;
           lvItem.iSubItem = 0;
           lvItem.pszText = "";
           iteration.Format("%d",i);
           TableView.InsertItem(&lvItem);
           TableView.SetItemText(i, 0, iteration);
           TableView.SetItemText(i, 1, sx[i]);
           TableView.SetItemText(i, 2, sy[i]);
           TableView.SetItemText(i, 3, sk1[i]);
           TableView.SetItemText(i, 4, sk2[i]);
           TableView.SetItemText(i, 5, sk3[i]);
           TableView.SetItemText(i, 6, sk4[i]);
      }
}
```

4.3 PARTIAL DIFFERENTIAL EQUATIONS

The general form of a partial differential equation is written as follows:

$$f\left[x, y, u(x, y), \frac{\partial u}{\partial x}, \frac{\partial u}{\partial y}, \frac{\partial^2 u}{\partial x^2}, \frac{\partial^2 u}{\partial y^2}, \frac{\partial^2 u}{\partial x \partial y}, \ldots\right] = 0 \tag{4.9}$$

In the above equation, $\partial u/\partial x$ is also written as u_x, $\partial^2 u/\partial x^2$ as u_{xx}, $\partial^2 u/(\partial x \partial y)$ as u_{xy}, and so on. An implicit first-order partial differential equation has the following form:

$$f(x, y, u_x, u_y) = g(x, y) \tag{4.10}$$

whereas an implicit second-order partial differential equation with parameters x, y, and $u(x, y)$ has the following general form:

$$f(x, y, u(x, y), u_x, u_y. u_{xx}, u_{yy}, u_{xy}) = g(x, y, u) \tag{4.11}$$

A second-order partial differential equation is said to be linear if it can expressed in the following form:

$$A(x, y)u_{xx} + B(x, y)u_{xy} + C(x, y)u_{yy} + D(x, y)u_x + E(x, y)u_y + F(x, y)u = G(x, y) \tag{4.12}$$

The linear equation in Equation (4.12) can further be classified as elliptic, parabolic, or hyperbolic according to the following rules:

If $B^2 - 4\,AC < 0$ then the equation is an elliptic PDE
If $B^2 - 4\,AC = 0$ then the equation is a parabolic PDE
If $B^2 - 4\,AC > 0$ then the equation is a hyperbolic PDE

A closed area or interval is a region bounded by one or more boundary points. The region inside these boundaries is governed by a partial differential equation that obeys the continuity and stability characteristics of the equation. It follows that a PDE problem with some given boundary values is called a *boundary value problem*. Some common PDE problems involving boundary values $a < x < b$ and $c < y < d$ are listed as follows:

Laplace equation: $\nabla^2 u = u_{xx} + u_{yy} = 0$, for $a < x < b$ and $c < y < d$
Poisson equation: $\nabla^2 u = u_{xx} + u_{yy} = g(x, y)$, for $a < x < b$ and $c < y < d$
Heat equation: $u_t - \alpha^2 u_{xx} = 0$, for $a < x < b$ and $t > 0$
Wave equation: $u_{tt} - \alpha^2 u_{xx} = 0$, for $a < x < b$ and $t > 0$

Poisson Equation: Finite Difference Method

A Poisson equation is an elliptic partial differential equation given by

$$\nabla^2 u = u_{xx} + u_{yy} = g(x, y) \tag{4.13}$$

for $a < x < b$ and $c < y < d$. The boundary value problem involving the Poisson equation consists of boundary conditions, given as follows:

$$u(a, y) = f_1(y)$$

$$u(b, y) = f_2(y)$$

$$u(x, c) = f_3(x)$$

$$u(x, d) = f_4(x)$$

The boundary value problem for the Poisson equation is characterized by a closed region bounded by the left, right, top, and bottom boundaries. The solution requires the bounded region to be discretized into several rectangular grids of equal widths. The boundaries in this problem are $x_0 < x < x_N$ and $y_0 < y < y_M$, where N is the number of intervals in the x-axis and M is the number of intervals in the y-axis. It can be seen that the number of horizontal intervals h and vertical intervals k are given by

$$h = \Delta x = \frac{x_N - x_0}{N}$$

$$k = \Delta y = \frac{y_M - y_0}{M}$$

We discuss a numerical method called the finite difference method for solving a boundary value problem involving the Poisson equation $u_{xx} + u_{yy} = g(x, y)$. The method consists of two steps, as shown in Figure 4.2. The first is the reduction of the Poisson equation into a system of linear equations by finding the finite difference formula from the equation. The second step is to solve this system of linear equations using a method such as Gauss elimination, which has been discussed in Chapter. 2

The *finite difference method* is an approximation method based on the discrete or finite form of the variables in an equation. For a function of one variable, the point (x, y) is approximated as (x_i, y_i), where i is the finite unit in the direction of x-axis. Similarly, a function of two variables $u(x, y)$ has its discrete form written as $g(x_i, y_j)$, where i and j are the finite units in the direction of the x and y axes, respectively.

In approximating the solution to the Poisson equation, we start with a scheme called the *central difference rules.* The one-dimensional form of the central difference rules for the function $y = f(x)$ is given as follows:

$$y_i' \approx \frac{y_{i+1} - y_{i-1}}{2h} \tag{4.14a}$$

$$y_i'' \approx \frac{y_{i+1} - 2y_i + y_{i-1}}{h^2} \tag{4.14b}$$

In Equations (4.14a) and (4.14b), (x_i, y_i) is the current point, (x_{i+1}, y_{i+1}) is its forward point, and (x_{i-1}, y_{i-1}) is its backward point. It is assumed that the intervals in

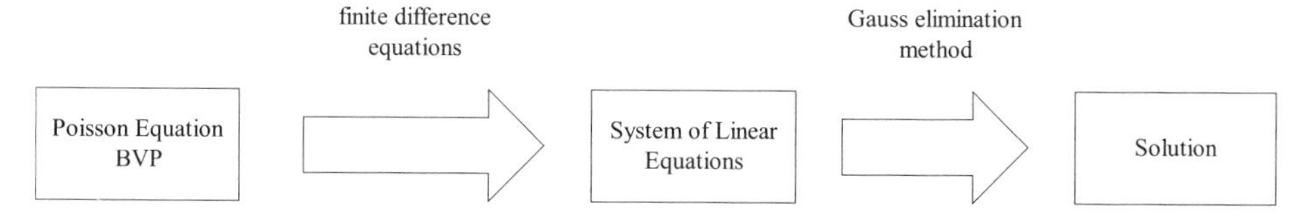

Figure 4.2 Two-step method for solving the Poisson equation boundary value problem.

the x-axis have an equal width given by $h = \Delta x$, as shown in Figure 4.3. It can be shown from the figure that $x_{i+1} = x_i + h$ and $x_{i-1} = x_i - h$.

Equations (4.14a) and (4.14b) are the finite approximations to the first and second derivatives of y as a function of one variable using the central difference rules. The corresponding approximations for a function of two variables $u(x, y)$ in two dimensions are given as follows:

$$u_x|_{(i,j)} \approx \frac{u_{i+1,j} - u_{i-1,j}}{2h} \tag{4.15a}$$

$$u_y|_{(i,j)} \approx \frac{u_{i,j+1} - u_{i,j-1}}{2k} \tag{4.15b}$$

$$u_{xx}|_{(i,j)} \approx \frac{u_{i+1,j} - 2u_{i,j} + u_{i-1,j}}{h^2} \tag{4.15c}$$

$$u_{yy}|_{(i,j)} \approx \frac{u_{i,j+1} - 2u_{i,j} + u_{i,j-1}}{k^2} \tag{4.15d}$$

In Equations (4.15a) to (4.15d), $u_{i,j}$ denotes $u(x_i, y_j)$, $u_{i+1,j}$ is $u(x_{i+1}, y_j)$ and so on. The finite points in the equations are shown in the grids in Figure 4.4.

The finite difference method has its roots in the central difference rules. For the Poisson equation, the finite difference model is obtained by replacing its partial derivatives with Equations (4.15c) and (4.15d), as follows:

$$u_{xx} + u_{yy} = g(x, y)$$

$$\frac{u_{i+1,j} - 2u_{i,j} + u_{i-1,j}}{h^2} + \frac{u_{i,j+1} - 2u_{i,j} + u_{i,j-1}}{k^2} = g(x_i, y_j)$$

■ Example 2

We discuss an example of a boundary value problem involving the Poisson function $u_{xx} + u_{yy} = g(x, y)$. The boundary values in this problem are

$u(0, y) = y \sin y$ and $u(0.8, y) = e^{-y}$, for $0 < y < 0.3$
$u(x, 0) = xe^{-x}$ and $u(x, 0.3) = 1 - x$, for $0 < x < 0.8$

Figure 4.5 shows the grids in the bounded region $0 < x < 0.8$ and $0 < y < 0.3$, having $h = \Delta x = 0.2$ and $k = \Delta y = 0.1$, respectively. The number of intervals for the

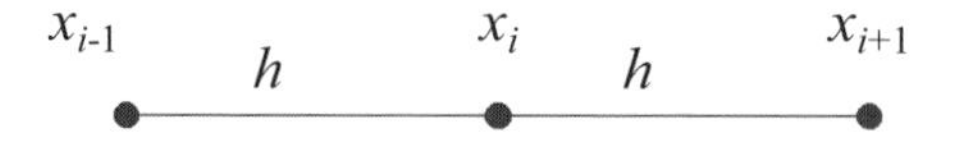

Figure 4.3 One-dimensional finite difference model.

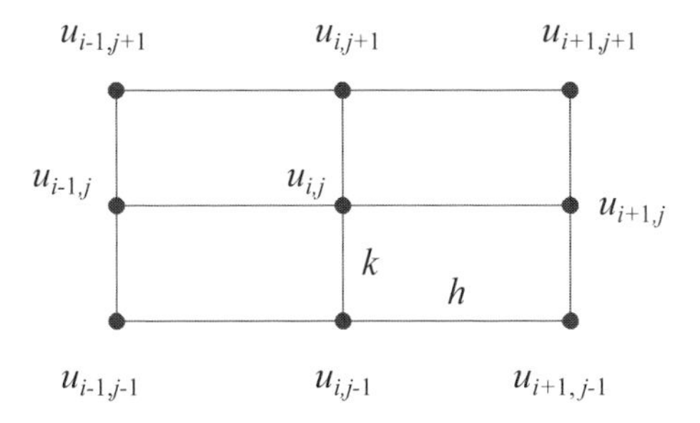

Figure 4.4 Two-dimensional finite difference model.

grids in the x and y axes are given by $N = (0.8 - 0)/0.1$ and $M = (0.3 - 0)/0.1 = 3$, respectively.

An element $u(x_i, y_j)$ in the grids is marked using the finite form simply as u_{ij}. The element $u_{0,0}$ has its home at the bottom-lefthand corner of the grids, with increasing i values in the direction of x-axis and increasing j values in the direction of the y-axis.

In Figure 4.5, the values $u_{0,1}$ and $u_{0,2}$ are determined from the left boundary condition $u(0, y) = y \sin y$, whereas $u_{4,1}$ and $u_{4,2}$ are determined from the right boundary condition $u(0.8, y) = e^{-y}$. Similarly, we obtain the values of $u_{1,0}$, $u_{2,0}$ and $u_{3.0}$ from the condition $u(x, 0) = xe^{-x}$, and $u_{1,3}$, $u_{2,3}$ and $u_{3,3}$ from $u(x, 0.3) = 1 - x$. For example, the value of $u_{2,0}$ is evaluated as follows:

$$u_{2,0} = u(x_2, y_0) = u(0.4, 0) = 0.4e^{-0.4} = 0.268$$

It is obvious from Figure 4.5 that there are six unknowns in the problem: $u_{1,1}$, $u_{2,1}$, $u_{3,1}$, $u_{1,2}$, $u_{2,2}$, and $u_{3,2}$. A system of linear equations of size 6×6 is needed to

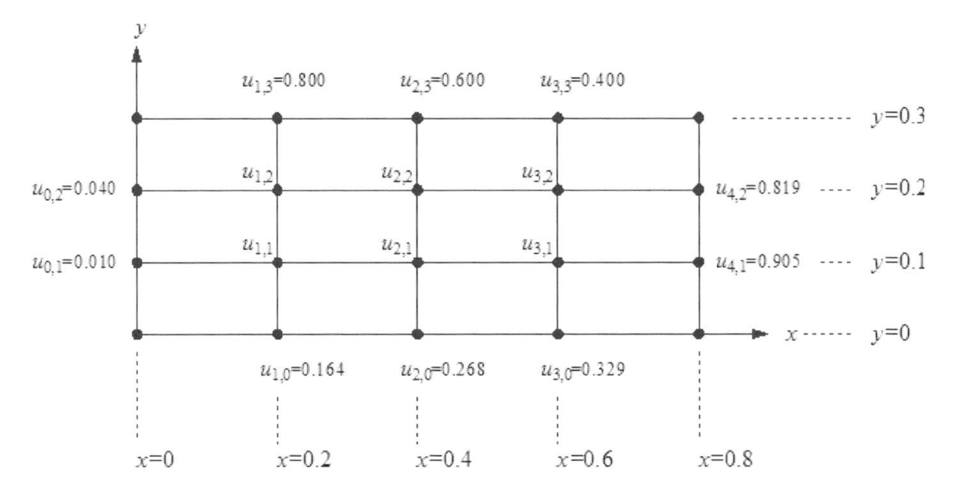

Figure 4.5 Rectangular grids for the Poisson equation.

find these unknowns. Therefore, it is important to derive the finite difference formula that will generate the system of linear equations. The finite difference formula is derived from the central difference rules for differentiation:

$$u_{xx} + u_{yy} = 10xy$$

Replacing each term in the above equation by Equations (4.16c) and (4.16d) we get

$$\frac{u_{i+1,j} - 2u_{i,j} + u_{i-1,j}}{h^2} + \frac{u_{i,j+1} - 2u_{i,j} + u_{i,j-1}}{k^2} = 10x_i y_j$$

Taking the common denominator simplifies the equation to

$$k^2 u_{i+1,j} - 2(k^2 + h^2)u_{i,j} + k^2 u_{i-1,j} + h^2 u_{i,j+1} + h^2 u_{i,j-1} = 10h^2 k^2 x_i y_j$$

Substituting the values of h and k, the equation becomes

$$0.01u_{i+1,j} - 2(0.001 + 0.04)u_{i,j} + 0.01u_{i-1,j} + 0.04u_{i,j+1} + 0.04u_{i,j-1} = 0.004x_i y_j$$

The final form of the finite difference equation is obtained by eliminating the decimal points in the above equation and rearranging the sequence of the subscripts. This is achieved by multiplying both sides of the above equation by 50 to produce

$$5u_{i+1,j} - 50u_{i,j} + 5u_{i-1,j} + 20u_{i,j+1} + 20u_{i,j-1} = 2x_i y_j \qquad (4.16)$$

Equation (4.16) is illustrated graphically as a molecule in Figure 4.6. This molecule acts like a template for the grids in Figure 4.5 for producing a system of linear equations. This step requires substituting the finite elements i and j in the unknowns in Figure 4.5 using Equation (4.16) to produce the following system of linear equations:

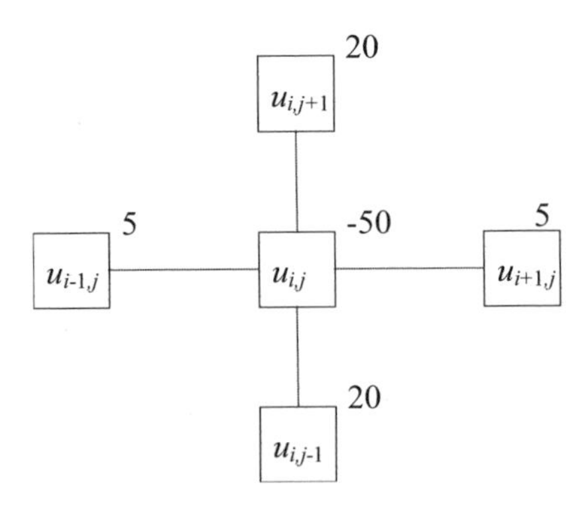

Figure 4.6 Molecular form of Equation (4.15).

$$i=1, j=1: \quad 5u_{2,1}-50u_{1,1}+5u_{0,1}+20u_{1,2}+20u_{1,0}=2(0.2)(0.1)$$
$$-50u_{1,1}+5u_{2,1}+20u_{1,2}=2(0.2)(0.1)-5(0.010)-20(0.164)=-3.290$$

$$i=2, j=1: \quad 5u_{3,1}-50u_{2,1}+5u_{1,1}+20u_{2,2}+20u_{2,0}=2(0.4)(0.1)$$
$$5u_{1,1}-50u_{2,1}+5u_{3,1}+20u_{2,2}=2(0.4)(0.1)-20(0.268)=-5.280$$

$$i=3, j=1: \quad 5u_{4,1}-50u_{3,1}+5u_{2,1}+20u_{3,2}+20u_{3,0}=2(0.6)(0.1)$$
$$5u_{2,1}-50u_{3,1}+20u_{3,2}=2(0.6)(0.1)-5(0.905)-20(0.329)=-10.985$$

$$i=1, j=2: \quad 5u_{2,2}-50u_{1,2}+5u_{0,2}+20u_{1,3}+20u_{1,1}=2(0.2)(0.2)$$
$$20u_{1,1}-50u_{1,2}+5u_{2,2}=2(0.2)(0.2)-5(0.040)-20(0.8)=-16.12$$

$$i=2, j=2: \quad 5u_{3,2}-50u_{2,2}+5u_{1,2}+20u_{2,3}+20u_{2,1}=2(0.4)(0.2)$$
$$20u_{2,1}+5u_{1,2}-50u_{2,2}+5u_{3,2}=2(0.4)(0.2)-20(0.600)=-11.840$$

$$i=3, j=2: \quad 5u_{4,2}-50u_{3,2}+5u_{2,2}+20u_{3,3}+20u_{3,1}=2(0.6)(0.2)$$
$$20u_{3,1}+5u_{2,2}-50u_{3,2}=2(0.6)(0.2)-5(0.819)-20(0.4)=-11.815$$

The above six equations are written in the matrix form, as follows:

$$\begin{bmatrix} -50 & 5 & 0 & 20 & 0 & 0 \\ 5 & -50 & 5 & 0 & 20 & 0 \\ 0 & 5 & -50 & 0 & 0 & 20 \\ 20 & 0 & 0 & -50 & 5 & 0 \\ 0 & 20 & 0 & 5 & -50 & 5 \\ 0 & 0 & 20 & 0 & 5 & -50 \end{bmatrix} \begin{bmatrix} u_{1,1} \\ u_{2,1} \\ u_{3,1} \\ u_{1,2} \\ u_{2,2} \\ u_{3,2} \end{bmatrix} = \begin{bmatrix} -3.290 \\ -5.280 \\ -10.985 \\ -16.120 \\ -11.840 \\ -11.815 \end{bmatrix} \tag{4.17}$$

The system of linear equations in Equation (4.17) can be solved using any suitable method such as the Gauss elimination method, LU factorization, and Gauss-Seidel iterative methods. We apply the Gauss elimination method to get the following results:

$$u(0.2, 0.1)=u_{1,1}=0.299$$
$$u(0.4, 0.1)=u_{2,1}=0.371$$
$$u(0.6, 0.1)=u_{3,1}=0.441$$
$$u(0.2, 0.2)=u_{1,2}=0.490$$
$$u(0.4, 0.2)=u_{2,2}=0.481$$
$$u(0.6, 0.1)=u_{3,2}=0.461$$

■

Code4B: Solving the Poisson Equation

We discuss the MFC approach for providing a user-friendly interface for solving the Poisson equation. The project is called Code4B and it consists of the files Code4B.h and Code4B.cpp. Figure 4.7 shows the output produced by this project from the same example discussed in the last section. It consists of edit boxes (white rectangles), static boxes (shaded rectangles), and the push button *Compute*.

In Code4B, input is made by the user in the edit boxes along the boundaries, whereas the output is produced in the static boxes when the push button *Compute* is

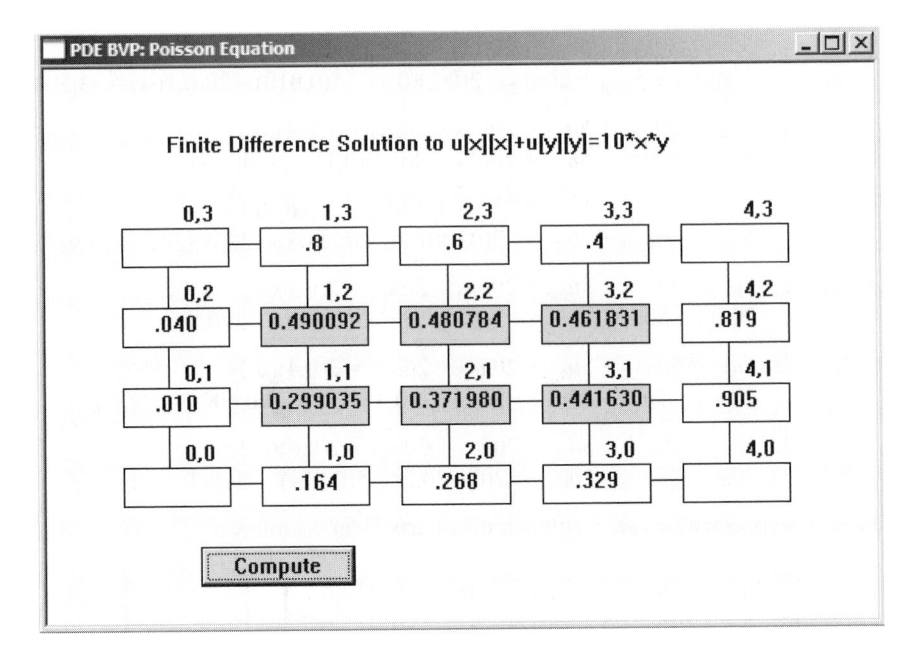

Figure 4.7 Output from Code4B.

clicked. Input in the form of edit boxes for the boundary values, instead of their functional form, provides better interaction and flexibility to the user. Also, as the finite difference molecule in Figure 4.6 has the shape of a star, its mapping on the grids ignores the corners. Therefore, the corner elements of the grids, $u_{0,0}$, $u_{0,3}$, $u_{4,3}$, and $u_{4,0}$, will not require input.

In Code4B.h, the only application class used is `CCode4B`. several global variables and objects are declared for use in this application, and they are summarized in Table 4.6.

Table 4.6 Variables and objects in Code4B

Variable/object	Type	Description
`u[i][j]`	`NODE`	Node u_{ij}
`a[i][j]`	`double`	Represents a_{ij} in the matrix A
`b[i]`	`double`	Represents b_i in the vector $\mathbf{b}$
`x[i]`	`double`	Represents x_i in the vector $\mathbf{x}$
`y[i]`	`double`	Represents y_i in the vector $\mathbf{y}$
`eu[i][j]`	`CEdit`	Edit box for the input u_{ij} at the boundaries
`su[i][j]`	`CStatic`	Static box for displaying the output u_{ij}
`bCompute`	`CButton`	Push button *Compute*
`h`	`double`	Interval size of the grids in the x-axis
`k`	`double`	Interval size of the grids in the y-axis

Several constants are declared as macros in Code4B.h. These constants together with the function $g(P, Q)$ which represents $g(x, y) = 10xy$ are defined using #define, as follows:

```
#define M 3                         // number of rows
#define N 4                         // number of columns
#define R (M-1)*(N-1)               // size of matrix A
#define IDC_COMPUTE 600             // control id for the push button
#define g(P,Q) ((double)10*P*Q)
```

In the above declarations, M and N are the constants representing the number of intervals in the x and y grids, respectively. There are (M+1)(N+1) elements in the grids. From this number, the number of unknowns is R=(M-1)(N-1), while the rest of the elements form the boundaries whose values are given.

The element u_{ij} in the grids is represented by the array u[i][j], declared using the structure NODE. We use a structure as the array is to be linked to two elements, as follows:

```
typedef struct
{
    CPoint Home;          // home position of u[i][j]
    double v;             // value of u[i][j]
} NODE;
NODE **u;
```

In the above declaration, the element u_{ij} is represented by u[i][j], declared as a double pointer. Its value is the linked element u[i][j].v and its position in the window is represented by u[i][j].Home.

The functions used in this application are summarized in Table 4.7. The constructor CCode4B() allocates memory for the class and creates the main window. This function also allocates memory dynamically for all the global arrays, as follows:

```
a=new double *[R+1];
b=new double [R+1];
x=new double [N+1];
```

Table 4.7 Member functions in Code4B

Function	Description
CCode4B()	Constructor of the class CCode4B. Creates the main window and child windows including the edit boxes, static boxes and push button.
~CCode4B()	Destructor of the class CCode4B
OnPaint()	Displays the opening message and initial output on the main window
OnCompute()	Reads the input, computes and displays the results on the static boxes
PGauss()	Solves the generated system of linear equations

```
y=new double [M+1];
eu=new CEdit *[N+1];
su=new CStatic *[N+1];
u=new NODE *[N+1];
for (i=0;i<=R;i++)
     a[i]=new double [R+1];
for (i=0;i<=N;i++)
{
     eu[i]=new CEdit [M+1];
     su[i]=new CStatic [M+1];
     u[i]=new NODE [M+1];
}
```

Several variables are also initialized in the constructor. This includes idc_eu and idc_su, which represent the control ids for the edit boxes and static boxes, respectively. Any integer value may be assigned as the control ids for their initial values.

```
k=0.1; h=0.2;
x[0]=0; y[0]=0;
idc_eu=500;
idc_su=700;
```

The constructor also computes the values of x[] and y[] in the grid lines. The last part of the constructor is the creation of the edit boxes, static boxes, and the push button *Compute*. In creating the edit boxes, the control ids represented by the variables idc_eu are incremented by one to distinguish one box from another. The same argument applies to the static boxes. The code is written as follows:

```
for (j=0;j<=M;j++)                          // row
{
     y[j]=y[0]+j*k;
     for (i=0;i<=N;i++)                     // column
     {
          x[i]=x[0]+i*h;
          u[i][j].Home=CPoint(50+i*90,250-j*50);
          if (i==0 || i==N || j==0 || j==M)
               eu[i][j].Create(WS_CHILD | WS_VISIBLE
               | WS_BORDER | SS_CENTER,
               CRect(u[i][j].Home,CSize(70,25)),this,idc_eu++);
          if (j>0 && j<M && i>0 && i<N)
               su[i][j].Create("",WS_CHILD | WS_VISIBLE
               | WS_BORDER | SS_CENTER,
               CRect(u[i][j].Home,CSize(70,25)),this,idc_su++);
     }
}
bCompute.Create("Compute",WS_CHILD | WS_VISIBLE | BS_DEFPUSH BUTTON,
        CRect(CPoint(100,300),CSize(100,25)),this, IDC_COMPUTE);
```

There are two events in this application. The first is the output in the main window, which is detected as WM_PAINT. The function OnPaint() responds to this event by displaying the opening message and prepares the initial setup on the main window. The initial display consists of grid lines for connecting the edit and static boxes. These lines are drawn as follows:

```
for (j=0;j<=M;j++)
{
     dc.MoveTo(u[0][j].Home+CPoint(10,10));
     dc.LineTo(u[N][j].Home+CPoint(10,10));
}
for (i=0;i<=N;i++)
{
     dc.MoveTo(u[i][0].Home+CPoint(30,10));
     dc.LineTo(u[i][M].Home+CPoint(30,10));
}
```

The second event is a click on the push button *Compute,* which is detected by BN_CLICKED. The function OnCompute() responds to this event by reading the input boundary values, computing the finite-difference formula to reduce the boundary value problem into a system of linear equations, calling the function PGauss() to solve the system of linear equations, and displaying the results on the static boxes.

The input data is read from the edit boxes using the array object eu[][], which is derived from the class CObject. The function GetWindowText() performs this task on the edit boxes along the boundary. The data is read as strings and they are immediately converted to the array u[][] using the C++ function atof().

```
for (j=0;j<=M;j++)
     for (i=0;i<=N;i++)
          if (i==0 || i==N || j==0 || j==M)
          {
               eu[i][j].GetWindowText(s);
               u[i][j].v=atof(s);
          }
```

The next task in OnCompute() is to form the arrays a[][] and b[] for the system of linear equations $A\mathbf{x} = \mathbf{b}$. This is achieved by substituting the given boundary values into the finite difference formula of Equation (4.16). Applying this procedure to the example problem, we obtain a system of linear equations as shown in Equation (4.17). This system of linear equations has the relationship expressed by the following equation:

$$\begin{bmatrix} -2(h^2+k^2) & k^2 & 0 & h^2 & 0 & 0 \\ k^2 & -2(h^2+k^2) & k^2 & 0 & h^2 & 0 \\ 0 & k^2 & -2(h^2+k^2) & 0 & 0 & h^2 \\ h^2 & 0 & 0 & -2(h^2+k^2) & k^2 & 0 \\ 0 & h^2 & 0 & k^2 & -2(h^2+k^2) & k^2 \\ 0 & 0 & h^2 & 0 & k^2 & -2(h^2+k^2) \end{bmatrix}$$

$$\times \begin{bmatrix} u_{1,1} \\ u_{2,1} \\ u_{3,1} \\ u_{1,2} \\ u_{2,2} \\ u_{2,3} \end{bmatrix} = \begin{bmatrix} h^2k^2g(x_1,y_1) - h^2u_{1,0} - k^2u_{0,1} \\ h^2k^2g(x_2,y_1) - h^2u_{2,0} \\ h^2k^2g(x_3,y_1) - h^2u_{3,0} - k^2u_{4,1} \\ h^2k^2g(x_1,y_2) - h^2u_{1,3} - k^2u_{0,2} \\ h^2k^2g(x_2,y_2) - h^2u_{2,3} \\ h^2k^2g(x_3,y_2) - h^2u_{3,3} - k^2u_{4,2} \end{bmatrix} \tag{4.18}$$

The following code fragments assign the elements in Equation (4.18) as the array $A = [a_{ij}]$:

```
for (j=1;j<=R;j++)
      for (i=1;i<=R;i++)
            a[i][j]=0;
for (j=1;j<=R;j++)
{
      if (j<=M+2)
            a[j][j+1]=a[j+1][j]=k*k;
      if (j>=1 && j<=M)
            a[j][j+M]=a[j+M][j]=h*h;
      for (i=1;i<=R;i++)
      {
            if (j==i)
                  a[i][j]=-2*(h*h+k*k);
            if (i==N && j==M)
                  a[i][j]=a[j][i]=0;
      }
}
```

Similarly, Equation (4.18) produces the code for vector **b**, as follows:

```
w=0;
for (j=0;j<=M-2;j++)
      for (i=0;i<=N-2;i++)
      {
            w++;
            b[w]=h*h*k*k*g(x[i+1],y[j+1]);
            if (j==0)
```

```
        {
             b[w] -= h*h*u[i+1][0].v;
             if (i==0)
                  b[w] -= k*k*u[0][1].v;
             if (i==N-2)
                  b[w] -= k*k*u[N][1].v;
        }
        if (j==M-2)
        {
             b[w] -= h*h*u[i+1][M].v;
             if (i==0)
                  b[w] -= k*k*u[0][M-1].v;
             if (i==N-2)
                  b[w] -= k*k*u[N][M-1].v;
        }
   }
```

Once the values of the arrays `a[][]` and `b[]` have been assigned, `OnCompute()` continues its operation by calling up the function `PGauss()` to solve the system of linear equations in Equation (4.18). `PGauss()` is basically the same function discussed earlier in Chapters 2 and 3. There is a slight modification to the code. The `PGauss()` function in the present application computes the system of linear equations using the Gauss elimination method to produce the solution as the matrix `u[][]`. The function is shown as follows:

```
void CCode4B::PGauss()              // compute the SLE
{
     int i,j,w;
     double m,Sum;
     for (w=1;w<=R-1;w++)           // row operations on a
         for (i=w+1;i<=R;i++)
          {
               m=a[i][w]/a[w][w];
               for (j=1;j<=R;j++)
                    a[i][j]=a[i][j]-m*a[w][j];
               b[i]=b[i]-m*b[w];
          }
     for (i=R;i>=1;i—)                // backstitutions on x
     {
               Sum=0;
          u[(i-1)%M+1][(i-1)/M+1].v=0;
          for (j=i;j<=R;j++)
               Sum +=a[i][j]*u[(j-1)%M+1][(j-1)/M+1].v;
          u[(i-1)%M+1][(i-1)/M+1].v=(b[i]-Sum)/a[i][i];
     }
}
```

The last step in the function `OnCompute()` is to display the results of the calculations. The solution from Equation (3.18) is obtained as u_{ij} using the array `u[][]`. These values are formatted as the `CString` objects `su[][]` and displayed in the static boxes using the MFC function `SetWindowText()`, as follows:

```
for (j=1;j<=M-1;j++)
     for (i=1;i<=N-1;i++)
     {
            s.Format("%lf",u[i][j].v);
            su[i][j].SetWindowText(s);
     }
```

Finally, the application on the finite difference method terminates by destroying all the arrays and the class `CCode4B`. This is done in the destructor, as follows:

```
CCode4B::~CCode4B()
{
      int i;
      for (i=0;i<=R;i++)
            delete a[i];
      for (i=0;i<=N;i++)
            delete eu[i],su[i],u[i];
      delete a,eu,su,u,b,x,y;
}
```

4.4 SUMMARY AND CONCLUSION

This chapter discusses the development of friendly interfaces for differential equation problems. We present two models for illustration, one for ordinary differential equations and another for partial differential equations. The ordinary differential problem is illustrated as an initial value problem of first order, and we solve this problem using the fourth-order Runge–Kutta method. The partial differential equation problem is the Poisson equation, which involves a closed domain. We apply the finite difference equation approach for solving this second problem.

We have discussed ways to present the problems and their solution visually. The first problem requires a large area for displaying the results. A list view window with scrollable options both horizontally and vertically is a suitable interface for this requirement. The method of presentation is very much similar to the bisection method for finding the root of an equation, discussed in Chapter 2. The second problem requires a dialog window that takes input from the user, solves the problem, and presents the solution in the same window. This is achieved in a manner similar to the problem of solving a system of linear equations, discussed in Chapter 2. Edit boxes, static boxes, and a push button are the resources used in this application.

Differential equations are indispensable for the effective modeling of real-world problems. The numerical solution provided in the model contributes to the real implementation of ideas for solving the problem. The real problem may consist of several modules for implementation. Each module may be developed as a set of differential equations, or a system, in the model. An effective model arises from the successful execution of each module, which may involve a tremendous amount computational power and complexity. Therefore, some nice structured methods for solving each problem are desirable in order for the whole system to function properly according to the plan.

BIBLIOGRAPHY

1. R. L. Burden and J. D. Faires, *Numerical Analysis,* Brooks Cole, 2000.
2. S. C. Chapra and R. Canale, *Numerical Methods for Engineers,* McGraw-Hill, 2001.
3. K. W. Morton and D. F. Mayers, *Numerical Solution of Partial Differential Equations,* Cambridge University Press, 1994.
4. W. H. Press and S. A. Teukolsky (Eds.), *Numerical Recipes in C++: The Art of Scientific Computing,* 2nd ed., Cambridge University Press, 2002.

CODE LISTINGS

Code4A: Runge–Kutta Method for ODE

```
// code4A.h
#include <afxcmn.h>
#include <math.h>
#define IDC_TABLEVIEW 300
#define nFIELDS 6
#define N 50
#define f(a,b) (pow(a,2)*cos(b)+1)

class CCode4A : public CFrameWnd
{
protected:
	double *x, *y, h;
	CString *sx, *sy;
	CString *sk1, *sk2, *sk3, *sk4;
public:
	CCode4A();
	~CCode4A();
	CListCtrl TableView;
	void CreateListColumns(), AddListItems(), RK4();
	afx_msg void OnPaint();
	DECLARE_MESSAGE_MAP();
};
```

```
class CMyWinApp : public CWinApp
{
public:
      virtual BOOL InitInstance();
};

// code4A: Runge-Kutta Mtd for IVP
#include "code4A.h"

CMyWinApp  MyApplication;

BOOL CMyWinApp::InitInstance()
{
     CCode4A* pFrame = new CCode4A;
     m_pMainWnd = pFrame;
     pFrame->ShowWindow(SW_SHOW);
     pFrame->UpdateWindow();
     return TRUE;
}

BEGIN_MESSAGE_MAP(CCode4A, CFrameWnd)
     ON_WM_PAINT()
END_MESSAGE_MAP()

CCode4A::CCode4A()
{
     x=new double [N+1];
     y=new double [N+1];
     sx=new CString [N+1];
     sy=new CString [N+1];
     sk1=new CString [N+1];
     sk2=new CString [N+1];
     sk3=new CString [N+1];
     sk4=new CString [N+1];
     x[0]=0; y[0]=-1; h=0.1;
     Create(NULL,"Runge-Kutta Method of Order 4");
     TableView.Create(WS_VISIBLE | WS_CHILD | WS_BORDER | LVS_REPORT
        | LVS_NOSORTHEADER,CRect(30,50, 550, 400), this, IDC_TABLEVIEW);
     CreateListColumns();
     RK4();
     AddListItems();
}

CCode4A::~CCode4A()
{
     delete x,y,sx,sy,sk1,sk2,sk3,sk4;
}

void CCode4A::OnPaint()
{
     CPaintDC dc(this);
     dc.TextOut(30,20,
          "Solutions to the problem dy/dx=pow(x,2)cos(y)+1, y(0)=-1.");
}
```

```
void CCode4A::CreateListColumns()
{
     char* column[nFIELDS+1]={"i","x","y","k1","k2","k3","k4"};
     int columnWidth[nFIELDS+1]={40,40,80,80,80,80,80};
     LV_COLUMN lvColumn;
     lvColumn.mask = LVCF_WIDTH | LVCF_TEXT | LVCF_FMT | LVCF_SUBITEM;
     lvColumn.fmt = LVCFMT_CENTER;
     lvColumn.cx = 85;
     for (int i=0;i<=nFIELDS;i++)
     {
          lvColumn.iSubItem = 0;
          lvColumn.pszText = column[i];
          TableView.InsertColumn(i, &lvColumn);
          TableView.SetColumnWidth(i,columnWidth[i]);
     }
}

void CCode4A::AddListItems()
{
     CString iteration;
     LV_ITEM lvItem;
     lvItem.mask = LVIF_TEXT | LVIF_STATE;
     lvItem.state = 0;
     lvItem.stateMask = 0;
     for (int i=0;i<=N;i++)
     {
          lvItem.iItem=i;
          lvItem.iSubItem = 0;
          lvItem.pszText = "";
          iteration.Format("%d",i);
          TableView.InsertItem(&lvItem);
          TableView.SetItemText(i, 0, iteration);
          TableView.SetItemText(i, 1, sx[i]);
          TableView.SetItemText(i, 2, sy[i]);
          TableView.SetItemText(i, 3, sk1[i]);
          TableView.SetItemText(i, 4, sk2[i]);
          TableView.SetItemText(i, 5, sk3[i]);
          TableView.SetItemText(i, 6, sk4[i]);
     }
}

void CCode4A::RK4()
{
     int i;
     double k1,k2,k3,k4;
     for (i=0;i<N;i++)
     {
          k1=h*f(x[i],y[i]);
          k2=h*f(x[i]+h/2,y[i]+k1/2);
          k3=h*f(x[i]+h/2,y[i]+k2/2);
          k4=h*f(x[i]+h,y[i]+k3);
          y[i+1]=y[i]+(k1+2*k2+2*k3+k4)/6;
          x[i+1]=x[0]+(double)(i+1)*h;         // increment for x
          // convert the values to strings
```

```
            sx[i].Format("%.1lf",x[i]);
            sy[i].Format("%lf",y[i]);
            sk1[i].Format("%lf",k1);
            sk2[i].Format("%lf",k2);
            sk3[i].Format("%lf",k3);
            sk4[i].Format("%lf",k4);
      }
}
```

Code4B: Poisson Equation Using the Finite Difference Method

```
// code4B.h
#include <afxwin.h>
#define M 3           // rows
#define N 4           // columns
#define R (M-1)*(N-1)
#define g(P,Q) ((double)10*P*Q)
#define IDC_COMPUTE 600

class CCode4B : public CFrameWnd
{
private:
      int idc_eu,idc_su;
      double h,k,**a,*b,*x,*y;
      CEdit **eu;
      CStatic **su;
      CButton bCompute;
      typedef struct
      {
            CPoint Home;
            double v;
      } NODE;
      NODE **u;
public:
      CCode4B();
      ~CCode4B();
      afx_msg void OnPaint();
      afx_msg void OnCompute();
      DECLARE_MESSAGE_MAP();
      void PGauss();
};

class CMyWinApp : public CWinApp
{
public:
      virtual BOOL InitInstance();
};

#include "code4B.h"

BOOL CMyWinApp::InitInstance()
{
      CCode4B* pFrame = new CCode4B;
      m_pMainWnd = pFrame;
```

```
    pFrame->ShowWindow(SW_SHOW);
    pFrame->UpdateWindow();
    return TRUE;
}
CMyWinApp  MyApplication;

BEGIN_MESSAGE_MAP(CCode4B,CFrameWnd)
     ON_WM_PAINT()
    ON_BN_CLICKED (IDC_COMPUTE,OnCompute)
END_MESSAGE_MAP()

CCode4B::CCode4B()
{
    int i,j;
    a=new double *[R+1];
    b=new double [R+1];
    x=new double [N+1];
    y=new double [M+1];
    eu=new CEdit *[N+1];
    su=new CStatic *[N+1];
    u=new NODE *[N+1];
    for (i=0;i<=R;i++)
        a[i]=new double [R+1];
    for (i=0;i<=N;i++)
    {
        eu[i]=new CEdit [M+1];
        su[i]=new CStatic [M+1];
        u[i]=new NODE [M+1];
    }
    Create(NULL,"PDE BVP: Poisson Equation");
    k=0.1; h=0.2;
    x[0]=0; y[0]=0;
    idc_eu=500;idc_su=700;
    for (j=0;j<=M;j++)// row
    {
        y[j]=y[0]+j*k;
        for (i=0;i<=N;i++)   // column
        {
            u[i][j].Home=CPoint(50+i*90,250-j*50);
            x[i]=x[0]+i*h;
            if (i==0 || i==N || j==0 || j==M)
                eu[i][j].Create(WS_CHILD| WS_VISIBLE| WS_BORDER| SS_CENTER,
                CRect(u[i][j].Home,CSize(70,25)),this,idc_eu++);
            if (j>0 && j<M && i>0 && i<N)
                su[i][j].Create("",WS_CHILD|WS_VISIBLE|WS_BORDER|SS_CENTER,
                CRect(u[i][j].Home,CSize(70,25)),this,idc_su++);
        }
    }
    bCompute.Create("Compute",WS_CHILD | WS_VISIBLE | BS_DEFPUSH BUTTON,
            CRect(CPoint(100,300),CSize(100,25)),this, IDC_COMPUTE);
}

CCode4B::~CCode4B()
{
    int i;
```

```
    for (i=0;i<=R;i++)
        delete a[i];
    for (i=0;i<=N;i++)
        delete eu[i],su[i],u[i];
    delete a,eu,su,u,b,x,y;
}

void CCode4B::OnPaint()
{
    CPaintDC dc(this);
    CString s;
    int i,j;
    for (j=0;j<=M;j++)
    {
        dc.MoveTo(u[0][j].Home+CPoint(10,10));
        dc.LineTo(u[N][j].Home+CPoint(10,10));
    }
    for (i=0;i<=N;i++)
    {
        dc.MoveTo(u[i][0].Home+CPoint(30,10));
        dc.LineTo(u[i][M].Home+CPoint(30,10));
    }
    for (j=0;j<=M;j++)
        for (i=0;i<=N;i++)
        {
            s.Format("%d,%d",i,j);
            dc.TextOut(u[i][j].Home.x+40,u[i][j].Home.y-17,s);
        }
    dc.TextOut(80,40,"Finite Difference Solution to u[x][x]+u[y][y]=10*x*y");
}

void CCode4B::OnCompute()
{
    CClientDC dc(this);
    CString s;
    int i,j,w;
    for (j=0;j<=M;j++)
        for (i=0;i<=N;i++)
            if (i==0 || i==N || j==0 || j==M)
            {
                eu[i][j].GetWindowText(s);
                u[i][j].v=atof(s);
            }
    for (j=1;j<=R;j++)
        for (i=1;i<=R;i++)
            a[i][j]=0;
    for (j=1;j<=R;j++)
    {
        if (j<=M+2)
            a[j][j+1]=a[j+1][j]=k*k;
        if (j>=1 && j<=M)
            a[j][j+M]=a[j+M][j]=h*h;
        for (i=1;i<=R;i++)
```

```
		{
			if (j==i)
				a[i][j]=-2*(h*h+k*k);
			if (i==N && j==M)
				a[i][j]=a[j][i]=0;
		}
	}
	w=0;
	for (j=0;j<=M-2;j++)
	{
		for (i=0;i<=N-2;i++)
		{
			w++;
			b[w]=h*h*k*k*g(x[i+1],y[j+1]);
			if (j==0)
			{
				b[w] -= h*h*u[i+1][0].v;
				if (i==0)
				   b[w] -= k*k*u[0][1].v;
				if (i==N-2)
				   b[w] -= k*k*u[N][1].v;
			}
			if (j==M-2)
			{
				b[w] -= h*h*u[i+1][M].v;
				if (i==0)
				   b[w] -= k*k*u[0][M-1].v;
				if (i==N-2)
				   b[w] -= k*k*u[N][M-1].v;
			}
		}
	}
	PGauss();
	for (j=1;j<=M-1;j++)
		for (i=1;i<=N-1;i++)
		{
			s.Format("%lf",u[i][j].v);
			su[i][j].SetWindowText(s);
		}
}

void CCode4B::PGauss()			// compute the SLE
{
	int i,j,w;
	double m,Sum;
	for (w=1;w<=R-1;w++)		// row operations on a
		for (i=w+1;i<=R;i++)
		{
			m=a[i][w]/a[w][w];
			for (j=1;j<=R;j++)
				a[i][j]=a[i][j]-m*a[w][j];
			b[i]=b[i]-m*b[w];
		}
	for (i=R;i>=1;i–)			// backstitutions on x
```

```
    {
        Sum=0;
        u[(i-1)%M+1][(i-1)/M+1].v=0;
        for (j=i;j<=R;j++)
            Sum +=a[i][j]*u[(j-1)%M+1][(j-1)/M+1].v;
        u[(i-1)%M+1][(i-1)/M+1].v=(b[i]-Sum)/a[i][i];
    }
}
```

CHAPTER 5

DRAWING CURVES

5.1 WINDOWS GRAPHICS REPRESENTATION

In general, it is easier for the brain to accept graphics than a series of text-oriented descriptions of a particular problem. This is because the graphical representation is natural and easily accepted by the human brain. A single graphical drawing maps directly to the human brain and is instantly recognized. A textual or numerical representation requires one extra step prior to its recognition: its transformation to the graphical form. This step produces an overhead that delays interpretation by the brain. Graphics provide a good visualization that contributes to a better understanding of a particular problem.

A curve is a visual representation of the relationship between two or more components. The simplest form of a curve in two-dimensional space is a single point. With two points, a straight line can be drawn to relate them. With three or more points, a curve can take its shape in many forms, depending on the points. In three-dimensional space, a plane represents a set of straight lines connecting three or more points. With more points in this dimension, we get a set of curves that form a surface.

This chapter discusses two projects involving the construction of several types of curves. The first is a discussion of techniques for plotting three types of curves: a polynomial, a polar curve, and a parametric curve. The second project is an interpolative method for constructing a cubic spline. This application allows the user to create the points by simply clicking the left button of the mouse. The spline is obtained by interpolating all these points. In plotting the curves, a conversion from the mathematical Cartesian coordinates to Windows needs to be considered. This is because the Cartesian system is based on real (floating) numbers, whereas the Windows system is based on integers.

Windows Coordinates System

Text or graphics drawing on the personal computer consists of raster operations based on the bitmap composition of pixels. A *pixel* is the smallest item that can be displayed on the screen. Several pixels form text characters such as “a,” “g,” and “%.” In the same way, a graphics primitive such as a line and a circle is formed by mapping the pixels according to its shape. A pixel consists of a unit on the screen formed from its row and column number. The size of a pixel is inversely proportional to the screen resolution. A large pixel makes the resolution low, and this produces low-quality drawings on the screen. A high screen resolution means that the size of each pixel is very small, and this produces very sharp drawings on the screen.

On a standard 800×600 display, the column of the window is measured from 0 to 799 eastward, and the row numbers from 0 to 599 southward, as shown in Figure 5.1. This mode of display produces $800 \times 600 = 480{,}000$ pixels. A better screen setup has more pixels distributed in the same size the screen, for example, $1{,}280 \times 800 = 1{,}204{,}000$. This setup produces finer pixels and a higher-resolution graphics display.

Mathematical curves are well presented on a computer. In Windows, a curve is obtained by activating a successive array of pixels. Each pixel represents a point corresponding to the coordinates (x, y) in mathematics. To draw a curve, the pixels are placed very close to each other so that they appear as a continuous curve. Another way of drawing a curve involves bringing the pen to the starting position, then drawing lines successively on the points that are placed close to each other.

In drawing curves on the computer screen, two coordinate systems must be considered. The first is the Windows coordinate system, in which only integer numbers corresponding to the column and row numbers of the pixels are supported. The sec-

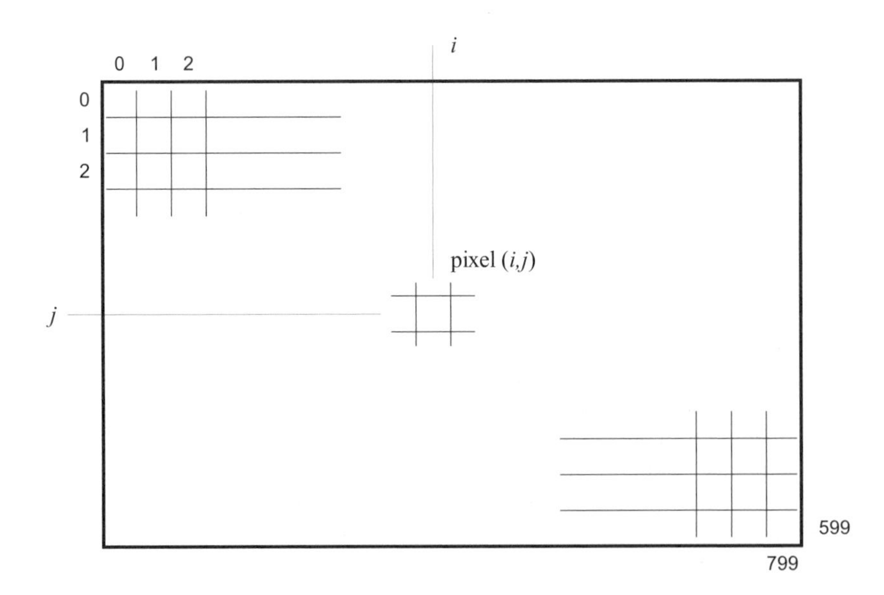

Figure 5.1 Windows coordinate system.

ond is the *Cartesian coordinate system,* which is a standard measuring system for all mathematical operations and applications. This coordinate system is shown in Figure 5.2, and it supports the floating point representation of numbers. The system has its origin at the intersection between the x and y axes, with x and y values increasing northward and eastward, respectively.

In order to draw a curve on the screen, it is necessary to convert the coordinates and their data types correctly from Windows to Cartesian, and vice versa. In general, a suitable conversion from Cartesian to Windows follows the following scheme:

$$\text{Windows } x \text{ value} = \text{Cartesian } x \text{ value} + x \text{ value of the origin} \quad (5.1a)$$

$$\text{Windows } y \text{ value} = y \text{ value of the origin} - \text{Cartesian } y \text{ value} \quad (5.1b)$$

Reversing Equations (5.1a) and (5.1b), the following equations convert the coordinates of a point from Windows to Cartesian:

$$\text{Cartesian } x \text{ value} = \text{Windows } x \text{ value} - x \text{ value of the origin} \quad (5.2a)$$

$$\text{Cartesian } y \text{ value} = y \text{ value of the origin} - \text{Windows } y \text{ value} \quad (5.2b)$$

5.2 MFC FUNCTIONS FOR DISPLAYING GRAPHICS

In MFC, dozens of useful functions are available for drawing. Some primitive tools for drawing graphical objects on the computer screen include pixels, lines, rectangles, circles, and polygons. Table 5.1 lists some of the most common functions for drawing graphics in Windows.

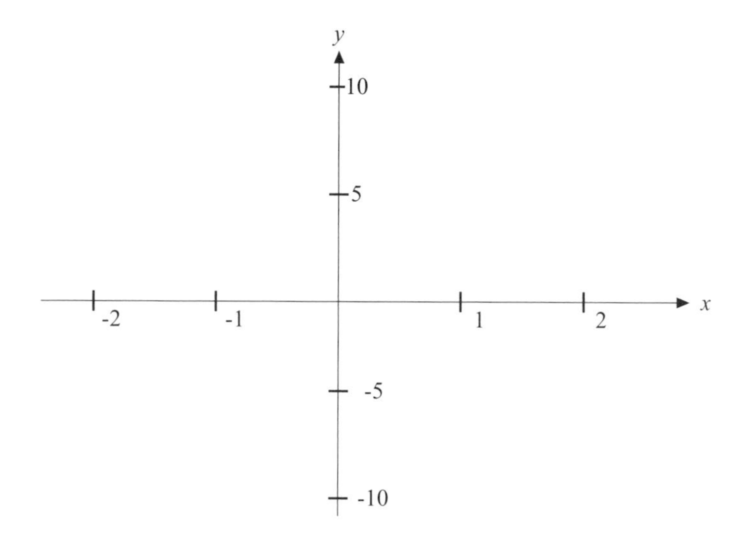

Figure 5.2 Cartesian coordinate system.

Table 5.1 Some common MFC functions for graphics

Function	Description
CPoint(*pt*)	Defines the window coordinates of the point *pt*
SetPixel(*pt*,*color*)	Displays a pixel with the color *color* at the point *pt*
GetPixel(*pt*,*color*)	Gets the pixel value at location *pt* and stores it as *color*
MoveTo(*pt*1)	Moves the pen to the location *pt*1
LineTo(*pt*2)	Draws a line to the location *pt*2
CRect(*a*,*b*,*c*,*d*)	Defines the rectangle bounded by the top-left coordinates (*a*,*b*) and the bottom-right coordinates (*c*,*d*)
Rectangle(*a*,*b*,*c*,*d*)	Displays a rectangle bounded by the top-left coordinates (*a*,*b*) and the bottom-right coordinates (*c*,*d*)
Ellipse(*a*,*b*,*c*,*d*)	Draws an ellipse specified by a rectangle with the top-left coordinates (*a*,*b*) and the bottom-right coordinates (*c*,*d*). If the rectangle is a square, then a circle is obtained
FillRect(*rc*,*brush*)	Fills the rectangle *rc* with a color defined by the brush *brush*
FillSolidRect(*rc*,*brush*)	Draws the rectangle *rc* using the color *brush*

Color Schemes. A pixel represents the smallest unit on the screen that can be displayed. A pixel has a value defined by the MFC function `RGB(r,g,b)`, defined as the composition of the red, green, and blue color components (see Table 5.2).

The color scheme employed in RGB(*r*,*g*,*b*) consists of the red, green, and blue components. Each component `r`, `g`, and `b` in the function is represented by a number from 0 to $2^8 - 1 = 255$, or 0000 0000 to 1111 1111 in binary and 00 to FF in hexadecimal. The value of 0 in each component is the darkest and this value becomes lighter as the number increases monotonically.

Windows interprets the entry in RGB (*r*, *g*, *b*) by placing *r* in the rightmost bits, followed by *g* in the middle, and *b* at the leftmost bits, as shown in Figure 5.3. This assignment of pixel values is illustrated in the figure using an example of a pixel with a value of `RGB(174,55,171)`. This value can also be written as `RGB(0xAE,0x37,0xAB)` in hexadecimal, and (10101011,00110111, 10101110) in binary.

Table 5.3 lists some of the primary colors obtained using `RGB()` by combining the `r`, `g`, and `b` components. In this table, yellow is obtained by combining the green and red components in the function `RGB()`. Similarly, cyan is produced from the combination of green and blue, and magenta is obtained from red and blue.

Table 5.2 Color manipulation function of a pixel

Function	Description
RGB(r,g,b)	The intensity of the pixel at the given location, which is composed of the red (`r`), green (`g`), and blue (`b`) components.

```
RGB(174,55,171) = RGB(0xAE,0x37,0xAB)
```

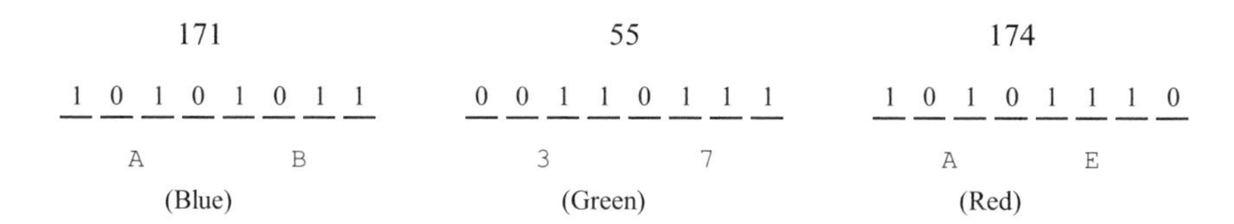

Figure 5.3 Pixel value of `RGB(174, 55, 171)`.

Figure 5.4 shows the relationship between the red, green, and blue color components in the form of a hypercube. These colors are obtained by referring to a coordinate system in which red and green are the planar axes, and blue is the vertical axis. The red, green, and blue components in the hypercube form an axis system with the colors at the corners.

Selecting an Object. A pen, brush, and other objects can be selected using the function `SelectObject()`. For example, the following code selects a gray pen by creating an object from the class `CPen`:

```
CPen penGray(PS_SOLID,1,RGB(100,100,100));
dc.SelectObject(&penGray);
```

Filling a Rectangular Area with a Color. A rectangular region is created by first creating an object from the class `CRect` to define its top-left and bottom-right coordinates. Another object, a brush, is created from the class `CBrush`. The rectangular area is then filled with the color `RGB(200,200,200)` using the function `FillRect`, as follows:

```
CRect rc=CRect(50,100,150,200);
CBrush grayBrush(RGB(200,200,200));
dc.FillRect(&rc,&grayBrush);
```

Table 5.3 Some common colors in `RGB()`

Function	Color
`RGB(255,0,0)`	Red
`RGB(0,255,0)`	Green
`RGB(0,0,255)`	Blue
`RGB(255,255,255)`	White
`RGB(0,0,0)`	Black
`RGB(255,255,0)`	Yellow
`RGB(0,255,255)`	Cyan
`RGB(255,0,255)`	Magenta

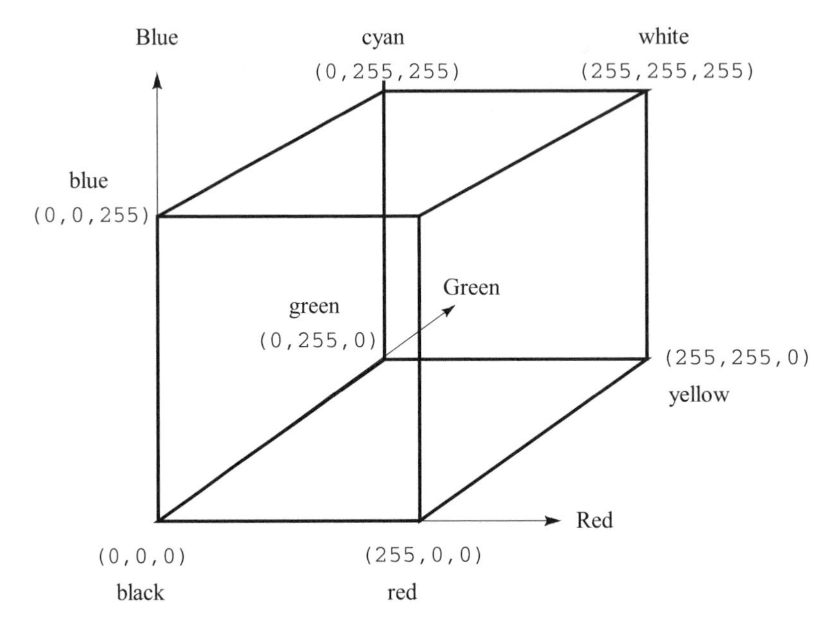

Figure 5.4 Hypercube showing the color combinations in the RGB system.

The same method can be used to erase an area on the screen where `rc` in the above example is the desired area. To erase the whole area in the window, the rectangular area in this case is the window. The area is obtained using the function `GetClientRect()`. The following code shows a way to erase the whole window:

```
CRect rc;
CBrush BgBrush(RGB(200,200,200));
GetClientRect(&rc);
dc.FillRect(&rc,&BgBrush);
```

Plotting a Point. The most basic way of assigning a pixel is the function `SetPixel`(*pt,color*). This function assigns the point *pt* with the color *color*. For example, the following code uses the device context object `dc` for plotting a point at the location (50,100):

```
dc.SetPixel(50,100);
```

In reverse, the function `GetPixel`(*pt,color*) gets the color value of the point *pt* on the screen and stores this value as *color*.

Drawing a Line. A line from *pt*1 to *pt*2 can be drawn by first moving the pen to *pt*1 using the function `MoveTo`(*pt*1) and dragging the pen to *pt*2 using

`LineTo`(*pt*2). For example, the following code draws a line from (50,100) to (120,200):

```
dc.MoveTo(50,100);
dc.LineTo(120,200);
```

Drawing an Object. An object such as a rectangle can be displayed on the window using an MFC function. For example, to draw a rectangle, an object from from the class `CRect` is first created:

```
CRect rc=CRect(50,100,150,200);
dc.Rectangle(&rc);
```

A rectangle also defines an area where other primitive objects such as an ellipse can be drawn. For example, the following code draws an ellipse with its top coordinates at (50,100) and bottom-right coordinates at (120,200):

```
CRect rc=CRect(50,100,150,200);
dc.Ellipse(rc);
```

A circle is an ellipse in which the rectangular box has equal width and height. Therefore, a circle can be drawn using the same function, `Ellipse()`.

A rectangle is selected for several purposes including drawing a box, erasing an area by filling it with a color, and selecting this area for updating. In operations such as changing the color of a rectangular area or erasing this area, a pen is not a good tool to use as it does not have features like filling or flooding the designated area. Instead, a brush object from the class `CBrush` offers a more practical choice.

5.3 DRAWING A CURVE

In mathematics, graphical objects include points, lines, curves, circles, and surfaces. Each of these objects is displayed to form the graphical visualization and illustration of a given problem. In this section, we discuss a method for drawing two-dimensional curves. In general, the rules for drawing a curve are governed by the fundamental rules of mathematics, having characteristics such as existence, continuity, convergence, and stability. Hence, if a curve does not exist at a point or an interval, then the program must identify this singularity and avoid drawing the curve at this point or interval.

A single-variable function is expressed as $y = f(x)$. A common function in this category is a *polynomial* of degree n, which has the following general form:

$$f(x) = a_0 + a_1 x + a_2 x^2 + \ldots + a_n x^n \tag{5.3}$$

In the above equation, a_i for $i = 0, 1, \ldots, n$ is a constant and n is a positive integer.

The function $y = f(x)$ is sometimes written in parametric form, with t as its parameter. In this case, the equation is written as $x(t)$ and $y(t)$ to denote each variable is a function of the parameter t. An example of a parametric curve is $x(t) = t$ and $y(t) = t^2$, which is the same as $y = x^2$. Another common parametric curve is the polar curve in which the parameter t is the angle of inclination. One such curve is $x(t) = r \cos t$ and $y(t) = r \sin t$, which represents the circle $x^2 + y^2 = r^2$.

Code5A: Mathematical Curves

In this section, we discuss some curve drawing techniques using the facilities in MFC. Three common types of curves are discussed: a polynomial, a polar curve, and a parametric curve. The project is called Code5A, and it consists of two files: Code5A.cpp and Code5A.h. Code5A has a rectangular menu in the form of shaded boxes. An item in the menu can be viewed simply by clicking the left button of the mouse on the respective box. Figures 5.5 shows an example of a curve generated from this application. The curve shown is a polynomial and it is obtained by clicking the *Polynomial* shaded box.

The project Code5A has one application class called `CCode5A`. The data structure in this application is described in Code5A.h. Table 5.4 describes the variables and objects in this data structure.

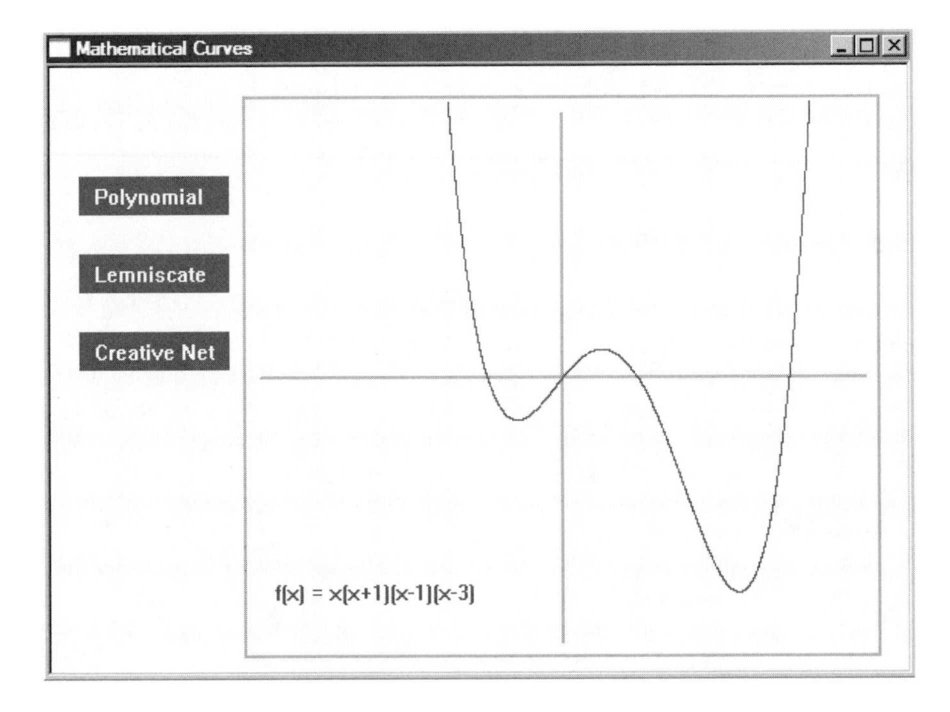

Figure 5.5 Output from Code5A.

Table 5.4 Variables/objects in Code5A

Variables/object	Type	Description
DrawRc	CRect	The rectangular area for drawing the curves
TopLeft	CPoint	Top-left coordinates of the drawing area
BottomRight	CPoint	Bottom-right coordinates of the drawing area
Origin	CPoint	The origin of the x and y axes
Pt1,Pt2,Pt3	CPoint	Home coordinates of the menu rectangles
rc1,rc2,rc3	CRect	First, second, and third menu rectangles
Color1,Color2	int	Color value, given by RGB(100,100,100) and RGB(170,170,170), respectively
MenuChoice	char	Character to respond to an update in the drawing area

The curve is displayed in a rectangular area bounded by the top-left and the bottom-right corners, TopLeft and BottomRight, respectively. The drawing area is defined by the object DrawRc. It has its Cartesian coordinates origin at Origin. Each menu item in the application is represented by a shaded rectangle with its top-left coordinates defined by Pt and the rectangular area by rc. For example, the item *Polynomial* has Pt1 as its top-left coordinates and rc1 as its rectangular object.

Table 5.5 describes several member functions from the class declared in Code5A.h. The event handlers are OnPaint() and OnLButtonDown(). The curves are drawn using the functions Polynomial(), Lemniscate(), and CreativeNet(), whereas DrawAxes() draws the x and y axes for the Cartesian coordinate system.

In the constructor, the global variables and objects are initialized as follows:

```
TopLeft=CPoint(130,20);
BottomRight=CPoint(550,380);
Origin.x=(TopLeft.x+BottomRight.x)/2;
Origin.y=(TopLeft.y+BottomRight.y)/2;
DrawRc=CRect(TopLeft.x+3,TopLeft.y+3,BottomRight.x-3,BottomRight.y-3);
pt1=CPoint(20,70);
rc1=CRect(pt1,pt1+CPoint(100,25));
pt2=CPoint(20,120);
rc2=CRect (pt2,pt2+CPoint(100,25));
pt3=CPoint(20,170);
rc3=CRect (pt3,pt3+CPoint(100,25));
Color1=RGB(100,100,100);
Color2=RGB(170,170,170);
```

The initial values are assigned mostly for producing the initial display in the main window. These include the position of the rectangular menu items, the drawing area, and the Cartesian origin for drawing the curves.

The initial display is invoked by the function OnPaint(), which responds to the message handler WM_PAINT. The main item in the main window is the drawing area for the curves, marked by the CRect object DrawRc. This rectangular area displays each of the three curves based on user's selection. The initial display also

Table 5.5 Member functions in Code5A

Function	Description
`CCode5A()`	The constructor
`~CCode5A()`	The destructor
`OnPaint()`	Produces the initial setup for the menus, drawing area, and an update area whenever `InvalidateRect()` is invoked
`OnLButtonDown()`	A response to the message handler `ON_WM_LBUTTONDOWN()`. The function takes input from the left mouse click as `pt` and checks if it is within one of the menu rectangles for an instant update using `InvalidateRect()`
`DrawAxes()`	Draw the Cartesian x and y axes, and the origin
`Polynomial()`	Draw a polynomial
`Lemniscate()`	Draw two polar curves
`CreativeNet()`	Draw a parametric curve called Creative Net

includes the three menu items in the form of shaded rectangles. The code fragment for the initial display is shown as follows:

```
CPaintDC dc(this);
CPen penDark(PS_SOLID,2,Color2);
dc.SelectObject(&penDark);
dc.Rectangle(CRect(CPoint(TopLeft),CPoint(BottomRight)));
dc.FillSolidRect(&rc1,Color1);
dc.FillSolidRect(&rc2,Color1);
dc.FillSolidRect(&rc3,Color1);
dc.SetTextColor(RGB(255,255,255));
dc.SetBkColor(Color1);
dc.TextOut (pt1.x+10,pt1.y+5, "Polynomial");
dc.TextOut (pt2.x+10,pt2.y+5, "Lemniscate");
dc.TextOut (pt3.x+10,pt3.y+5, "Creative Net");
```

A character variable called `MenuChoice` is used to store the user's choice from the menu. An item in the menu is selected by clicking the mouse's left button on the respective rectangle. The click is an event that is detected by the message handler `WM_LBUTTONDOWN`. A click on one of the shaded boxes activates the function `OnLButtonDown()`, which assigns one of the characters "P," "L," or "C" to the variable `MenuChoice`. This active variable is passed to `OnPaint()` through the update function `InvalidateRect()`. `OnPaint()` responds to this function by calling one of the functions `Polynomial()`, `Lemniscate()`, or `CreativeNet()`, as follows:

```
switch(MenuChoice)
{
        case 'P':
                Polynomial(); break;
```

```
        case 'L':
                Lemniscate(); break;
        case 'C':
                CreativeNet(); break;
}
```

The function `OnLButtonDown()` responds to the event invoked through the user's click on one of the three rectangular menu boxes. This function has the argument `pt`, which returns the Windows coordinates at the point of click. It is through the function `PtInRect()` that the choice from the user is determined. The code fragment for this function is written as follows:

```
void CCode5A::OnLButtonDown(UINT nFlags,CPoint pt)
{
        if (rc1.PtInRect(pt))
        {
                MenuChoice = 'P';
                InvalidateRect(DrawRc);
        }
        if (rc2.PtInRect(pt))
        {
                MenuChoice = 'L';
                InvalidateRect(DrawRc);
        }
        if (rc3.PtInRect(pt))
        {
                MenuChoice = 'C';
                InvalidateRect(DrawRc);
        }
        if (!rc1.PtInRect(pt) && !rc2.PtInRect(pt)
                                        && !rc3.PtInRect(pt))
                MenuChoice = 'X';
}
```

In the above event handler, `InvalidateRect`(*rc*) is a powerful function that brings the control instantly to `OnPaint()` for updating a rectangular area marked by its argument *rc*, which is a `CRect` object. Related to this function is a function called `Invalidate()` that updates the whole area instead of a portion of the main window. Both, `InvalidateRect()` and `Invalidate()` are called from a client function, outside of `OnPaint()`.

The above code assigns the active variable `MenuChoice` with one of the characters, "P," "L," or "C" if `pt` is inside one the three menu boxes marked by `rc1`, `rc2`, and `rc3`, respectively. Each of these three choices calls up the function `InvalidateRect()`, which updates a rectangular area in the window marked by `DrawRc` by transferring the control to the function `OnPaint()`. If `pt` is not inside any of the three boxes, `MenuChoice` will be assigned with the value of "X," which will not be doing anything.

The character variable MenuChoice activates the area of the screen marked by DrawRc. This is done through the function InvalidateRect(), which transfers control to OnPaint(), bringing along the value assigned to the variable MenuChoice. This character variable determines a curve to be displayed in the drawing area, either a polynomial, a lemniscate, or Creative Net.

Drawing a Polynomial. A polynomial is displayed by the function Polynomial() through the option MenuChoice='P'. Figure 5.5 shows an output showing this option. It consists of a display of the polynomial $y = f(x) = x(x + 1)(x - 1)(x - 3)$ for $-2 \le x \le 4$. The function is written as follows:

```
void CCode5A::Polynomial()
{
        CClientDC dc(this);
        double X,Y;
        CPoint pt;
        DrawAxes();

        // draw the curve, where X is magnified 50 times, Y 20 times
        X=-2;
        while (X<=4)
        {
                Y=X*(X+1)*(X-1)*(X-3);
                pt.x=(int)(50*X);
                pt.y=(int)(20*Y);
                pt.x += Origin.x; pt.y=Origin.y-pt.y;
                if (DrawRc.PtInRect(pt))
                        dc.SetPixel(pt,Color1);
                X += 0.001;
        }
        dc.SetTextColor(Color1);
        dc.TextOut(TopLeft.x+20,BottomRight.y-50,"f(x) = x(x+1)(x-1)(x-3)");
}
```

In the above function, pt is a CPoint object used for plotting the points in the curve. The curve is drawn by plotting the points successively close to each other in such a way that they appear as one continuous curve. X and Y are two local variables of type double, representing the Cartesian coordinates (x, y). They are not to be confused with the elements x and y (both in lower case) linked from the objects pt. The value of X is converted to pt.x for mapping in Windows using Equation (5.1a). Since the increment of X is too small to be viewed in the window, its value is magnified by some large number, for example, 50, to make the pixels visible on the screen. In the same way, the value of Y is magnified 20 times and mapped as pt.y on the window using Equation (5.1b).

To make sure that the curve is drawn wholly inside the drawing area DrawRc only, the following conditional test is performed:

```
if (DrawRc.PtInRect(pt))
        dc.SetPixel(pt,Color1);
```

The Boolean function `PtInRect(pt)` with the parameter `pt` checks for the point of click of the `CPoint` object `pt`. If this point lies inside the rectangle `DrawRc`, then it returns TRUE (1); otherwise it returns FALSE (0).

Drawing a Lemniscate. The second item in the menu is the lemniscate, which is a curve based on polar coordinates. This curve is displayed when the rectangular box *Lemniscate* is clicked, as shown in Figure 5.6.

A polar curve is a curve expressed in the form of $r = f(t)$, where $x = r \cos t$ and $y = r \sin t$. In this function, r is the length of the curve from the origin and t is its angle of inclination. A polar representation is a suitable way for representing a curve instead of the normal form $y = f(x)$ in cases where y is not a function of x in some parts of the domain (see Figure 5.7).

The lemniscates $r = \sin kt$ and $r \cos kt$, where k is a constant, are special types of polar curves that are petal shaped. In general, the following rules apply to the shape of this type of curve:

If k is even then $r = \sin kt$ and $r \cos kt$ each have $2k$ petals.
If k is odd then $r = \sin kt$ and $r \cos kt$ each have k petals.

Figure 5.6 shows two polar curves, $r = \sin 3t$ and $r = \cos 8t$, for $0 \leq t \leq 8$. The curve $r = \sin 3t$ is shown as the darker curve. The first curve is drawn on the win-

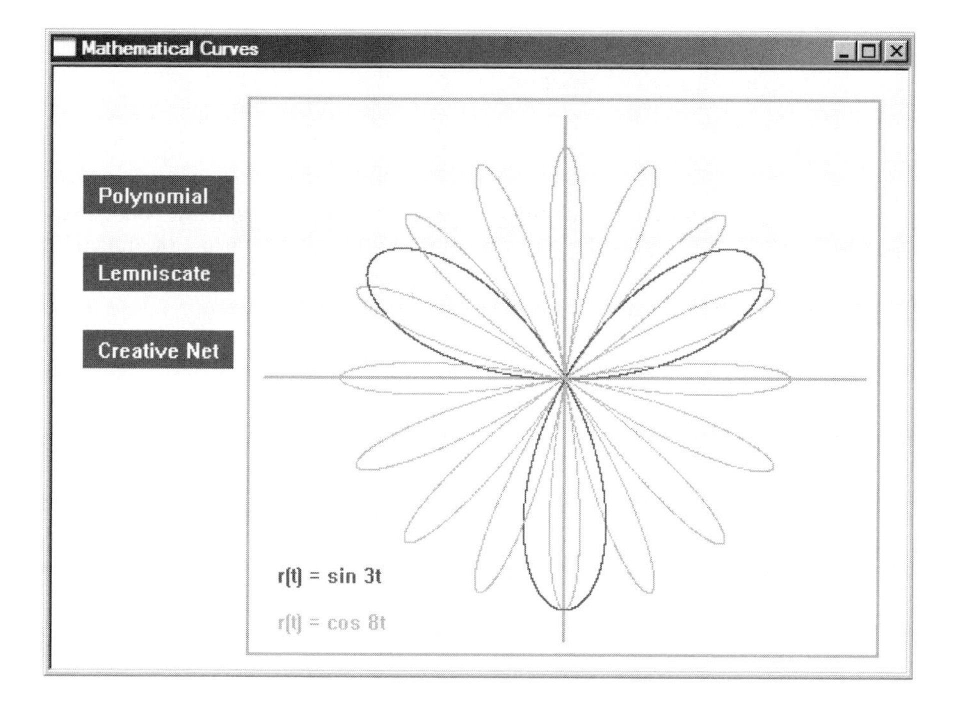

Figure 5.6 The lemniscate output from Code5A.

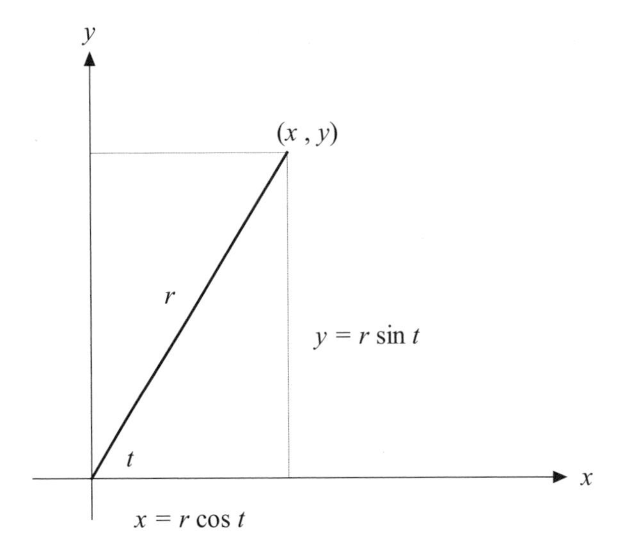

Figure 5.7 Polar representation of a point.

dow using the function `r1=sin(3*t)`, whereas the second function is `r2=cos(8*t)`. The points in the first curve are mapped on the window with their x and y values magnified 150 times to make them visible, as follows:

```
pt.x=(int)(150*r1*cos(t));
pt.y=(int)(150*r1*sin(t));
pt.x += Origin.x;
pt.y=Origin.y-pt.y;
```

The second curve is also mapped with the same magnification scale. In both cases, the `CPoint` object `pt` is used to draw the points on the curves using the function `SetPixel()`. The curves are drawn by iterating the local variable t from 0 to 8. The full code for this function is shown below:

```
void CCode5A::Lemniscate()
{
        CClientDC dc(this);
        double r1,r2,t;
        CPoint pt;
        DrawAxes();

        // draw the curves where x,y are all magnified 150 times
        t=0;
        while (t<=8)
        {
                r1=sin(3*t);
                pt.x=(int)(150*r1*cos(t));
```

```
            pt.y=(int)(150*r1*sin(t));
            pt.x += Origin.x;
            pt.y=Origin.y-pt.y;
            if (DrawRc.PtInRect(pt))
                    dc.SetPixel(pt,Color1);

            r2=cos(8*t);
            pt.x=(int)(150*r2*cos(t));
            pt.y=(int)(150*r2*sin(t));
            pt.x += Origin.x;
            pt.y=Origin.y-pt.y;
            if (DrawRc.PtInRect(pt))
                    dc.SetPixel(pt,Color2);
            t += .001;
        }
        dc.SetTextColor(Color1);
        dc.TextOut(TopLeft.x+20,BottomRight.y-60,"r(t) = sin 3t");
        dc.SetTextColor(Color2);
        dc.TextOut(TopLeft.x+20,BottomRight.y-30,"r(t) = cos 8t");
}
```

Drawing Creative Net. The last item in the menu generates a curve called *Creative Net,* which is a parametric curve with t as the parameter, given as follows:

$$x(t) = \sin 0.99t + 2 \cos 3.01t$$

$$y(t) = \cos 1.01t - 0.1 \sin 15.03t$$

The curve is plotted when the function `CreativeNet()` is called up. Creative Net is not a standard curve as it is produced by taking a sum of terms from some combination of sine and cosine functions. A different shape of the curve can be obtained by changing the coefficients of t in the above equations.

Figure 5.8 shows Creative Net when plotted in the interval $-150 \leq t \leq 150$. In `CreativeNet()`, (x, y) values are represented as `(pt.x,pt.y)`. The conversion from the Cartesian coordinate system to Windows involves the magnification of the x and y values by a factor 60 and 120, respectively, as follows:

```
x=sin(0.99*t)+2*cos(3.01*t);
y=cos(1.01*t)-0.1*sin(15.03*t);
pt.x=(int)(60*x);
pt.y=(int)(120*y);
```

We adopt a different strategy for drawing this curve. This time, the line drawing function `LineTo()` is used to draw the curve. The curve is drawn by first moving the pen to the origin using the function `MoveTo()`. The drawing starts by continuously iterating the x and y values using the function `LineTo()`. As the points are

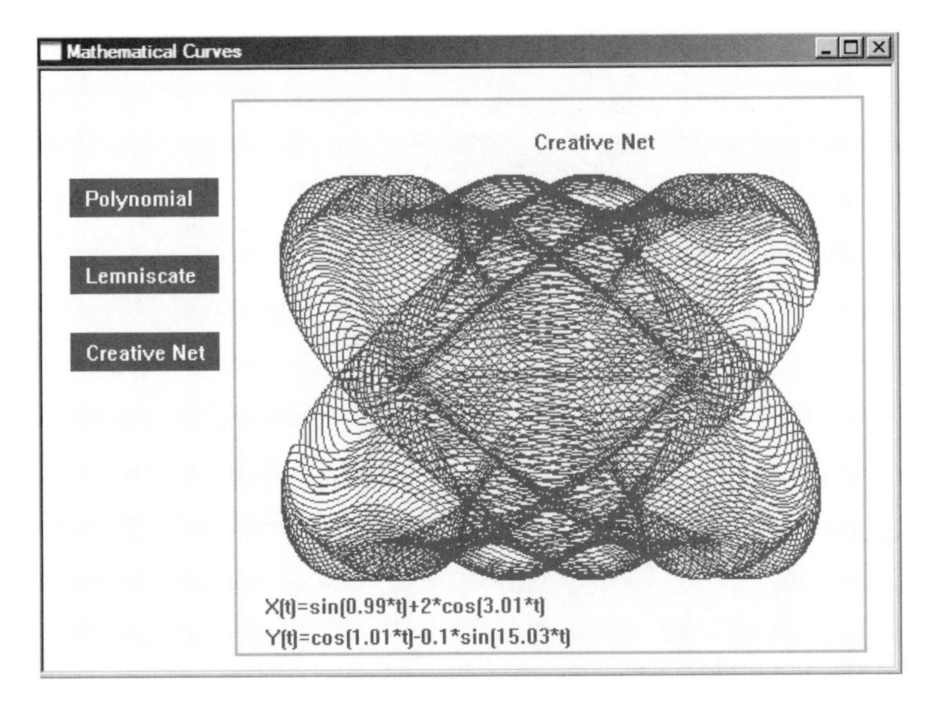

Figure 5.8 Creative Net output from Code5A.

very close to each other on the screen, the straight lines between them appear as one continuous curve. In this way, we obtain a smooth curve occupying the region, as shown in Figure 5.8.

With today's fast computers, each drawing is completed within a few microseconds. To see the line-drawing action, the plot can be paused for several milliseconds at each iteration using the function `Sleep(0)`. This extra statement is included to slow down the curve drawing process. `CreativeNet()` is written as follows:

```
void CCode5A::CreativeNet()
{
        CClientDC dc(this);
        double X,Y,t;
        CPen penDark(PS_SOLID,1,Color1);
        CPoint pt;

        dc.SetTextColor(Color1);
        dc.TextOut(TopLeft.x+200,TopLeft.y+20,"Creative Net");
        dc.TextOut(TopLeft.x+20,BottomRight.y-40,
              "X(t)=sin(0.99*t)+2*cos(3.01*t)");
        dc.TextOut(TopLeft.x+20,BottomRight.y-20,
              "Y(t)=cos(1.01*t)-0.1*sin(15.03*t)");
```

```
        dc.SelectObject(&penDark);
        t=-150;
        X=sin(0.99*t)+2*cos(3.01*t); Y=cos(1.01*t)-0.1*sin(15.03*t);
        pt.x=(int)(60*X); pt.y=(int)(120*Y);
        pt.x += Origin.x; pt.y=Origin.y-pt.y;

        dc.MoveTo(pt);
        while (t<=150)
        {
              X=sin(0.99*t)+2*cos(3.01*t);
              Y=cos(1.01*t)-0.1*sin(15.03*t);
              pt.x=(int)(60*X);
              pt.y=(int)(120*Y);
              pt.x += Origin.x; pt.y = Origin.y-pt.y;
              if (DrawRc.PtInRect(pt))
                    dc.LineTo(pt.x,pt.y);
              ::Sleep(0);
              t += 0.005;
        }
}
```

5.4 CUBIC SPLINE INTERPOLATION

The previous application shows several ways of plotting a curve using MFC functions when the curve function is given. It is also possible to draw a curve when its function is not given. In this case, the curve can be generated from a set of points either through interpolation or approximation. In *interpolation,* the curve is drawn so that it passes through all the given points. In *approximation,* the curve may pass through some or none of the points. A relationship is derived from these terminologies: interpolation is a form of approximation in which the curve passes through all the given points. A curve drawn through interpolation or approximation is a generalization of its close relationship to the points.

Curve interpolation can be performed numerically using many classical techniques such as the Lagrange, Newton, and Hermite methods, which produce polynomials. In approximation, several methods based on the least-squares method are applied. The least-squares method performs approximation by minimizing the sum of the squares of the errors to produce low-degree polynomials. Interested readers can refer to Burden and Faires [1] for details on these techniques.

An interesting curve that finds a lot of applications in mathematics is the spline. A *spline* is a set of piecewise continuous functions obtained through the interpolation of several points. A spline of degree n is defined as follows:

$$S_j(x) = a_j + b_j(x - x_j) + c_j(x - x_j)^2 + \ldots + d_j(x - x_j)^n \quad (5.4)$$

where a_j, b_j, c_j, and d_j are constants, for $0 \leq j \leq m - 1$. This spline interpolates $m + 1$ points (x_j, y_j) for $j = 0, 1, \ldots, m$ and it consists of m segments, $S_j(x)$ for $j = 0, 1,$

..., $m - 1$. A spline of degree two is called a *quadratic spline,* whereas a third degree spline is a *cubic spline*. Figure 5.9 shows a spline formed from four points having three segments.

The classical Lagrange and Newton methods for interpolation produce a polynomial whose degree is determined by the number of interpolating points. For n points, these methods produce a polynomial of degree $n - 1$. As the degree of the polynomial in these methods is dependent on the number of interpolating points, the higher the number of points, the greater is the degree of the polynomial. Therefore, these methods of interpolation are not suitable for applications requiring smooth curves, as a high-degree polynomial means there are many extremum points in the curve.

A spline is formed independently of the number of interpolating points. It is possible to obtain a smooth curve using a low-degree polynomial even if the number of interpolating points is large. For this reason, a spline is preferred over most other interpolating techniques for producing low-degree polynomials.

A derivative of the cubic spline called B-Spline is often used to help in the design of smooth surfaces such as the bodies of ships, cars, and aircraft. In contrast to the interpolating spline, B-spline is obtained by approximating the curve in the convex hull formed from a set of points. Another form is the Bezier curve, which is widely used for the same purposes.

In this section, we discuss a cubic spline and its interpolative technique for drawing the curve. A cubic spline consists of a set of piecewise polynomials of degree three, or $n = 3$, in Equation (5.4). A segment $S_j(x)$ in a cubic spline is obtained by interpolating three successive points. Therefore, a cubic spline can be constructed from a minimum of three points.

In order to have a smooth curve at each connecting point, an interpolating point must be continuous both from left and right. In addition, continuity must also be observed in terms of the first and second derivatives at each interpolating point in the curve. The full properties of a cubic spline are listed as follows:

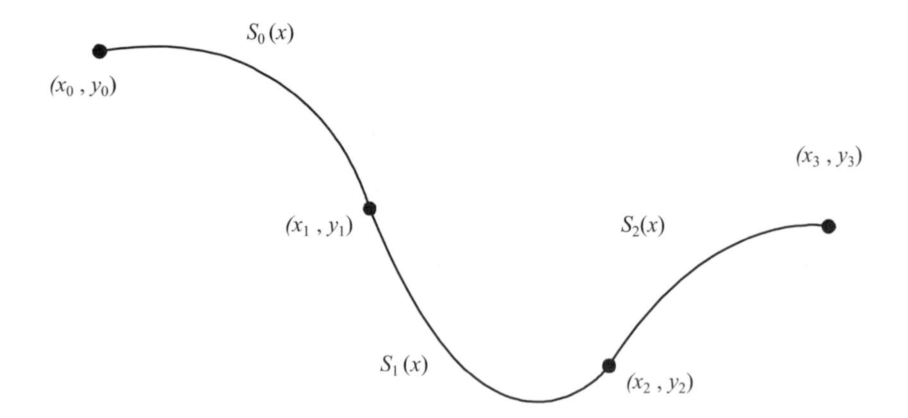

Figure 5.9 A cubic spline formed from three segments.

1. $S(x_j) = y_j$, for $j = 0, 1, 2, \ldots, m$
2. $S_{j+1}(x_{j+1}) = S_j(x_{j+1})$, for $j = 0, 1, 2, \ldots, m-2$
3. $S'_{j+1}(x_{j+1}) = S'_j(x_{j+1})$, for $j = 0, 1, 2, \ldots, m-2$
4. $S''_{j+1}(x_{j+1}) = S''_j(x_{j+1})$, for $j = 0, 1, 2, \ldots, m-2$
5. $S''(x_0) = S''(x_m) = 0$

The following algorithm constructs a cubic spline for interpolating a set of points (x_i, y_i), for $i = 0, 1, 2, \ldots, m$:

for $j = 0$ to m
  let $a_j = y_j$;
endfor
for $i = 0$ to $m - 1$
  let $h_i = x_{i+1} - x_i$;
endfor
for $i = 1$ to $m - 1$
  compute $\alpha_i = \dfrac{3[a_{i+1}h_{i-1} - a_i(h_i + h_{i-1}) + a_{i-1}h_i]}{h_i h_{i-1}}$;
endfor
let $l_0 = 1$, $\mu_0 = 0$ and $z_0 = 0$;
for $i = 1$ to $m - 1$
  compute $l_i = 2(h_i + h_{i-1}) - h_{i-1}\mu_{i-1}$;
  compute $\mu_i = \dfrac{h_i}{l_i}$;
  compute $z_i = \dfrac{\alpha_i - h_{i-1}z_{i-1}}{l_i}$;
endfor
let $l_m = 1$, $z_m = c_m = 0$;
for $j = m - 1$ downto 0
  compute $c_j = z_j - \mu_j c_{j+1}$
  compute $b_j = \dfrac{a_{j+1} - a_j}{h_j} - \dfrac{h_j(c_{j+1} - 2c_j)}{3}$
  compute $d_j = \dfrac{c_{j+1} - c_j}{3h_j}$
endfor

Code5B: Constructing a Cubic Spline

The project Code5B implements the above algorithm. The interpolating points are represented by the coordinates `(X[],Y[])`. The function `ComputeCSpline()` in this project reads these values and produces the cubic spline based on the above algorithm. The results are the arrays `a[j]`, `b[j]`, `c[j]`, and `d[j]` that represent a_j, b_j, c_j, and d_j, respectively in Equation (5.4).

We discuss a friendly interface for constructing a cubic spline in Code5B based on the above algorithm. Figure 5.10 shows an output from this project with a cubic spline having eight segments, with $m = 8$, obtained from nine interpolating points. The program allows the user to choose the points by clicking the mouse's left button on any point in the drawing area. The clicked points are indexed from left to right, and their coordinates are shown on the top-right-hand side of the window. A maximum of 10 points are allowed in this application but this number can be increased or decreased by changing the macro in the header file. The cubic spline is displayed when the user clicks the push button *Plot Curve*. Besides the curve, the output also includes the coefficient values a_j, b_j, c_j, and d_j of the segments $S_j(x)$ in Equation (5.4), shown in the right half of the window.

Code5B has two files: Code5B.h and Code5B.cpp. The class used in this application is `CCode5B` and it is derived from the MFC class `CFrameWnd`. The constants used in this application are the maximum number of clicked points `M` and the push-button id `IDC_PUSHBUTTON`. The global objects and variables in this project are shown in Table 5.6.

In constructing a cubic spline, we first need to find the values of `a[j]`, `b[j]`, `c[j]`, and `d[j]` in Equation (5.4). These values are the coefficients values of $S_j(x)$ as described in Equation (5.4). The interpolating points in this problem are `X[j]` and `Y[j]` for `j=0,1,...,m`.

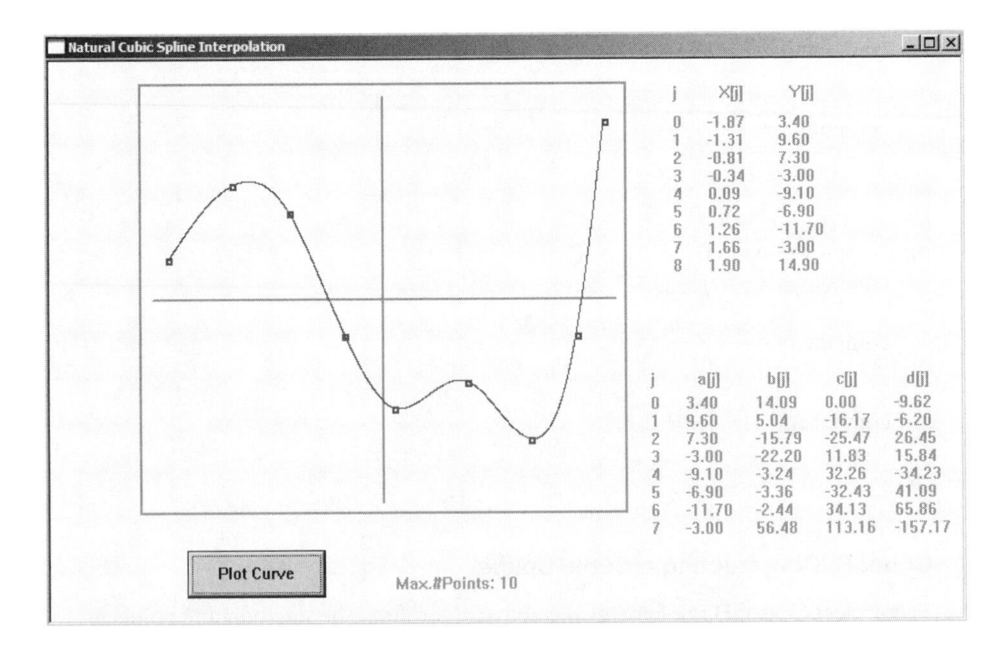

Figure 5.10 Cubic spline output from Code5B.

Table 5.6 Global objects and variables in Code5B

Variables/objects	Type	Description
`TopLeft, BottomRight`	`CPoint`	Top-left and bottom-right coordinates of the drawing area
`a[j],b[j],c[j],d[j]`	`double`	The coefficients a_j, b_j, c_j, and d_j in $S_j(x)$
`X[j],Y[j]`	`double`	The interpolating points (x_j, y_j)
`m`	`int`	Number of interpolated points
`Pushbutton`	`CButton`	`CButton` object that activates `OnClickCalc()`
`PushbuttonRect`	`CRect`	`CRect` object for the push button
`Color1, Color2`	`int`	Colors for both text and graphics
`Push buttonClick`	`int`	The flag for checking the number of left mouse clicks

The class `CCode5B` has member functions as listed in Table 5.7. It includes three event handlers, `OnPaint()`, `OnLButtonDown()`, and `OnClick-Calc()`. Another function, `ComputeCSpline()`, computes the spline from the interpolated points.

The constructor `CCode5B()` constructs the class, creates the window, and initializes all the global variables in the application. They are shown as follows:

```
a=new double [M+1];
b=new double [M+1];
c=new double [M+1];
d=new double [M+1];
X=new double [M+1];
Y=new double [M+1];
TopLeft=CPoint(80,20);
BottomRight=CPoint(500,380);
Origin=CPoint((TopLeft.x+BottomRight.x)/2,(TopLeft.y+BottomRight.y)/2);
m=0; Push buttonClick=0;
Color1=RGB(100,100,100);
Color2=RGB(150,150,150);
```

In the above initializations, the origin of the Cartesian coordinates is a `CPoint` object called `Origin`. This point has the Cartesian coordinates (0,0) and Windows coordinates as given by Equations (5.1a) and (5.1b). For displaying text and graphics in the window, two gray-scale colors `Color1` and `Color2` are used. The push-button child window called *Plot Curve* is created as follows:

```
Push buttonRc=CRect(120,410,240,450);
Push button.Create("Plot Curve",WS_CHILD | WS_VISIBLE
       | BS_DEFPUSH BUTTON, Push buttonRc,this, IDC_PUSH BUTTON);
```

The initial display event is handled by the function `OnPaint()`. The display in

Table 5.7 Member functions in Code5B

Function	Description
`CCode5B()`	The constructor of the class. This function also initializes the values of several variables.
`OnPaint()`	Provides the initial display in response to the message handler `WM_PAINT`
`OnLButtonDown()`	Responds to the message handler `WM_LBUTTONDOWN`, which records the points clicked using the left mouse and displays small rectangles to indicate the points
`OnClickCalc()`	Responds to the push button and message handler `BN_CLICKED`. This function reads the input from the interpolated points, sends these values to `ComputeCSpline()` for calculation, and displays the spline.
`ComputeCSpline()`	Computes the spline from the given interpolated points

the main window consists of a rectangular area for drawing the curve and some messages, as follows:

```
void CCode5B::OnPaint()
{
	CPaintDC dc(this);
	CPen penGray(PS_SOLID,2,Color2);
	dc.SelectObject(&penGray);
	dc.Rectangle(CRect(CPoint(TopLeft),CPoint(BottomRight)));

	// draw the axes
	dc.MoveTo(Origin.x,TopLeft.y+10);
	dc.LineTo(Origin.x,BottomRight.y-10);
	dc.MoveTo(TopLeft.x+10,Origin.y);
	dc.LineTo(BottomRight.x-10,Origin.y);

	// plot the input values
	dc.SetTextColor(Color2);
	dc.TextOut(BottomRight.x-200,BottomRight.y+50,"Max.#Points: 10");
	dc.TextOut(BottomRight.x+40,TopLeft.y,"i");
	dc.TextOut(BottomRight.x+80,TopLeft.y,"X[i]");
	dc.TextOut(BottomRight.x+140,TopLeft.y,"Y[i]");
}
```

The event-handling function `OnLButtonDown()` responds to the mouse's left button click inside the drawing. The drawing area, in this case, is a rectangular region denoted by the `CRect` object `DrawRc`. Code5B allows the user to create the interpolating points by clicking the mouse's left button within the drawing area. This clicked point is read as `pt` and is converted to the local object `Point`. A conversion to the Cartesian coordinates is then performed on this point using Equations (5.1a) and (5.1b). The number of clicked points is represented by the variable `m`, and this number is less than or equals to maximum allowed, `M`. At each mouse click inside

DrawRc, the value of m increases by one to denote the current number of points. As in the previous application, we check the validity of this point for plotting inside the rectangle DrawRc using the function PtInRect(pt)). If rc is inside DrawRc, then this point is immediately assigned to the Cartesian arrays X[] and Y[].

The function OnLButtonDown() is written as follows:

```
void CCode5B::OnLButtonDown (UINT nFlags, CPoint pt)
{
        CPoint Point;
        CClientDC dc(this);
        CString s;
        CRect rc=CRect(CPoint(TopLeft),CPoint(BottomRight));
        CPen penGray(PS_SOLID,2,Color1);
        dc.SelectObject(&penGray);
        dc.SetTextColor(Color2);
        if (rc.PtInRect(pt))
               if (m<=M)
               {
                       Point=pt;
                       X[m]=(double)(Point.x-Origin.x)/100;
                       Y[m]=(double)(Origin.y-Point.y)/10;
                       dc.Rectangle(pt.x,pt.y,pt.x+5,pt.y+5);
                       s.Format("%d",m);
                       dc.TextOut(BottomRight.x+40,TopLeft.y+25+15*m,s);
                       s.Format("%.2lf",X[m]);
                       dc.TextOut(BottomRight.x+70,TopLeft.y+25+15*m,s);
                       s.Format("%.2lf",Y[m]);
                       dc.TextOut(BottomRight.x+130,TopLeft.y+25+15*m,s);
                       m++;
               }
}
```

The function OnClickCalc() responds to the click on the push button *Plot Curve*. This event is detected by BN_CLICKED in the message map. OnClick-Calc() calls up the function ComputeCSpline() to compute the interpolating cubic spline and plots this curve on the window, as follows:

```
void CCode5B::OnClickCalc()
{
        CClientDC dc(this);
        CString s;
        CPoint Point;
        CRect DrawRc=CRect(TopLeft.x+5,TopLeft.y+5,
                           BottomRight.x-5,BottomRight.y-5);
        double xC,yC;

        Push buttonClick++;
        if (Push buttonClick==1)
               m--;
        ComputeCSpline();                // compute the spline

        dc.SetTextColor(Color2);
```

```
        dc.TextOut(BottomRight.x+20,BottomRight.y-120,"j");
        dc.TextOut(BottomRight.x+60,BottomRight.y-120,"a[j]");
        dc.TextOut(BottomRight.x+120,BottomRight.y-120,"b[j]");
        dc.TextOut(BottomRight.x+180,BottomRight.y-120,"c[j]");
        dc.TextOut(BottomRight.x+240,BottomRight.y-120,"d[j]");

        // draw the curve here
        dc.SetTextColor(Color2);
        for (int j=0;j<=m-1;j++)
        {
               xC=X[j];
               while (xC<=X[j+1])
               {
                      yC=a[j]+b[j]*(xC-X[j])
                            +c[j]*pow(xC-X[j],2)+d[j]*pow(xC-X[j],3);
                      Point.x = (int)(100*xC)+Origin.x;
                      Point.y=Origin.y-(int)(10*yC);
                      if (DrawRc.PtInRect(Point))
                            dc.SetPixel(CPoint(Point),Color1);
                      xC += 0.001;
               }
               s.Format("%d",j);
               dc.TextOut(BottomRight.x+20,BottomRight.y-100+15*j,s);
               s.Format("%.2lf",a[j]);
               dc.TextOut(BottomRight.x+50,BottomRight.y-100+15*j,s);
               s.Format("%.2lf",b[j]);
               dc.TextOut(BottomRight.x+110,BottomRight.y-100+15*j,s);
               s.Format("%.2lf",c[j]);
               dc.TextOut(BottomRight.x+170,BottomRight.y-100+15*j,s);
               s.Format("%.2lf",d[j]);
               dc.TextOut(BottomRight.x+230,BottomRight.y-100+15*j,s);
        }
}
```

In `OnClickCalc()`, each Cartesian point to be plotted is represented by the coordinates `(xC,yC)`. The values of `(xC,yC)` are initially obtained from `(X[j],Y[j])` at each end of the segment $S_j(x)$. These coordinates are converted to the Windows coordinates `Point.x` and `Point.y` using Equations (5.1a) and (5.1b), respectively. It is not necessary to use an array for `Point` in this case as the coordinates are just temporary variables needed only for plotting the points on the curve.

The last function in this application is `ComputeCSpline()`. This function gets input from the interpolating points that have been converted to the Cartesian coordinates earlier in the function `OnLButtonClicked()`. The function `ComputeCSpline()` produces the values of `a[j]`, `b[j]`, `c[j]`, and `d[j]` for Equation (5.4) to generate the cubic spline based on the cubic spline algorithm discussed earlier. The following code fragment shows the function:

```
void CCode5B::ComputeCSpline()
{
       double *h, *l, *mu, *z, *alpha;
```

```
        h=new double [M+1];
        l=new double [M+1];
        mu=new double [M+1];
        z=new double [M+1];
        alpha=new double [M+1];
        for (int j=0;j<=m;j++)
                a[j] = Y[j];
        for (j=0;j<=m-1;j++)
                h[j] = X[j+1] - X[j];
        for (j=1;j<=m-1;j++)
                alpha[j] = 3*(a[j+1]*h[j-1]-a[j]
                        *(h[j]+h[j-1])+a[j-1]*h[j])/(h[j]*h[j-1]);

        l[0]=1; mu[0]=0; z[0]=0;
        for (j=1;j<=m-1;j++)
        {
                l[j]=2*(h[j]+h[j-1])-h[j-1]*mu[j-1];
                mu[j]=h[j]/l[j];
                z[j]=(alpha[j]-h[j-1]*z[j-1])/l[j];
        }

        l[m]=1; z[m]=0; c[m]=0;
        for (j=m-1;j>=0;j—)
        {
                c[j]=z[j]-mu[j]*c[j+1];
                b[j]=(a[j+1]-a[j])/h[j]-h[j]*(c[j+1]+2*c[j])/3;
                d[j]=(c[j+1]-c[j])/(3*h[j]);
        }
        delete h,l,mu,z,alpha;
}
```

5.5 SUMMARY AND CONCLUSION

Curve drawing is an important component in numerical simulation and visualization. A curve represents the visual relationship between its parameters, which is more acceptable to humans compared to a series of numbers. It is through a curve that the concepts and principles underlying the given problem are better understood.

We have discussed two approaches for constructing curves in this chapter. The first method presents the fundamental approach to drawing three types of curves, that is, through their functions. The method involves a series of iterations on the parameters in the functions to produce a polynomial, lemniscate, and a parametric curve called Creative Net. The second approach is about constructing a cubic spline. This method is more challenging as the function is not given. Instead, a set of (x, y) values is read from the user input and the curve is constructed by interpolating these points. In the cubic spline approach, input comes in the form of the mouse's left clicks. The user places the points visually in the window by clicking the mouse. The spline is generated when a push button in the window is clicked.

This provides a friendly interface to the problem that benefits the user and makes the problem easy to understand.

Curve construction finds applications in many areas of computer aided designs. Cubic spline is a fundamental topic for constructing low-degree curves, such as B-splines and Bezier curves. These curves have applications in the design of objects with smooth surfaces, such as the bodies of aircraft, ships, and cars. A good understanding on this topic will definitely pave the way for exploring these areas further.

BIBLIOGRAPHY

1. R. L. Burden and J. D. Faires, *Numerical Analysis,* Brooks Cole, 2000.
2. S. C. Chapra and R. Canale, *Numerical Methods for Engineers,* McGraw-Hill, 2001.
3. G. D. Knott, *Interpolating Cubic Splines (Systems and Control),* Birkhauser Verlag, 2000.
4. H. Spath, *One Dimensional Spline Interpolation Algorithms,* AK Peters Ltd., 1995.
5. E. Cohen, R. F. Riesenfeld, and G. Elber, *Geometric Modeling with Splines: An Introduction,* AK Peters Ltd, 2001.

CODE LISTINGS

Code5A: Mathematical Curves

```
// code5A.h
#include <afxwin.h>
#include <math.h>

class CCode5A : public CFrameWnd
{
private:
       CPoint TopLeft, BottomRight, Origin;
       CPoint pt1,pt2,pt3;
       CRect DrawRc, rc1, rc2, rc3;
       int Color1,Color2;
       char MenuChoice;
public:
       CCode5A();
       ~CCode5A()                    {}
       afx_msg void OnPaint();
       afx_msg void OnLButtonDown(UINT nFlags,CPoint pt);
       void Polynomial(),Lemniscate(),CreativeNet(),DrawAxes();
       DECLARE_MESSAGE_MAP();
};

class CMyWinApp : public CWinApp
{
public:
  virtual BOOL InitInstance();
};
```

```
// code5A.cpp
#include "chap5A.h"
BOOL CMyWinApp::InitInstance()
{
        CCode5A* pFrame = new CCode5A;
        m_pMainWnd = pFrame;
        pFrame->ShowWindow(SW_SHOW);
        pFrame->UpdateWindow();
        return TRUE;
}

CMyWinApp  MyApplication;

BEGIN_MESSAGE_MAP(CCode5A, CFrameWnd)
        ON_WM_PAINT()
        ON_WM_LBUTTONDOWN()
END_MESSAGE_MAP()

CCode5A::CCode5A()
{
        Create(NULL,"Mathematical Curves",
               WS_OVERLAPPEDWINDOW,CRect(20,30,600,450));
        TopLeft=CPoint(130,20); BottomRight=CPoint(550,380);
        Origin.x=(TopLeft.x+BottomRight.x)/2;
        Origin.y=(TopLeft.y+BottomRight.y)/2;
        DrawRc=CRect(TopLeft.x+3,TopLeft.y+3,BottomRight.x-3,BottomRight.y-3);
        pt1=CPoint(20,70);
        rc1=CRect(pt1,pt1+CPoint(100,25));
        pt2=CPoint(20,120);
        rc2=CRect (pt2,pt2+CPoint(100,25));
        pt3=CPoint(20,170);
        rc3=CRect (pt3,pt3+CPoint(100,25));
        Color1=RGB(100,100,100);
        Color2=RGB(170,170,170);
}

void CCode5A::OnPaint()
{
        CPaintDC dc(this);
        CPen penDark(PS_SOLID,2,Color2);
        dc.SelectObject(&penDark);
        dc.Rectangle(CRect(CPoint(TopLeft),CPoint(BottomRight)));
        dc.FillSolidRect(&rc1,Color1);
        dc.FillSolidRect(&rc2,Color1);
        dc.FillSolidRect(&rc3,Color1);
        dc.SetTextColor(RGB(255,255,255));
        dc.SetBkColor(Color1);
        dc.TextOut (pt1.x+10,pt1.y+5, "Polynomial");
        dc.TextOut (pt2.x+10,pt2.y+5, "Lemniscate");
        dc.TextOut (pt3.x+10,pt3.y+5, "Creative Net");
        switch(MenuChoice)
        {
               case 'P':
                      CCode5A::Polynomial();break;
```

```
            case 'L':
                    CCode5A::Lemniscate();break;
            case 'C':
                    CCode5A::CreativeNet();break;
        }
}

void CCode5A::OnLButtonDown(UINT nFlags,CPoint pt)
{
        if (rc1.PtInRect(pt))
        {
                MenuChoice = 'P';
                InvalidateRect(DrawRc);
        }
        if (rc2.PtInRect(pt))
        {
                MenuChoice = 'L';
                InvalidateRect(DrawRc);
        }
        if (rc3.PtInRect(pt))
        {
                MenuChoice = 'C';
                InvalidateRect(DrawRc);
        }
        if (!rc1.PtInRect(pt) && !rc2.PtInRect(pt)
                                    && !rc3.PtInRect(pt))
                MenuChoice = 'X';
}

void CCode5A::DrawAxes()
{
        CClientDC dc(this);
        CPen penGray(PS_SOLID,2,Color2);
        dc.SelectObject(&penGray);
        dc.MoveTo(Origin.x,TopLeft.y+10);
        dc.LineTo(Origin.x,BottomRight.y-10);
        dc.MoveTo(TopLeft.x+10,Origin.y);
        dc.LineTo(BottomRight.x-10,Origin.y);
}

void CCode5A::Polynomial()
{
        CClientDC dc(this);
        double X,Y;
        CPoint pt;
        DrawAxes();

        // draw the curve, where X is magnified 50 times, Y 20 times
        X=-2;
        while (X<=4)
        {
                Y=X*(X+1)*(X-1)*(X-3);
                pt.x=(int)(50*X);
```

```
            pt.y=(int)(20*Y);
            pt.x += Origin.x; pt.y=Origin.y-pt.y;
            if (DrawRc.PtInRect(pt))
                  dc.SetPixel(pt,Color1);
            X += 0.001;
      }
      dc.SetTextColor(Color1);
      dc.TextOut(TopLeft.x+20,BottomRight.y-50,"f(x) = x(x+1)(x-1)(x-3)");
}

void CCode5A::Lemniscate()
{
      CClientDC dc(this);
      double r1,r2,t;
      CPoint pt;
      DrawAxes();

      // draw the curves where x,y are all magnified 150 times
      t=0;
      while (t<=8)
      {
            r1=sin(3*t);
            pt.x=(int)(150*r1*cos(t)); pt.y=(int)(150*r1*sin(t));
            pt.x += Origin.x; pt.y=Origin.y-pt.y;
            if (DrawRc.PtInRect(pt))
                  dc.SetPixel(pt,Color1);

            r2=cos(8*t);
            pt.x=(int)(150*r2*cos(t)); pt.y=(int)(150*r2*sin(t));
            pt.x += Origin.x; pt.y=Origin.y-pt.y;
            if (DrawRc.PtInRect(pt))
                  dc.SetPixel(pt,Color2);
            t += .001;
      }
      dc.SetTextColor(Color1);
      dc.TextOut(TopLeft.x+20,BottomRight.y-60,"r(t) = sin 3t");
      dc.SetTextColor(Color2);
      dc.TextOut(TopLeft.x+20,BottomRight.y-30,"r(t) = cos 8t");
}

void CCode5A::CreativeNet()
{
      CClientDC dc(this);
      double X,Y,t;
      CPen penDark(PS_SOLID,1,Color1);
      CPoint pt;

      dc.SetTextColor(Color1);
      dc.TextOut(TopLeft.x+200,TopLeft.y+20,"Creative Net");
      dc.TextOut(TopLeft.x+20,BottomRight.y-40,
            "X(t)=sin(0.99*t)+2*cos(3.01*t)");
      dc.TextOut(TopLeft.x+20,BottomRight.y-20,
            "Y(t)=cos(1.01*t)-0.1*sin(15.03*t)");
```

```
        dc.SelectObject(&penDark);
        t=-150;
        X=sin(0.99*t)+2*cos(3.01*t); Y=cos(1.01*t)-0.1*sin(15.03*t);
        pt.x=(int)(60*X); pt.y=(int)(120*Y);
        pt.x += Origin.x; pt.y=Origin.y-pt.y;

        dc.MoveTo(pt);
        while (t<=150)
        {
                X=sin(0.99*t)+2*cos(3.01*t);
                Y=cos(1.01*t)-0.1*sin(15.03*t);
                pt.x=(int)(60*X);
                pt.y=(int)(120*Y);
                pt.x += Origin.x; pt.y = Origin.y-pt.y;
                if (DrawRc.PtInRect(pt))
                        dc.LineTo(pt.x,pt.y);
                ::Sleep(0);
                t += 0.005;
        }
}
```

Code5B: Natural Cubic Spline

```
// code5B.h
#include <afxwin.h>
#include <math.h>
#define M 10
#define IDC_PUSH BUTTON 301

class CCode5B : public CFrameWnd
{
private:
        CPoint TopLeft, BottomRight, Origin;
        double *a,*b,*c,*d;
        double *X,*Y;
        int m, Push buttonClick;
        int Color1,Color2;
        CButton Push button;
        CRect Push buttonRect;
public:
        CCode5B();
        ~CCode5B();
        afx_msg void OnPaint();
        afx_msg void OnLButtonDown (UINT nFlags, CPoint pt);
        afx_msg void OnClickCalc();
        void ComputeCSpline();
        DECLARE_MESSAGE_MAP();
};

class CMyWinApp : public CWinApp
{
public:
  virtual BOOL InitInstance();
};
```

```
// Natural Cubic Spline Interpolation
#include "chap5B.h"
BOOL CMyWinApp::InitInstance()
{
        CCode5B* pFrame = new CCode5B;
        m_pMainWnd = pFrame;
        pFrame->ShowWindow(SW_SHOW);
        pFrame->UpdateWindow();
        return TRUE;
}

CMyWinApp  MyApplication;

BEGIN_MESSAGE_MAP(CCode5B, CFrameWnd)
        ON_WM_PAINT()
        ON_WM_LBUTTONDOWN()
    ON_BN_CLICKED  (IDC_PUSH BUTTON,OnClickCalc)
END_MESSAGE_MAP()

CCode5B::CCode5B()
{
        a=new double [M+1];
        b=new double [M+1];
        c=new double [M+1];
        d=new double [M+1];
        X=new double [M+1];
        Y=new double [M+1];
        Create(NULL,"Natural Cubic Spline Interpolation",
               WS_OVERLAPPEDWINDOW,CRect(0,0,800,500));
        Push buttonRect=CRect(120,410,240,450);
        Push button.Create("Plot Curve",WS_CHILD | WS_VISIBLE
               | BS_DEFPUSH BUTTON, Push buttonRect,this, IDC_PUSH BUTTON);

        TopLeft=CPoint(80,20); BottomRight=CPoint(500,380);
        Origin=CPoint((TopLeft.x+BottomRight.x)/2,(TopLeft.y+BottomRight.y)/2);
        m=0; Push buttonClick=0;
        Color1=RGB(100,100,100);
        Color2=RGB(150,150,150);
}

CCode5B::~CCode5B()
{
        delete a,b,c,d,X,Y;
}

void CCode5B::OnPaint()
{
        CPaintDC dc(this);
        CPen penGray(PS_SOLID,2,Color2);
        dc.SelectObject(&penGray);
        dc.Rectangle(CRect(CPoint(TopLeft),CPoint(BottomRight)));

        // draw the axes
        dc.MoveTo(Origin.x,TopLeft.y+10);
        dc.LineTo(Origin.x,BottomRight.y-10);
```

```
        dc.MoveTo(TopLeft.x+10,Origin.y);
        dc.LineTo(BottomRight.x-10,Origin.y);
        // plot the input values
        dc.SetTextColor(Color2);
        dc.TextOut(BottomRight.x-200,BottomRight.y+50,"Max.#Points: 10");
        dc.TextOut(BottomRight.x+40,TopLeft.y,"j");
        dc.TextOut(BottomRight.x+80,TopLeft.y,"X[j]");
        dc.TextOut(BottomRight.x+140,TopLeft.y,"Y[j]");
}

void CCode5B::OnLButtonDown (UINT nFlags, CPoint pt)
{
        CPoint Point;
        CClientDC dc(this);
        CString s;
        CRect rc=CRect(CPoint(TopLeft),CPoint(BottomRight));
        CPen penGray(PS_SOLID,2,Color1);
        dc.SelectObject(&penGray);
        dc.SetTextColor(Color2);
        if (rc.PtInRect(pt))
               if (m<=M)
               {
                      Point=pt;
                      X[m]=(double)(Point.x-Origin.x)/100;
                      Y[m]=(double)(Origin.y-Point.y)/10;
                      dc.Rectangle(pt.x,pt.y,pt.x+5,pt.y+5);
                      s.Format("%d",m);
                      dc.TextOut(BottomRight.x+40,TopLeft.y+25+15*m,s);
                      s.Format("%.2lf",X[m]);
                      dc.TextOut(BottomRight.x+70,TopLeft.y+25+15*m,s);
                      s.Format("%.2lf",Y[m]);
                      dc.TextOut(BottomRight.x+130,TopLeft.y+25+15*m,s);
                      m++;
               }
}

void CCode5B::OnClickCalc()
{
        CClientDC dc(this);
        CString s;
        CPoint Point;
        CRect DrawRc=CRect(TopLeft.x+5,TopLeft.y+5,
                           BottomRight.x-5,BottomRight.y-5);
        double xC,yC;

        Push buttonClick++;
        if (Push buttonClick==1)
               m—;
        ComputeCSpline();                    // computes the spline

        dc.SetTextColor(Color2);
        dc.TextOut(BottomRight.x+20,BottomRight.y-120,"j");
        dc.TextOut(BottomRight.x+60,BottomRight.y-120,"a[j]");
        dc.TextOut(BottomRight.x+120,BottomRight.y-120,"b[j]");
```

```
        dc.TextOut(BottomRight.x+180,BottomRight.y-120,"c[j]");
        dc.TextOut(BottomRight.x+240,BottomRight.y-120,"d[j]");
        // draw the curve here
        dc.SetTextColor(Color2);
        for (int j=0;j<=m-1;j++)
        {
               xC=X[j];
               while (xC<=X[j+1])
               {
                      yC=a[j]+b[j]*(xC-X[j])
                             +c[j]*pow(xC-X[j],2)+d[j]*pow(xC-X[j],3);
                      Point.x = (int)(100*xC)+Origin.x;
                      Point.y=Origin.y-(int)(10*yC);
                      if (DrawRc.PtInRect(Point))
                             dc.SetPixel(CPoint(Point),Color1);
                      xC += 0.001;
               }
               s.Format("%d",j);
               dc.TextOut(BottomRight.x+20,BottomRight.y-100+15*j,s);
               s.Format("%.2lf",a[j]);
               dc.TextOut(BottomRight.x+50,BottomRight.y-100+15*j,s);
               s.Format("%.2lf",b[j]);
               dc.TextOut(BottomRight.x+110,BottomRight.y-100+15*j,s);
               s.Format("%.2lf",c[j]);
               dc.TextOut(BottomRight.x+170,BottomRight.y-100+15*j,s);
               s.Format("%.2lf",d[j]);
               dc.TextOut(BottomRight.x+230,BottomRight.y-100+15*j,s);
        }
}

void CCode5B::ComputeCSpline()
{
        double *h, *l, *mu, *z, *alpha;
        h=new double [M+1];
        l=new double [M+1];
        mu=new double [M+1];
        z=new double [M+1];
        alpha=new double [M+1];
        for (int j=0;j<=m;j++)
               a[j] = Y[j];
        for (j=0;j<=m-1;j++)
               h[j] = X[j+1] - X[j];
        for (j=1;j<=m-1;j++)
               alpha[j] = 3*(a[j+1]*h[j-1]-a[j]
                      *(h[j]+h[j-1])+a[j-1]*h[j])/(h[j]*h[j-1]);

        l[0]=1; mu[0]=0; z[0]=0;
        for (j=1;j<=m-1;j++)
        {
               l[j]=2*(h[j]+h[j-1])-h[j-1]*mu[j-1];
               mu[j]=h[j]/l[j];
               z[j]=(alpha[j]-h[j-1]*z[j-1])/l[j];
        }
```

```
    l[m]=1; z[m]=0; c[m]=0;
    for (j=m-1;j>=0;j—)
    {
          c[j]=z[j]-mu[j]*c[j+1];
          b[j]=(a[j+1]-a[j])/h[j]-h[j]*(c[j+1]+2*c[j])/3;
          d[j]=(c[j+1]-c[j])/(3*h[j]);
    }
    delete h,l,mu,z,alpha;
}
```

CHAPTER 6

WORKING WITH IMAGES

6.1 HANDLING IMAGES

Images are represented in Windows as digital objects stored either in the computer memory or as files. An image is a graphical object displayed in Windows as a graphics device interface (GDI) object. As discussed earlier, GDI consists of functions from the `CDC` and `CGdiObject` classes to manage graphics within a Windows application. This graphical standard has a set of device-independent routines that can be used to draw, print, display, and store images and other graphical objects based on a device context. GDI objects include pens, brushes, fonts, palettes, and bitmaps. An image may be created internally using some drawing or painting tools. In most cases, an image is created from external sources such as scanners, digital cameras, and downloaded files. An image can be modified, edited, or deleted using tools provided in the device context, including pens, brushes, fonts, and color palettes.

An image is stored and displayed in Windows using several different file formats. Some of the most common formats are bitmap, JPEG, and GIF. The difference between these format lie mostly in their way of storing data and on factors like data compression, portability, and application. A bitmap image is represented by tiny units called pixels. This form of image is normally large in size as it is normally not compressed. JPEG and GIF file formats, on the other hand, are represented as compressed files, or files smaller than what they actually are.

JPEG stands for *Joint Photographic Experts Group* and is a lossy compression method standardized by the International Standards Organization (ISO) in 1990. Each JPEG file is small in size as it is stored in a compressed format, which results in some loss in quality. Size reduction is accomplished by grouping the data into several squares and applying the discrete cosine transform method for turning these squares into a curve. The file keeps track of changes when updates are made.

Through this process, some not-so-critical data is lost to achieve compression. For example, compression may degrade or distort the original colors of the image. The JPEG standard is widely used for storing large-size images or photographs based on 24 bits of data and 16.7 million colors. The format is suitable for storing still images with continuous tones, as in high-quality photographs and images. JPEG does not support animation.

GIF, or *Graphics Interchange Format,* is suitable for images with large areas of flat color. The format compresses an image without losing any data from the original image. This is done through a compression method that stores important keys of the data in a hash table. However, the format supports a palette of less than 256 colors in an 8-bit format. GIF is suitable for storing icons, clip art, buttons, and images for animation. Unlike JPEG, GIFF is quicker to load, making it suitable for moving applications such as graphics animation.

The JPEG and GIF file formats are more open and portable as they can be read from different computer systems, including Unix and Apple Macintosh. The bitmap format, on the other hand, is confined to the Microsoft Windows system only. However, there are many tools available in the market to convert a bitmap into other file formats, and vice versa.

In this chapter, we discuss some fundamental concepts in using MFC for manipulating images. Two examples are presented. The first example shows how a color image is converted into several monotone scales. The second example explores a fundamental area of study in image processing involving the edge-detection problem.

6.2 BITMAP FILE FORMAT

A *bitmap* is a native file format for the Microsoft Windows environment for displaying images based on a rectangular mesh of cells called pixels. To the computer, a bitmap represents a drawing surface in the computer memory that can be manipulated by the device context in the window. A bitmap is a form of raster data storage with almost no compression. As a result, a bitmap file occupies a large storage area in the disk, which makes it inferior to the JPEG and GIF formats in terms of the size of data storage.

There are two types of bitmaps. The first is the *device-dependent bitmap,* or DDB. A DDB does not have a color palette and, therefore, is dependent on devices such as the screen for colors. DDBs are used mostly for transferring the information of an image between the memory and the screen. This facility is especially useful in applications requiring quick graphics redraws and updates, such as in graphics animation. The DDB format is seldom stored in a disk.

The second type is the *device-independent bitmap,* or DIB. The DIB includes a color palette that is stored with the BMP file extension. The DIB format is more relevant to Windows as it is not dependent on specific devices. A DIB file consists of a file header, a bitmap header, a bitmap color table, and the data. Unlike a DDB, a DIB operation usually involves storage in a disk.

Raster Operations Involving Bit Shifting

We discuss a method for manipulating the colors of an image by transforming the pixels of a colored image into red, green, blue, and black/white monotone scales. The transformation is made possible through a series of operations on the pixels called *bit shifting*. Bit shifting is a mathematical operation for moving some bits from a given string to another location within the string.

In Windows, a bitmap is represented by 24 bits of data, representing a total of 16,777,126 colors. An image is formed from the rectangular composition of pixels. Each pixel is represented by a string consisting of 24 bits of binary digits in the MFC function `RGB()`. The first eight bits in the string starting from the right form the red component. This is followed by eight bits of green in the middle, and the remaining bits make up blue.

Figure 6.1 shows a pixel in the shaded square having its value defined as `RGB(174,55,171)`, represented as a 24-bit string. The figure also shows the corresponding values in hexadecimals: AB for 171, 37 for 55, and AE for 174.

The monotone scales for red, green, and blue are easily obtained by blanking the other two color components, as follows:

Red	`RGB(r,0,0)`
Green	`RGB(0,g,0)`
Blue	`RGB(0,0,b)`

A grayscale color is obtained by setting `r=g=b` in `RGB(r,g,b)`. A solid black color is obtained by setting `r=g=b=0`, whereas `r=g=b=255` produces white.

A raster operation involving bitmaps combines one or two source pixels to pro-

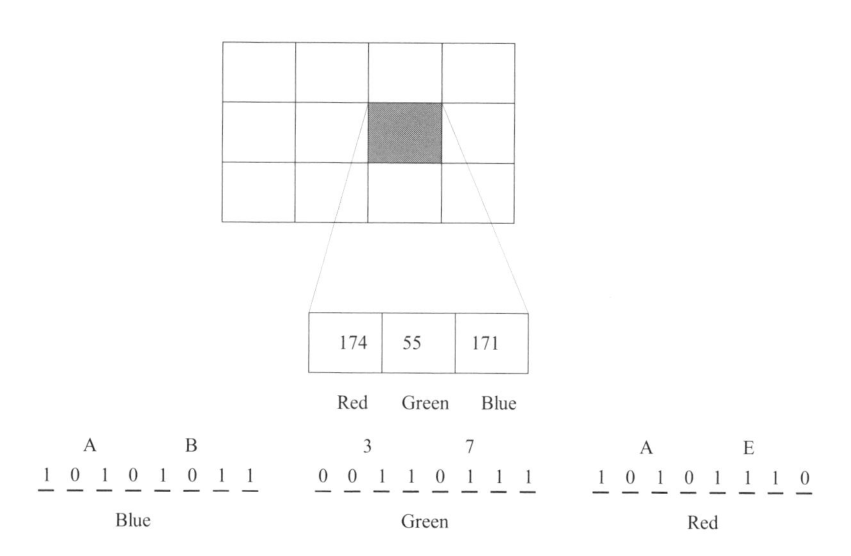

Figure 6.1 A pixel represented as a 24-bit string.

duce a destination pixel. Bit shifting is a form of raster operation that transforms the original color of a given pixel to a different color. Bits in a string can be shifted from the left or right. Left shifting is denoted as <<, whereas right shifting uses the symbol >>. In general, right shifting causes the string to lose its value as the vacated bits on the right are filled with zeros. Left shifting, on the other hand, may result in a gain or a loss in value for the original pixel.

Figure 6.2 (a) and (b) shows left and right shift operations on a 8-bit string. In Figure 6.2 (a), a right shift of 3 bits causes the bits 10101 occupying the first five places from the left to be shifted three places to the right. The new string has a binary value of 00010101, or 21 in decimal. The vacated three places from the left in this string are all filled with zeros. On the other hand, left shifting of 3 bits as shown in Figure 6.2 (b) produces a new binary value of 01110000, or 112 in decimal. In this case, the three places from the right are all filled with zeros.

Figure 6.3 shows an example of bit shifting on a pixel that has an original value of `RGB(181, 55, 174)`. We call the red, green and blue components of the pixel simply `r`, `g`, and `b`, respectively. The red component of a pixel occupies the first eight bits from the right. Therefore, a pixel can be converted into its red monotone scale by performing right shifting on the given pixel by 16 bits. A right shift of 16 bits is performed that results in a new value given by `r=RBG(174,0,0)`.

In a similar manner, the green component is obtained by right-shifting `r` by 8 bits to produce `r=RBG(0,174,0)`, as shown in the figure. Finally, we obtain the blue component of the pixel by right-shifting another 8 bits to `g` to get `b=RBG(0,0,174)`. The red, green, and blue components of the bitmap image are displayed at the bottom of this figure.

The operation extends to finding the grayscale version of the bitmap image, represented as `bw`. A grayscale image consists of pixels whose red, green, and blue components have equal values. This objective is achieved simply by adding the three components obtained earlier from the bit operations, as follows:

```
bw=r+g+b
  =RGB(174,0,0)+RGB(0,174,0)+RGB(0,0,174)
  =RGB(174,174,174)
```

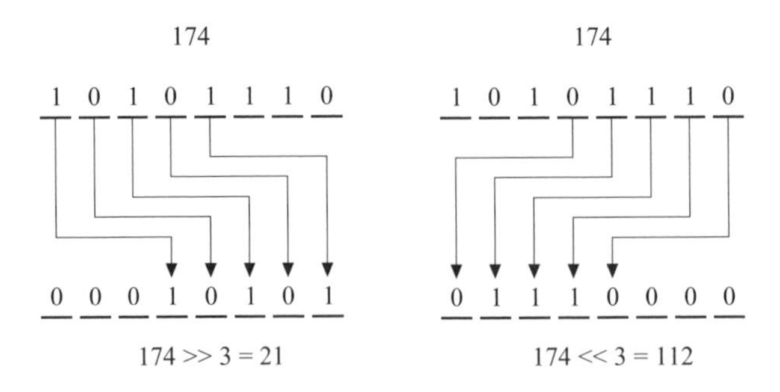

Figure 6.2 (a) Right shifting. (b) Left shifting.

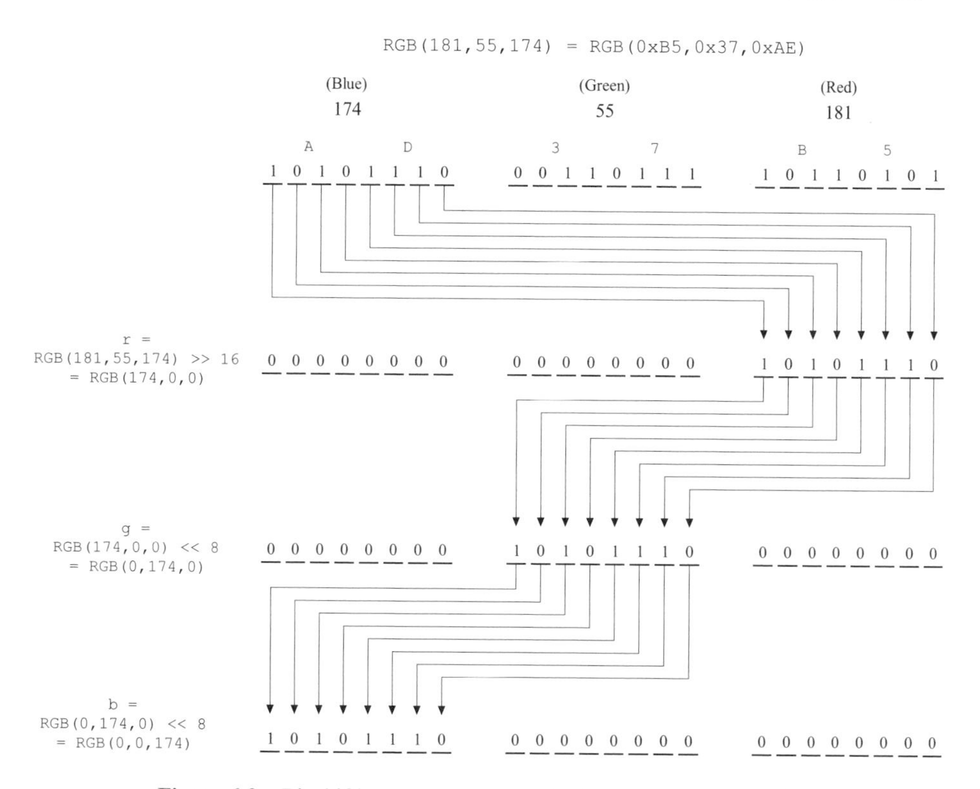

Figure 6.3 Bit shifting operations on a pixel into red, green, and blue.

Code6A: Demonstrating Bit Shifting

We discuss a way to manipulate an image by shifting the bits from the pixels of an image. Figure 6.4 shows the output from the project Code6A, including a color image (top left, shown here as a grayscale image), its extract copy (top right), and its manipulations in the form of bit shifting (the rest). The extract copy is obtained by copying a rectangular area from the original image as an array and pasting this array to its location. The red monotone scale of the image is obtained by right shifting the array by 16 bits. A similar operation involving shifting the red image to the left by 8 bits produces a green monotone image. A blue image is obtained by shifting the green image to the left by another 8 bits. Finally, the red, green, and blue images are then combined to produce a grayscale image.

The project Code6A consists of the files Code6A.cpp, Code6A.h, and Code6A.rc. The latter is a resource file that is needed in order to include a bitmap file in the project. A resource file also includes resources such as edit boxes, static boxes, dialog buttons, images, and list view windows. The global variables used in the application are listed in Table 6.1. Table 6.2 lists all the member functions used in this application.

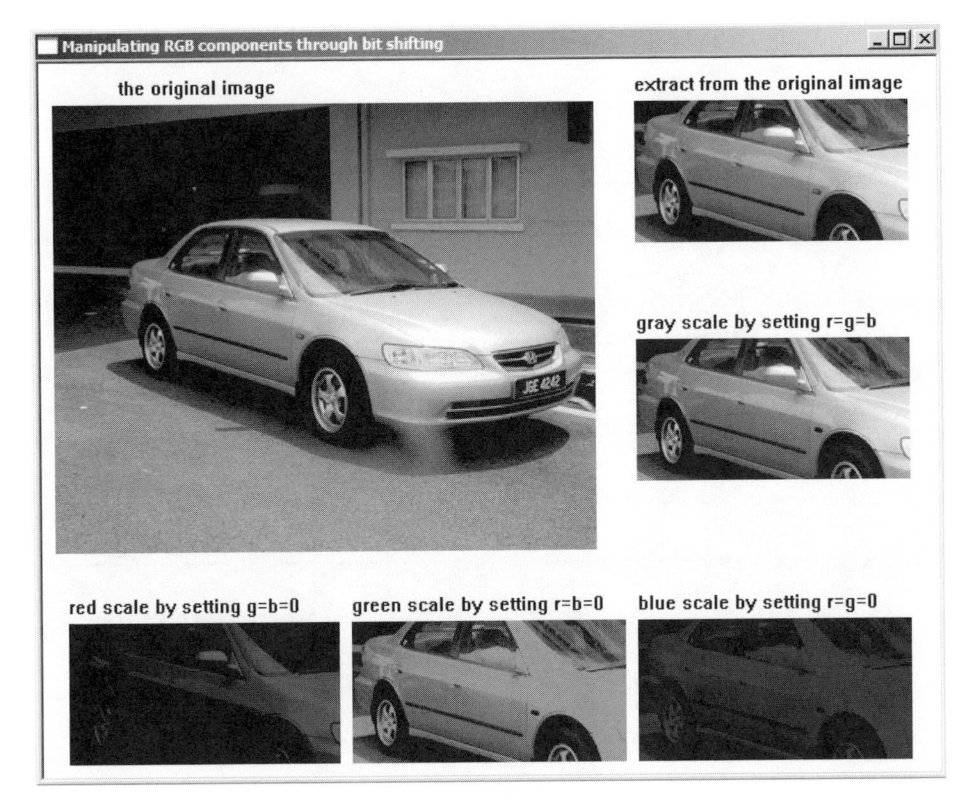

Figure 6.4 Transformation of a color image (original image, shown here as a grayscale image) into red, green, blue, and gray monotone scales.

The raster operation for displaying an image involves moving the bitmaps from the image to the memory device context currently in use. A *memory device context* is an object created from the class `CDC` that holds the image before displaying it on the screen. In this case, a Windows device context from `CPaintDC` is needed for the display. As the formats used in the memory device context and the device context may not be the same, an object is created to make these formats compatible with each other. Once the bitmap has been successfully loaded into the

Table 6.1 Global variables and objects in Code6A

Variable/object	Type	Description
`f[i][j]`	`int`	The pixel value at location `(i,j)` of the image, where `i` is the column and `j` is the row
`Home`	`CPoint`	Home or the top-left coordinates of the original image
`sImage`	`CSize`	Size of the sample image
`MyImage`	`CBitmap`	The sample image

Table 6.2 Functions used in Code6A

Function	Description
`CCode6A()`	The constructor of the class
`~CCode6A()`	The destructor of the class
`OnPaint()`	Provides the initial display in response to the message handler `ON_WM_PAINT`. The function also displays some instructions for using the program.
`OnLButtonDown()`	Responds to the message handler `ON_WM_LBUTTONDOWN`, which records the points clicked using the left mouse, and displays small rectangles to indicate the points

device context, it can be displayed at any position in the window through the device context.

Several steps are needed to accomplish this task. First, a resource file called Code6B.rc is created to enable an object to handle the image. This is achieved by clicking the mouse's right button on *Resource Files*. A menu appears, as shown in Figure 6.5. Choose *Add* and another submenu appears. Choose *Add Resources* from this submenu.

The image is marked by clicking the mouse's right button on *Bitmap* and assigned with an id inside the resource file according to the steps shown in Figure 6.6.

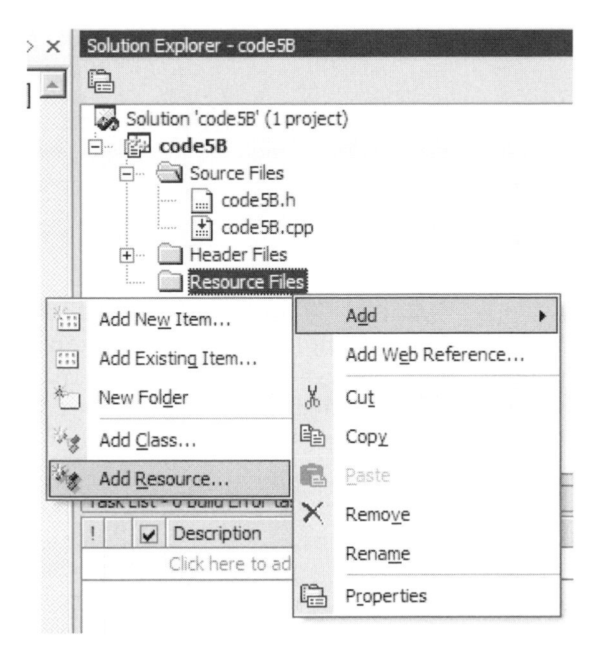

Figure 6.5 Creating the resource file Code6B.rc.

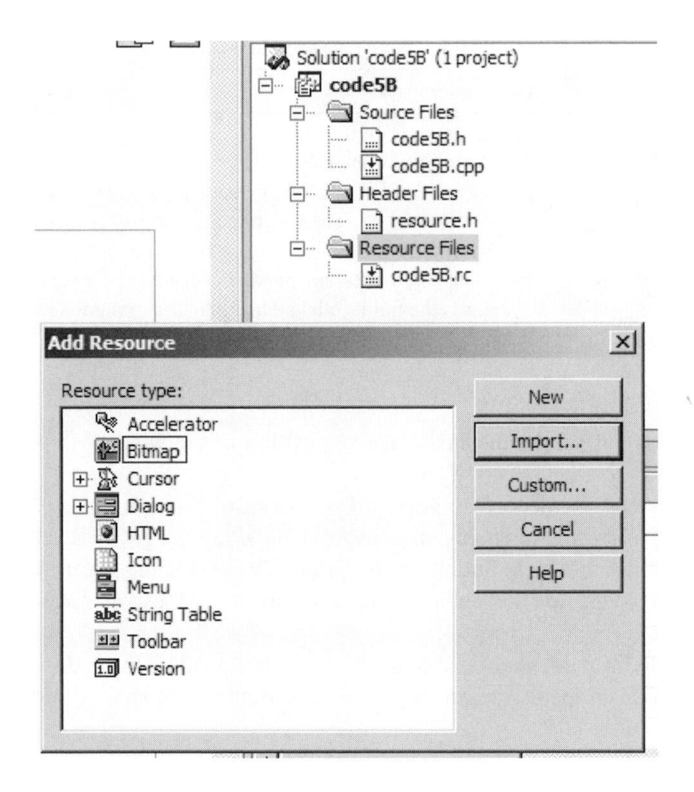

Figure 6.6 Importing the bitmap file as `IDB_BITMAP1`.

Another object is `MyImage`, which is created from the class `CBitmap` to represent the image. The image is is assigned with the id `IDB_BITMAP1` and loaded into the memory using the function `LoadBitmap()`. This is done as follows:

```
CBitmap MyImage;
MyImage.LoadBitmap(IDB_BITMAP1)
```

The next step is to link the image with the class device context, `CDC`, before displaying it in the main window through the function `OnPaint()`. The linking process involves moving the bitmaps from the memory device context to the device context in `CPaintDC`, as follows:

```
CPaintDC dc(this);
CDC memDC;
```

An object from the memory device context is created to make it compatible with the device context `dc`, derived from the class `CPaintDC`, as shown in the following code fragments:

```
memDC.CreateCompatibleDC(&dc);
memDC.SelectObject(&MyImage);
```

Finally, the image is displayed on the screen using the function `BitBlt()` through the device context, as follows:

```
dc.BitBlt(Home.x,Home.y,400,320,&memDC,0,0,SRCCOPY);
```

The function `BitBlt()` represents a raster operation that takes eight parameters as input, as follows:

```
void BilBlt(a,b,c,d,e,f,g,h);
```

where
`a` = `x` coordinate of the top-left position of the image
`b` = `h` coordinate of the top-left position of the image
`c` = width of the image
`d` = height of the image
`e` = a pointer to the `dc` that has the image
`f` = `x` coordinate of the top-left position of the bitmap
`g` = `y` coordinate of the top-left position of the bitmap
`h` = raster operation option

The `h` parameter above is the raster option in the mapping, as shown in Table 6.3. A pixel can be mapped onto the screen using the fundamental logical operations in the binary system: OR, AND, and XOR operations. Table 6.4 shows some common MFC functions for manipulating a bitmap in the CDC class.

Bit shifting is performed in the function `OnLButtonDown()`, which responds to the mouse's left-button click. Several operations are performed to manipulate the image. First, a smaller size of the image is copied as the array `f[i][j]` using the function `GetPixel()`, where `i` and `j` are the row and column numbers, respectively. This function reads the value of the pixel at the given location and returns this value as an integer. The pixels are immediately transferred to the new locations on the right using `SetPixel()`, as follows:

```
f[i][j]=dc.GetPixel(i+Home.x+50,j+Home.y+100);
dc.SetPixel(440+i,Home.y+j,f[i][j]);
```

The red monotone image is obtained by shifting the values of the pixels by 16 bits to the right, as follows:

```
r= f[i][j] >> 16;
dc.SetPixel(20+i,400+j,r);
```

Table 6.4 Some common MFC functions for manipulating a bitmap in CDC

Function	Description
`LoadBitmap()`	Load the bitmap into the memory
`StretchBlt()`	Copy and display a bitmap from the memory to the window, allowing some degree of stretching and compression of the original image
`CreateCompatibleDC()`	Make the image compatible between the memory device context and the Windows device context
`GetPixel()`	Read a value from the pixel
`SetPixel()`	Write a value on the pixel

Table 6.3 Some raster options for displaying the bitmap using `BitBlt`

Function	Description
`SRCCPY`	Copies the source bitmap to its destination
`SRCPAINT`	Applies OR in combining the source and destination pixels
`SRCERASE`	Inverts the destination bitmap and applies AND with the source pixels
`SRCAND`	Applies AND in combining the source and destination pixels
`SRCINVERT`	Applies XOR in combining the source and destination pixels

Shifting the red pixels to the left by eight bits produces green pixels, as follows:

```
g = r << 8;
dc.SetPixel(230+i,400+j,g);
```

Similarly, blue pixels are obtained by shifting the green pixels to left by eight bits:

```
b = g << 8;
dc.SetPixel(440+i,400+j,b);
```

Finally, we get the gray components by combining the red, green, and blue components, as follows:

```
bw=r+g+b;
dc.SetPixel(440+i,200+j,bw);
```

6.3 EDGE-DETECTION PROBLEM

Image processing is an area of study for manipulating and modifying images. In most cases, image processing deals with methods for improving the quality of an image. For example, the quality of a bad image taken from a digital camera can be improved by reducing the noise and sharpening the edges. Image processing also includes image modification. An image may be stretched, magnified, warped, rotated, dithered, and so on, so that it appears more appealing.

Another important application of image processing is in recognizing the objects in an image, which may help in decision making. During the Bosnian war in the mid-1990s, for example, photos taken from American military aircraft were analyzed to identify mass graves in certain parts of the country. A photo of a geographical area taken from an aircraft thousands of feet above the ground may not be clear enough to identify some or all the objects in it. The quality of this image can be improved by applying several techniques in image processing. The improved image will then describe the area better, making it possible to recognize the objects in the image.

Image processing involves techniques for image enhancement, restoration, and smoothing by moving, copying, deleting, and modifying the contents of the pixels in the image. Image enhancement is a technique for improving the quality of an image. For example, a blurred image can be improved by sharpening the edges and removing the unwanted noise in the image. Image restoration involves recovering the missing bits and color values in an image. Smoothing is a technique for reducing noise in an image.

Edge detection is one of the most fundamental operations in image processing. Basically, edge detection is a technique for finding the edges of an image that separate pixels of high intensity from low intensity or, in other words, finding the binary form of the image. The *edges* of an image consist of boundary lines separating pixels of high intensity in the image from those of low intensity. The separation scheme between the high- and low-intensity pixels for producing the edges is based on a preset threshold value.

Edge detection is carried out through a linear operation called *convolution,* or sometimes called filtering. Convolution involves an operator called *kernel* that adds the product of the convolution coefficients to the pixel's neighbor values. This operation has the effect of filtering the image from unwanted noise, besides sharpening the edges.

A pixel $f(x, y)$ changes its value to $f'(x, y)$ through a convolution kernel K, as shown in the following equation:

$$f'(x, y) = f(x, y)K \tag{6.1}$$

In the above equation, K represents the method of the convolution. Equation (6.1) represents a linear transformation from $f(x, y)$ to $f'(x, y)$. The new value $f'(x, y)$ is compared to a threshold value. The binary assignment follows from this comparison: if the new value is greater than the threshold, then 1 (or 0) is assigned to the pixel. Otherwise, the pixel is assigned with 0 (or 1). By mapping all the pixels in the image according to this rule, a binary version of the image is obtained that clearly shows the edges in the image.

Some of the most commonly applied convolution methods for edge detection include the Roberts, Prewitt, Sobel, Canny, and Laplacian methods. In this section, we focus our discussion on the Sobel and Laplacian methods.

Sobel Filtering Method

The Sobel method is one the most popular digital filtering methods for detecting the edges of an image. The method is based on a one-dimensional approximation to gradient for smoothing. In real life, a Sobel filter is used in applications such as medical imaging. One typical use is in counting the number of human blood cells. In this application, an image is acquired by scanning or by inserting a tiny camera into the body.

A Sobel filter consists of two convolution kernels for detecting the horizontal and vertical changes in an image. The kernels make up the gradient vector $S = [S_x \backslash S_y]$, where S_x is the horizontal kernel and S_y is the vertical kernel. The horizontal kernel S_x is used for smoothing along the direction of the x-axis, and the vertical kernel S_y applies to smoothing along the y-axis.

The horizontal kernel S_y is produced by applying the first derivative using the central difference approximation rule to produce the following matrix:

$$S_y = \begin{bmatrix} 1 & 2 & 1 \\ 0 & 0 & 0 \\ -1 & -2 & -1 \end{bmatrix} \tag{6.1}$$

The vertical kernel is produced in the same manner, as follows:

$$S_x = \begin{bmatrix} 1 & 0 & -1 \\ 2 & 0 & -2 \\ 1 & 0 & -1 \end{bmatrix} \tag{6.2}$$

The two kernels can be used to compute the magnitude of the gradient, which is the filter value, and the direction of the edges in the image. The magnitude of the gradient is determined as follows:

$$|S| = \sqrt{S_x^2 + S_y^2} \tag{6.3}$$

The above operation, which involves a square root, is computationally expensive. A more practical approximation has the following form:

$$|S| = |S_x| + |S_y| \tag{6.4}$$

The Sobel filter also gives the direction of the edges in the image as follows:

$$\theta = \tan^{-1} \frac{S_y}{S_x} \tag{6.5}$$

Figure 6.7 shows an approximation map for a 3 × 3 neighborhood of an image. A pixel at (i, j) with its value given by $f_{i,j}$ is updated to $f'_{i,j}$ using the Sobel method based on the values of its eight neighboring pixels, as shown in the figure. The following equation performs this operation:

$f_{i-1,j+1}$	$f_{i,j+1}$	$f_{i+1,j+1}$
$f_{i-1,j}$	$f_{i,j}$	$f_{i+1,j}$
$f_{i-1,j-1}$	$f_{i,j-1}$	$f_{i+1,j-1}$

Figure 6.7 Pixel update at (i, j).

$$f'_{i,j} = \sum_{h=1}^{n} \sum_{k=1}^{n} f_{i+h-2,j+k-2} S_{h,k} \tag{6.6}$$

In the above equation, n is the neighborhood size, which is 3 in this case, and $S_{h,k}$ is an element in the kernel matrix. The pixels are updated using the kernels by scanning the image starting from left to right of the top row, then repeating the procedure for the middle and bottom rows.

Laplacian Filtering Method

The Laplacian method is a second-order differential equation operator based on the Laplace equation, given as follows:

$$\nabla^2 f = \frac{\partial^2 f}{\partial x^2} + \frac{\partial^2 f}{\partial y^2} = 0 \tag{6.7}$$

A numerical approximation to the above equation is obtained from the central-difference rules given by

$$\nabla^2 f = \frac{f_{i+1,j} - 2f_{i,j} + f_{i-1,j}}{h^2} + \frac{f_{i,j+1} - 2f_{i,j} + f_{i,j-1}}{k^2} \tag{6.8}$$

where h and k are constants. A simple Laplacian kernel over a 3 × 3 neighborhood is obtained by rearranging the coefficients of Equation (6.8) and setting $h = k = 1$, as follows:

$$L = \begin{bmatrix} 0 & 0 & 0 \\ 1 & -2 & 1 \\ 0 & 0 & 0 \end{bmatrix} + \begin{bmatrix} 0 & 1 & 0 \\ 0 & -2 & 0 \\ 0 & 1 & 0 \end{bmatrix} = \begin{bmatrix} 0 & 1 & 0 \\ 1 & -4 & 1 \\ 0 & 1 & 0 \end{bmatrix} \tag{6.9}$$

Another form of Laplacian kernel for the 3 × 3 neighborhood is given below:

$$L = \begin{bmatrix} -1 & -1 & -1 \\ -1 & 8 & -1 \\ -1 & -1 & -1 \end{bmatrix} \tag{6.10}$$

Laplacian method is isotropic in the x and y directions as it is independent of these axes. The method is also invariant to the 90 degree rotation. The Laplacian method of edge detection is simple to implement by starting the scan from the top row in the direction from left to right. A scan of the pixel at (i, j) having its initial value f_{ij} using $L_{h,k}$ from Equations (6.9) or (6.10) produces a new value f'_{ij}, as follows:

$$f'_{ij} = \sum_{h=1}^{n} \sum_{k=1}^{n} f_{i+h-2,j+k-2} L_{h,k} \tag{6.11}$$

The Laplacian method has the disadvantage of being very sensitive to noise, which affects the output. Therefore, it is seldom used on its own. Normally, the method is used in line with a technique called Laplacian of Gaussian (LoG) which handles noise in a more effective way.

Code6B: Detecting the Edges of an Image

We discuss a model for detecting the edges of a bitmap image using the Sobel and Laplacian methods. The project is called Code6B and it consists of the files Code6B.h, Code6B.cpp, and Code6B.rc. Figure 6.8 shows the output from the project. It consists of the original black and white image displayed at the left and three smaller images at the right. Convolutions using the Sobel and Laplacian methods are performed when the mouse's left button is clicked. The top image on the

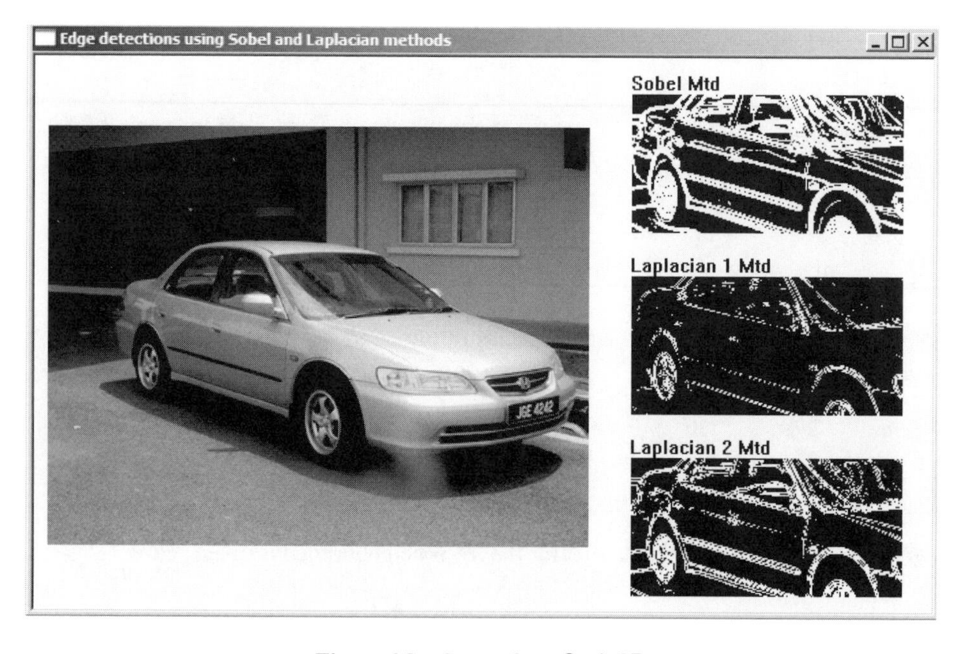

Figure 6.8 Output from Code6B.

right shows the results from convolution using the Sobel method. The middle image is produced from the Laplacian filter using Equation (6.9), and the bottom image is produced from Equation (6.10).

The application class used in Code6B is CCode6B and it has a few global variables and objects, as listed in Table 6.5. The bitmap image is defined in the resource file with the id IDB_BITMAP1. It is represented in the memory as the object bmp. The image has its top-left-hand coordinates defined by the object Home and its size declared as sImage. The pixel value f_{ij} at (i, j) which displays the image is represented by f[i][j].

Table 6.6 describes the global functions in this application, OnPaint() and OnLButtonDown(), which respond to the events detected by WM_PAINT and WM_LBUTTONDOWN, respectively.

The constructor in this project allocates memory for the array f[][] and loads the image into the computer memory. The image is loaded as the bitmap object bmp using the function LoadBitmap().

The original image is displayed in the main window using the function OnPaint(). To display the image, a memory device context called memDC is created to hold the image and makes it compatible with the Windows device context called dc. The image is then displayed using the function BitBlt(). The following routine shows this task:

```
CDC memDC;
memDC.CreateCompatibleDC(&dc);
memDC.SelectObject(&bmp);
dc.BitBlt(Home.x,Home.y,400,300,&memDC,0,0,SRCCOPY);
```

The array f[][] is an extract copy of the original image. Its values are assigned by reading the pixel values in the rectangular defined by sImage using the function GetPixel(). From the array, the threshold value of the edges Threshold is computed simply by taking the mean of the pixel values, as follows:

```
for (int j=0;j<=sImage.cy;j++)
     for (int i=0;i<=sImage.cx;i++)
     {
          f[i][j]=dc.GetPixel(Home.x+i+50,Home.y+j+100);
          Threshold += (double)f[i][j];
     }
Threshold /= (double)(N*N);
```

Convolutions using the Sobel and Laplacian methods are performed in OnLButtonDown(). The Sobel method updates a pixel at (i, j) by masking the pixels horizontally and vertically. Horizontal and vertical maskings on this pixel are performed using Equations (6.1) and (6.2), respectively, by referring to its six neighboring pixels. Convolution is obtained by combining the two results Sx and Sy as S using Equation (6.4). The operation is performed as follows:

Table 6.5 Global variables and objects in Code6B

Variable/object	Type	Description
`f[i][j]`	`int`	Array representing the pixels of the image
`Threshold`	`double`	Threshold value of the edges that separates pixels of high intensity from low intensity
`bmp`	`CBitmap`	Object representing the image
`sImage`	`CSize`	Size of the image
`Home`	`CPoint`	Top-left coordinates of the original image in the window

```
Sy=f[i-1][j+1]-f[i+1][j+1]+2*f[i-1][j]     // horizontal kernel
          -2*f[i+1][j]+f[i-1][j-1]-f[i+1][j-1];
Sx=-f[i-1][j+1]-2*f[i][j+1]-f[i+1][j+1]    // vertical kernel
          +f[i-1][j-1]+2*f[i][j-1]+f[i+1][j-1];
S=abs(Sy)+abs(Sx);
```

The edges of the image are drawn as white lines. An edge is formed if the convolution result S is greater than the threshold value, as follows:

```
if (S<=(int)Threshold)
     S=RGB(0,0,0);
else
     S=RGB(255,255,255);
dc.SetPixel(440+i,Home.y-25+j,S);
```

The Laplacian method is applied in a similar manner. Convolution using Equation (6.9) as the kernel is implemented as follows:

```
L=-f[i][j+1]-f[i-1][j]+4*f[i][j]-f[i+1][j]-f[i][j-1]; // kernel 1
L=abs(L);
if (L<=(int)Threshold)
     L=RGB(0,0,0);
else
     L=RGB(255,255,255);
dc.SetPixel(440+i,Home.y+5+sImage.cy+j,L);
```

Table 6.6 Global functions in Code6B

Function	Description
`OnPaint()`	Function for the initial display in the window
`OnLButtonDown()`	Function that responds to the message handler `ON_WM_LBUTTONDOWN` for the mouse's left-click event

Another kernel, using Equation (6.10), is written as follows:

```
L=-f[i-1][j+1]-f[i][j+1]-f[i+1][j+1]          // kernel 2
          -f[i-1][j]+8*f[i][j]-f[i+1][j]
          -f[i-1][j-1]-f[i][j-1]-f[i+1][j-1];
L=abs(L);
if (L<=(int)Threshold)
     L=RGB(0,0,0);
else
     L=RGB(255,255,255);
    dc.SetPixel(440+i,Home.y+35+2*sImage.cy+j,L);
```

6.4 SUMMARY AND CONCLUSION

In this chapter, we discussed two small examples of using MFC involving images. The first application involves manipulating the pixels of an image for producing their monotone color scales. This is done through a technique called bit shifting that alters the values of the pixels. The second application involves two elementary methods for finding the edges of an image. The problem is called edge detection. The two methods are the first-order Sobel filtering method and the second-order Laplacian method.

Applications involving images are used in many areas of study, including multimedia, image processing, signal processing, visualizations, and numerical methods. There are many ways an image can be applied in these areas of study. For example, the edge-detection application discussed in this chapter can easily be extended to include things like object recognition. It is necessary to identify primitive objects like lines and circles as they lead to recognizing characters and patterns in the image. C++ with MFC will definitely be useful as a programming tool for developing solutions to these problems.

BIBLIOGRAPHY

1. R. C. Gonzalez and R. Woods, *Digital Image Processing,* 2nd ed., Prentice-Hall, 2002.
2. M. Seul, L. O'Gorman, and M. J. Sammon, *Practical Algorithms for Image Analysis: Descriptions, Example and Code,* Cambridge University Press, 2000.
3. A. Bovik (Ed.), *Handbook of Image and Video Processing,* Academic Press, 2000.

CODE LISTINGS

Code6A: Working with Colors

```
// code6A.h
#include <afxwin.h>
```

```
#include "resource.h"
#define N 500

class CCode6A : public CFrameWnd
{
private:
     int **f;
     CSize sImage;
     CPoint Home;
     CBitmap MyImage;
public:
     CCode6A();
     ~CCode6A();
     afx_msg void OnLButtonDown(UINT,CPoint);
     afx_msg void OnPaint();
     DECLARE_MESSAGE_MAP()
};

class CMyWinApp : public CWinApp
{
public:
  BOOL InitInstance();
};

// code6A.cpp: Working with colors
#include "code6A.h"

CMyWinApp MyApplication;

BOOL CMyWinApp::InitInstance()
{
     m_pMainWnd = new CCode6A();
     m_pMainWnd->ShowWindow(m_nCmdShow);
     m_pMainWnd->UpdateWindow();
     return TRUE;
}

BEGIN_MESSAGE_MAP(CCode6A, CFrameWnd)
     ON_WM_PAINT()
     ON_WM_LBUTTONDOWN()
END_MESSAGE_MAP()

CCode6A::CCode6A()
{
     f=new int *[N+1];
     for (int i=0;i<=N;i++)
          f[i]=new int [N+1];
     Home=CPoint(10,30);
     sImage=CSize(200,100);
     Create(NULL, "Manipulating RGB components through bit shifting",
          WS_OVERLAPPEDWINDOW,CRect(0,0,700,600));
     MyImage.LoadBitmap(IDB_BITMAP1);
}
```

```
CCode6A::~CCode6A()
{
     for (int i=0;i<=N;i++)
          delete f[i];
     delete f;
}

afx_msg void CCode6A::OnPaint()
{
     CPaintDC dc(this);
     CDC memDC;
     memDC.CreateCompatibleDC(&dc);
     memDC.SelectObject(&MyImage);
     dc.BitBlt(Home.x,Home.y,400,320,&memDC,0,0,SRCCOPY);
     dc.TextOut(Home.x+50,Home.y-20,"the original image");
     dc.TextOut(440,50,"press left button to see the rest");
}

afx_msg void CCode6A::OnLButtonDown (UINT nFlags,CPoint pt)
{
     CClientDC dc(this);
     int r,g,b,bw;
     dc.TextOut(440,Home.y-20,"extract from the original image");
     dc.TextOut(440,180,"gray scale by setting r=g=b");
     dc.TextOut(20,380,"red scale by setting g=b=0");
     dc.TextOut(230,380,"green scale by setting r=b=0");
     dc.TextOut(440,380,"blue scale by setting r=g=0");
     for (int j=0;j<=sImage.cy;j++)
          for (int i=0;i<=sImage.cx;i++)
          {
               f[i][j]=dc.GetPixel(Home.x+i+50,Home.y+j+100);
               dc.SetPixel(440+i,Home.y+j,f[i][j]);
               r= f[i][j] >> 16;
               dc.SetPixel(20+i,400+j,r);
               g = r << 8;
               dc.SetPixel(230+i,400+j,g);
               b = g << 8;
               dc.SetPixel(440+i,400+j,b);
               bw=r+g+b;
               dc.SetPixel(440+i,200+j,bw);
          }
}
```

Code6B: Edge Detection Problem

```
// code6B.h
#include <afxwin.h>
#include <math.h>
#include "resource.h"
#define N 200

class CCode6B : public CFrameWnd
{
```

```
public:
     CBitmap bmp;
     CPoint Home;
     CSize sImage;
     double threshold;
     int **f;
public:
     CCode6B();
     afx_msg void OnLButtonDown(UINT, CPoint);
     afx_msg void OnPaint();
     DECLARE_MESSAGE_MAP()
};

class CMyWinApp : public CWinApp
{
public:
  BOOL InitInstance();
};

// code6B.cpp: Edge detection of an image
#include "code6B.h"

CMyWinApp MyApplication;

BOOL CMyWinApp::InitInstance()
{
     m_pMainWnd = new CCode6B();
     m_pMainWnd->ShowWindow(m_nCmdShow);
     m_pMainWnd->UpdateWindow();
     return TRUE;
}

BEGIN_MESSAGE_MAP(CCode6B, CFrameWnd)
     ON_WM_LBUTTONDOWN()
     ON_WM_PAINT()
END_MESSAGE_MAP()

CCode6B::CCode6B()
{
     f=new int *[N+1];
     for (int i=0;i<=N;i++)
          f[i]=new int [N+1];
     Home=CPoint(10,50);
     sImage=CSize(200,100);
     threshold=0;
     Create(NULL,"Edge detections using Sobel and Laplacian methods",
          WS_OVERLAPPEDWINDOW,CRect(0,0,700,600));
     bmp.LoadBitmap(IDB_BITMAP1);
}

void CCode6B::OnPaint()
{
     CPaintDC dc(this);
     CDC memDC;
     memDC.CreateCompatibleDC(&dc);
```

```
	memDC.SelectObject(&bmp);
	dc.BitBlt(Home.x,Home.y,400,300,&memDC,0,0,SRCCOPY);
	for (int j=0;j<=sImage.cy;j++)
		for (int i=0;i<=sImage.cx;i++)
		{
			f[i][j]=dc.GetPixel(Home.x+i+50,Home.y+j+100);
			threshold += (double)f[i][j];
		}
	threshold /= (double)(N*N);
}

afx_msg void CCode6B::OnLButtonDown(UINT flags, CPoint pt)
{
	CClientDC dc(this);
	int Sy,Sx,S,L;

	dc.TextOut(440,Home.y-40,"Sobel Mtd");
	dc.TextOut(440,Home.y-10+sImage.cy,"Laplacian 1 Mtd");
	dc.TextOut(440,Home.y+20+2*sImage.cy,"Laplacian 2 Mtd");
	for (int j=1;j<=sImage.cy-1;j++)
		for (int i=1;i<=sImage.cx-1;i++)
		{
			// Sobel masking for detecting the edges
			Sy=f[i-1][j+1]-f[i+1][j+1]+2*f[i-1][j]   // horizontal kernel
				-2*f[i+1][j]+f[i-1][j-1]-f[i+1][j-1];
			Sx=-f[i-1][j+1]-2*f[i][j+1]-f[i+1][j+1]  // vertical kernel
				+f[i-1][j-1]+2*f[i][j-1]+f[i+1][j-1];
			S=abs(Sy)+abs(Sx);
			if (S<=(int)threshold)
				S=RGB(0,0,0);
			else
				S=RGB(255,255,255);
			dc.SetPixel(440+i,Home.y-25+j,S);

			// Laplacian masking for detecting the edges
			L=-f[i][j+1]-f[i-1][j]
					+4*f[i][j]-f[i+1][j]-f[i][j-1]; // kernel 1
			L=abs(L);
			if (L<=(int)threshold)
				L=RGB(0,0,0);
			else
				L=RGB(255,255,255);
			dc.SetPixel(440+i,Home.y+5+sImage.cy+j,L);

			L=-f[i-1][j+1]-f[i][j+1]-f[i+1][j+1]       // kernel 2
				-f[i-1][j]+8*f[i][j]-f[i+1][j]
				-f[i-1][j-1]-f[i][j-1]-f[i+1][j-1];
			L=abs(L);
			if (L<=(int)threshold)
				L=RGB(0,0,0);
			else
				L=RGB(255,255,255);
			dc.SetPixel(440+i,Home.y+35+2*sImage.cy+j,L);
		}
}
```

CHAPTER 7

VISUALIZING A GRAPH

7.1 ELEMENTARY GRAPH CONCEPTS

Graph theory is a branch of mathematics that deals with the study of graphs and their properties and applications. The importance of graph theory is reflected in the way a given problem is reduced to the form of a graph before its solution is obtained. Several applications in the real world can be expressed as graphs. For example, a tanker transporting fuel from the source (depot) to several destinations in the city will have to follow routes with the shortest paths in order to save cost and time. In this case, the source and the destinations make up the nodes of the graph, whereas their delivery routes make up a problem of finding the shortest paths in graph theory. Some other applications of graph theory include problems of optimization such as scheduling, placement, routing, and inventory and resource management.

A graph G consists of a set of *vertices*, V, and a set of *edges*, or *links*, E. The set of vertices having N vertices is denoted as $V = \{v_i\}$, for $i = 1, 2, \ldots, N$. The set of edges is symbolized by $E = \{e_{ij}\}$, where i and j are the pair of nodes with a direct link e_{ij} in the graph. A link e_{ij} is said to exist between a pair of nodes (v_i, v_j) if the two nodes are adjacent to each other.

Figure 7.1 shows a graph G with $V=\{v_k\}$ for $k = 1, 2, \ldots, 8$ and $E = \{e_{ij}\}$ for $i, j = 1, 2, \ldots, 8$; e_{ij} has a nonzero value if a link exists and 0 otherwise. A link in the graph may have a value called its *weight*. A weight may represent the length, duration, or any other cost associated with the two adjacent nodes. A graph with all its links having weights is called a *weighted graph*. Figure 7.1 shows a weighted graph with eight nodes and 11 links. A weighted graph may represent a network of cities, where the nodes are the cities and the links are the roads linking the cities. The weights in the links, in this case, may represent their road distances or the time taken for road travel between the cities. The weights in the edges e_{ij} and e_{ji} may or may not have the same values. For example, although the road distances from i to j are

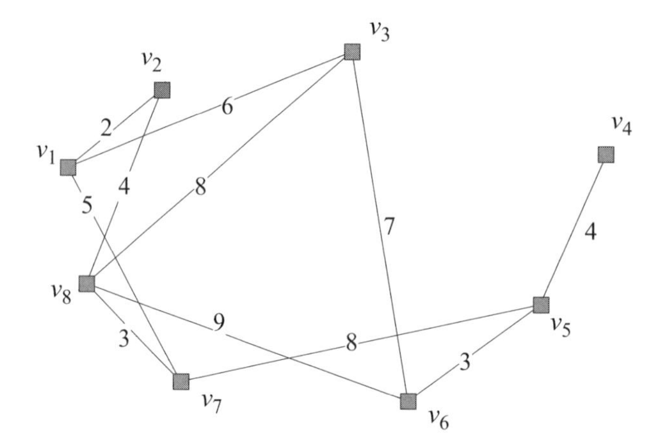

Figure 7.1 A weighted graph with eight nodes and 11 links.

the same as from j to i, the time taken to travel may not be the same as a result of different road and traffic conditions between them.

To understand a graph, several fundamental properties are discussed. A graph G' in which all nodes and links are contained inside another graph G is said to be a *subgraph* of that graph, denoted as $G' \subseteq G$. In Figure 7.1, G', with $V' = \{v_5, v_6, v_7\}$ and $E' = \{e_{56}, e_{57}\}$, is a subgraph of G. A graph in which any node can find a path to any other node in the graph is called a *connected graph*. Otherwise, the graph is disconnected and, in this case, it consists of more than one component. A link in the graph that causes the graph to be disconnected is called a *bridge*. In a similar manner, a node is called an *articulation point* if its removal causes the graph to be disconnected. In Figure 7.1, the connected graph has one bridge, e_{45}, with v_5 as the only articulation point.

A node that has a physical link to its neighbor is said to be *adjacent* to that node. The adjacency information of all the nodes in the graph is stored as the *adjacency matrix*. In this matrix, row number i corresponds to the node number, whereas the column number j is its adjacency status relative to that node. The entry of 1 in row i and column j indicates nodes i and j are adjacent, otherwise they are not. The adjacency matrix of the graph in Figure 7.1 is given as follows:

$$
\begin{array}{cc}
\textit{Adjacency Matrix} & \textit{Degree} \\
\begin{bmatrix}
0 & 1 & 1 & 0 & 0 & 0 & 1 & 0 \\
1 & 0 & 0 & 0 & 0 & 0 & 0 & 1 \\
1 & 0 & 0 & 0 & 0 & 1 & 0 & 1 \\
0 & 0 & 0 & 0 & 1 & 0 & 0 & 0 \\
0 & 0 & 0 & 1 & 0 & 1 & 1 & 0 \\
0 & 0 & 1 & 0 & 1 & 0 & 0 & 1 \\
1 & 0 & 0 & 0 & 1 & 0 & 0 & 1 \\
0 & 1 & 1 & 0 & 0 & 1 & 1 & 0
\end{bmatrix}
&
\begin{array}{c}
3 \\ 2 \\ 3 \\ 1 \\ 3 \\ 3 \\ 3 \\ 4
\end{array}
\end{array}
$$

Related to the graph adjacency issue, the *degree of a node* in a graph is defined as the number of edges that pass through the given node. The degree of a node can also be determined from the adjacency matrix simply by counting the number of 1's in the node row. The *degree of a graph* is the maximum of the degrees of the nodes. From Figure 7.1, v_1 has a degree of three since it is adjacent to v_2, v_3, and v_7. The graph in this figure has a degree of four as its maximum degree is determined by v_8.

A *path* in a graph is a successive series of links from a source node to its destination node. The path is called a *cycle* if it starts at a node and ends at the same node. A graph is said to be *cyclic* if it has a cycle, otherwise it is *acyclic*. A path in a graph that visits all the edges in the graph exactly once is called an *Eulerian path*. Related to this terminology is the *Hamiltonian path,* which is a path that visits every node in the graph exactly once. An interesting application related to the Eulerian path is the well-known *traveling salesman problem* (TSP), which may be applied for solving several difficult combinatorial optimization problems. TSP involves finding a Hamiltonian path in the graph in which the source and destination nodes are the same, and whose sum of the links is minimum.

A connected graph is said to be *bipartite* if the nodes can be partitioned into two subgraphs in such a way that no two nodes in the subgraph are adjacent to each other. Figure 7.2 shows a bipartite graph (left) as the nodes can be partitioned into two bins (right). Similarly, if the graph can be partitioned into three disjoint subgraphs, then the graph is *tripartite*. In general, an m-partite graph consists of m disjoint subgraphs of that graph.

One useful application of the m-partite graph is in the *node coloring problem.* One version of the node coloring problem can be stated as finding the least number of colors so that no two adjacent nodes in the graph share the same color. In this problem, it is easy to verify that an m-partite graph requires m colors to make sure that every node in a partition has its own distinct color. Another version is the edge

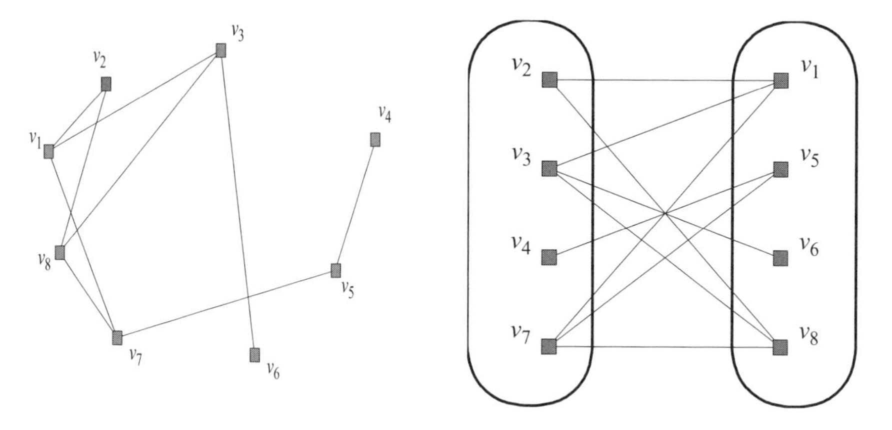

Figure 7.2 A bipartite graph.

coloring problem, which is about finding the least number of colors in such a way that all the edges originating from a node have distinct colors. The coloring problem has many applications; for example, the *channel assignment problem,* which is about assigning radio frequencies to mobile users in a wireless cellular telephone network. In this problem, the frequencies assigned to the mobile users in a cell must be well separated in order to avoid electromagnetic interference. In this case, the frequencies are colors and the network is the graph. We will discuss this problem further in Chapter 11.

A link that has a direction is called a *directed link.* The direction in an edge is shown as an arrow, where the tail is the source node and the head is the destination node. The directed edges between pairs of nodes in the graph show the partial order on a sequence of job activities in the graph. In this scenario, the node at the tail of a directed edge has a job that needs to be completed before the job at the head can start. A graph with directed edges is called a *directed graph.* If all the edges have weights, then the graph is called a *weighted directed graph,* as shown in Figure 7.3. In this figure, e_{28} is a directed link from v_2 to v_8 with a weight of 4 units. This link is not the same as e_{82}, which, in this case, does not exist as there is no directed link in this path.

Some other properties of a graph are described briefly here. A graph in which all the nodes are adjacent to each other is called a *complete graph.* A subgraph that is also a complete graph is called a *clique* to the graph. The maximum clique of a graph is then a clique that has the most number of nodes in the graph. It can be shown that the subgraph $\{v_3, v_6, v_8\}$ in Figure 7.1 is the *maximum clique* of the graph. Related to the maximum clique problem is the *maximum independent set* problem. An *independent set* of a graph is a subgraph that has nonadjacent nodes only. The maximum independent set is then an independent set that has the most

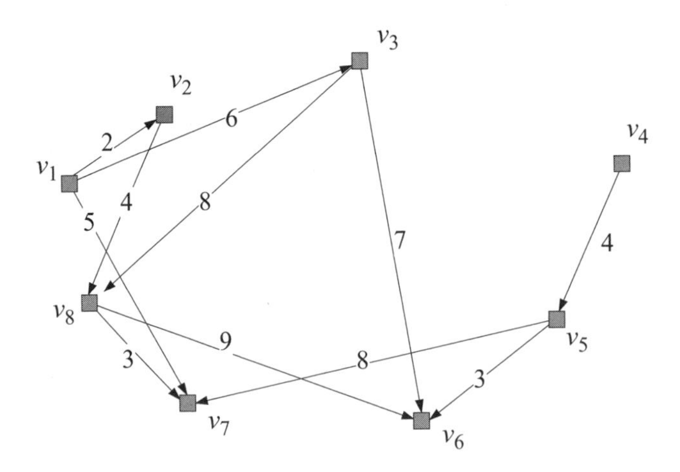

Figure 7.3 A weighted directed graph.

number of nodes in the graph. In Figure 7.1, $\{v_2, v_3, v_4\}$ and $\{v_1, v_4, v_8\}$ are two independent sets of the graph, whereas the maximum independent set is given by $\{v_2, v_3, v_4, v_6, v_7\}$. Two graphs are said to be *isomorphic* if there is a direct mapping between the nodes in the two graphs in such a way that the neighbors of the nodes in the first graph are also the corresponding neighbors of the nodes in the second graph.

This chapter presents two models involving the creation of a graph and its application. The first is a visualization model that allows the graph to be constructed using a mouse. The second is an application in graph theory for finding the minimum spanning tree of a graph using Prim's algorithm. The two models have been designed in such a way that they share a common platform in the form of a user-friendly interface for drawing the graph.

7.2 GRAPH VISUALIZATION MODEL

A visual display of a graph is very helpful in the study of its features and characteristics. Several tools in MFC are available for providing an interface that allows the user to construct a graph visually in Windows. The easiest way to do this is with the use of a mouse. A mouse provides full navigational capability for the window on the screen. A Windows user can easily plot a point by clicking the mouse button. However, it is a challenge for the programmer to provide this facility for the benefit of his/her users. In Visual C++, this feature is available as the language itself is object-oriented. In one of its powerful features, MFC supports the creation of several events related to this requirement through the resources available in its library. In this section, we discuss a method of drawing a graph using these resources.

Code7A: Drawing a Graph

We discuss a method for constructing a graph using a mouse. Figure 7.4 shows our tool for drawing a graph. In the figure, a node in the graph is created as a small rectangular box by clicking the left button of the mouse inside the drawing area. A link between a pair of nodes is formed by clicking the right button on the two pairing nodes consecutively. In addition, each time a link between two nodes is created, the program immediately updates the adjacency matrix and the degree of the nodes in the graph. In the adjacency matrix, an entry has a value of 1 if a link between the pairing nodes exists, and 0 otherwise.

Our tool for visualizing a graph is Code7A. The program allows the user to draw a graph having a maximum of N nodes, and $N \times N$ links. The macro N is set to be 10 initially, using the `#define` facility. It is, therefore, easy to support a higher number of nodes in the graph by changing this value. The program provides an easy interface for drawing a graph. The display consists of a drawing area, some simple instructions, and an output area that shows the adjacency matrix and the degree of

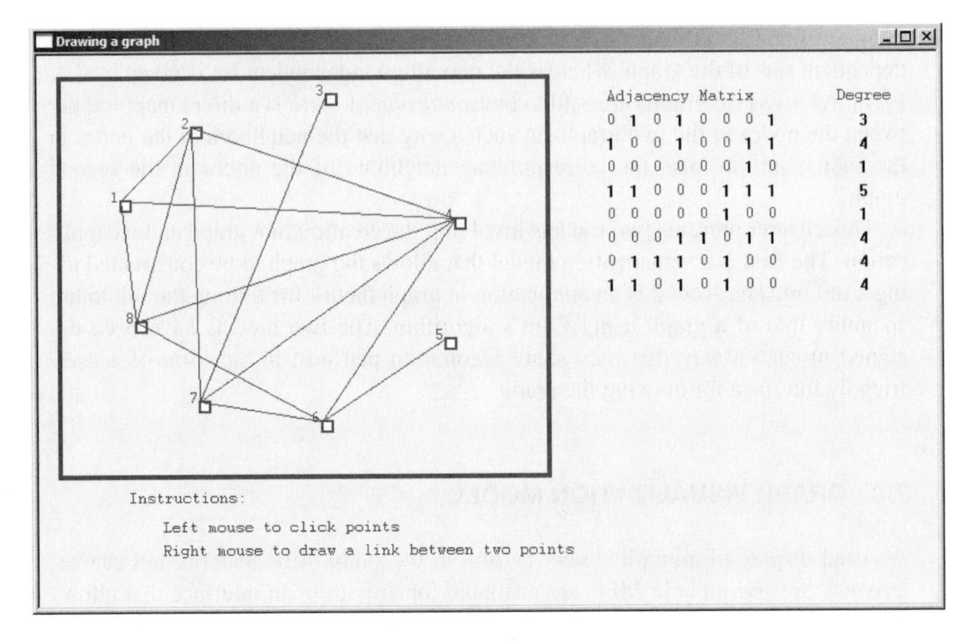

Figure 7.4 A user-friendly way of drawing a graph in Code7A.

the nodes in the graph. All the graphical objects, text, and background colors have been set in grayscale.

The Code7A project consists of the files Code7A.h and Code7A.cpp. The events associated with this project are the initial display, left-button click, and right-button click of the mouse, as shown in Figure 7.5. These events are mapped as `WM_PAINT`, `WM_LBUTTONDOWN` and `WM_RBUTTONDOWN`, respectively. In response to `WM_PAINT`, the function `OnPaint()` provides the initial display of the window, which consists of the drawing area, an instruction area, and an output area. The function `OnLButtonDown()` responds to `WM_LBUTTONDOWN` by drawing a node of the graph each time the left button of the mouse is clicked. Similarly, the function `OnRButtonDown()` responds to `WM_RBUTTONDOWN` by drawing an edge between any two nodes clicked consecutively using the mouse's right button.

Several variables, defined inside Code7A.h, are used for setting up the initial display on the screen. These variables are described briefly in Table 7.1.

Several variables and objects are used to represent the items in the graph. A node v_i is declared in the structure `NODE` and is represented as `v[i]`, whereas the link between the nodes v_i and v_j, denoted by e_{ij}, is represented by the array `e[i][j]`. This array has a value of 1 if v_i and v_j are adjacent, and 0 otherwise.

The structure `NODE` defines all the features of a node in the graph, namely, the node number, its position on the screen, and its representation as a rectangular box. For example, the degree of v_4 is represented as `v[4].degree`. The contents of the structure are shown as follows:

```
typedef struct
{
    CPoint Home;        // node coordinates in the window
    CRect Box;          // node representation as a box
    int degree;         // degree of the node
} NODE;
NODE *v;                // declares the  nodes as an array
```

The variables and objects in the structure NODE are briefly described in Table 7.2.

The main task of the constructor CCode7A() is to allocate memory for the class CCode7A. It is also in the constructor that the main window is created and

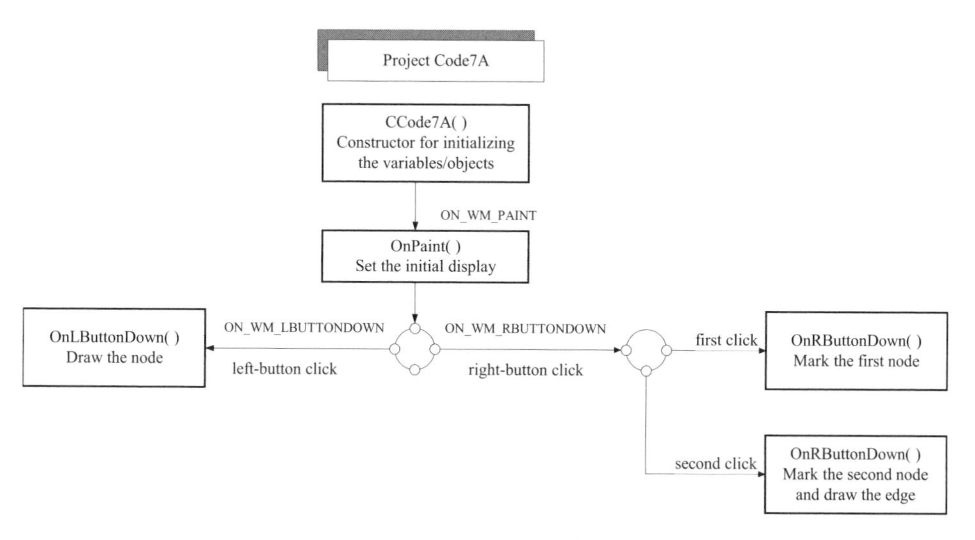

Figure 7.5 The events in Code7A.

Table 7.1 Display setup of Code7A

Variable/object	Class	Description
TopLeft	CPoint	Top-left point of the drawing area
BottomRight	CPoint	Bottom-right point of the drawing area
InstText	CPoint	Starting text coordinates for the instructions
MatText	CPoint	Starting text coordinates of the adjacency matrix
TextGap	int	Horizontal spacing between the elements in the adjacency matrix
FontCourier	CFont	Font used in the whole program
BoxSize	CSize	Size of the clicked point (node) in the drawing area
TextColor	int	Standard text color
BgColor	int	Standard background color

Table 7.2 Variables in the graph

Variable/object	Class	Description
v[i]		The node v_i
v[i].Home	CPoint	v_i coordinates on the main window
v[i].Box	CRect	v_i representation as a rectangular box
v[i].degree	int	Degree of v_i
e[i][j]	int	The weighted link e_{ij}
Pt1,Pt2	int	The first and second points clicked consecutively using the left button of the mouse for establishing a link
nv	int	Number of nodes in the graph
nLinks	int	Number of links in the graph
RButtonFlag	int	A flag to indicate if the right-button click on a node rectangle is the first (1) or second node (2) in the link

several initial values of variables and objects of the entire program are set. The role of the constructor is outlined in Figure 7.6.

In the constructor, the memory for the arrays `v[]` and `e[][]`, declared from the structure NODE and the type `int`, respectively, are allocated as follows:

```
int i,j;
v=new NODE [N+1];
e=new int *[N+1];
for (i=0;i<=N;i++)
     e[i]=new int [N+1];
```

In the main window, the drawing area of the graph is bounded by the `CPoint` objects, `TopLeft` and `BottomRight`, which represent the top-left and bottom-right corners, respectively. The constructor initializes the positions of the drawing area by assigning values to `TopLeft` and `BottomRight`, as follows:

```
TopLeft=CPoint(20,20); BottomRight=CPoint(450,360);
```

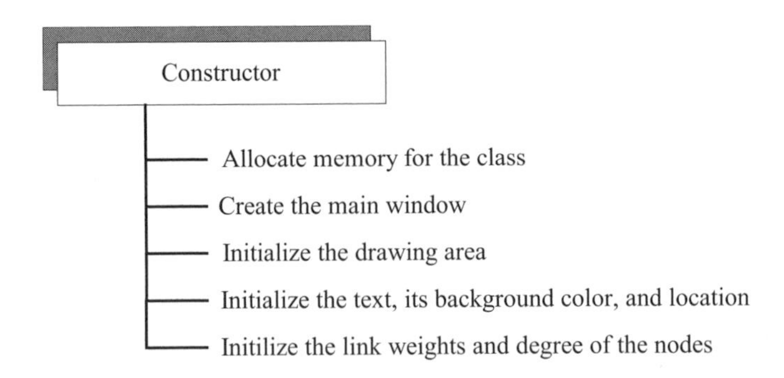

Figure 7.6 Contents of the constructor `CCode7A()`.

In addition, most of the settings for the display, for example, the location and size of the instruction area, the background color, and the standard font for the text messages, are also initialized in the constructor. Text messages are displayed using the Courier font of size 7. As each standard unit in this font requires 10 pixels for drawing, the first parameter in the function `CreatePointFont()` is set to 60 to represent a font size of 7. The initial display is realized through the following code:

```
MatText=CPoint(500,30); InstText=CPoint(80,370);
FontCourier.CreatePointFont(60,"Courier");
BgColor=RGB(240,240,240); TextGap=20; BoxSize=CSize(10,10);
TextColor=RGB(100,100,100);
```

Initializations also include the number of nodes and links in the graph, and the flag value of the mouse's right button, as follows:

```
nv=0; nLink=0; RButtonFlag=0;
```

It is necessary to assign the weights of the links with some remote values such as 99 to denote their initial values. The number 99 could mean the links between the pairs of nodes have not been established yet. A number such as 9 may not serve as a good representation as it could mean the real weight of the link. Also, as the degree of a node depends on the number of links on that node, it is a good idea to initialize its value to 0. These two quantities, the weight of the links and the degree of nodes, need to be initialized as their values will be referred to in subsequent updates later in the program. The following code fragment performs this task:

```
for (i=1;i<=N;i++)
{
      v[i].degree=0;
      for (j=1;j<=N;j++)
            e[i][j]=e[j][i]=99;
}
```

There are three events involved in the program, namely, the initial display (`WM_PAINT`), the left mouse click (`WM_LBUTTONDOWN`), and the right mouse click (`WM_RBUTTONDOWN`). The events are mapped as follows:

```
BEGIN_MESSAGE_MAP(CCode7A,CFrameWnd)
     ON_WM_PAINT()
     ON_WM_LBUTTONDOWN()
     ON_WM_RBUTTONDOWN()
END_MESSAGE_MAP()
```

The initial display is set using the `OnPaint()` function. As mentioned earlier, the display consists of an area for drawing the graph, an area for displaying the instructions, and an output area that displays the adjacency matrix and the degree of each node. The activities in this function are outlined in Figure 7.7.

In the function `OnPaint()`, the main window is first cleared using the background color, defined as `BgColor`, as follows:

```
CRect rc;
CBrush BgBrush(BgColor);
GetClientRect(&rc);
dc.FillRect(&rc,&BgBrush);
```

The above code erases the main window and fills it with the background color `BgColor`. The task starts by creating a brush called `BgBrush` using this color. The statement `GetClientRect(&rc)` reads the rectangular area of the main window and assigns this area as the `CRect` object `rc`. It follows that the fill function `FillRect(&rc,&BgBrush)` erases the area marked by `rc` using the `BgBrush` brush.

The rest of the code in `OnPaint()` displays the initial settings of the main window, as follows:

```
CPen penDrawingBox(PS_SOLID,4,RGB(100,100,100));
dc.SelectObject(penDrawingBox);
dc.SelectStockObject(HOLLOW_BRUSH);
rc=CRect(TopLeft,BottomRight);
dc.Rectangle(rc);

dc.SelectObject(FontCourier);
dc.SetTextColor(TextColor); dc.SetBkColor(BgColor);
dc.TextOut(InstText.x,InstText.y,"Instructions:");
dc.TextOut(InstText.x+30,InstText.y+25,
     "Left mouse to click points");
dc.TextOut(InstText.x+30,InstText.y+45,
     "Right mouse to draw a link between two points");
dc.TextOut(MatText.x,MatText.y,"Adjacency Matrix");
dc.TextOut(MatText.x+TextGap*N,MatText.y,"Degree");
```

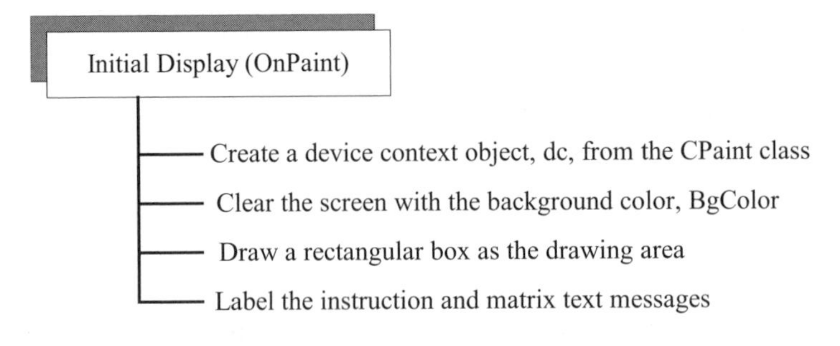

Figure 7.7 Contents of `OnPaint()`.

The second event is the left-button click of the mouse event, which creates the nodes of the graph. Essentially, a click in the drawing area updates the graph by adding a new node and displays the node on the screen. This increases the number of nodes, `nv`, in the graph by one. The new node occupies the coordinates `v[nv].Home` on the screen. In addition, the event also updates the adjacency graph with zero entries of the row and column of the new node. The contents of the left-click button operation are outlined in Figure 7.8.

The left-button click is detected by `WM_LBUTTONDOWN` and handled by the function `OnLButtonDown()`. This function prepares to draw small rectangular boxes as the nodes of the graph. The following code prepares the initial settings of the drawing pen and text styles in this function:

```
CClientDC dc(this);
CString s;
CPen penGray(PS_SOLID,2,TextColor);
dc.SelectObject(penGray);

dc.SelectObject(FontCourier);
dc.SetTextColor(TextColor);
dc.SetBkColor(BgColor);
```

The following code in `OnLButtonDown()` draws the nodes in the graph:

```
if (CRect(TopLeft,BottomRight).PtInRect(pt))
     if (nv<=N)
     {
          nv++;
          v[nv].Home=pt;
          v[nv].Box=CRect(CPoint(pt),CSize(BoxSize));
          dc.Rectangle(v[nv].Box);
          s.Format("%d",nv);
          dc.TextOut(v[nv].Home.x-10,v[nv].Home.y-10,s);
          for (int i=1;i<=nv;i++)
          {
               dc.TextOut(MatText.x+TextGap*(i-1),
                    MatText.y+TextGap+TextGap*(nv-1),"0");
               dc.TextOut(MatText.x+TextGap*(nv-1),
                    MatText.y+TextGap+TextGap*(i-1),"0");
               s.Format("%d",v[nv].degree);
               dc.TextOut(MatText.x+TextGap+TextGap*N,
                    MatText.y+TextGap+TextGap*(nv-1),s);
          }
     }
```

The function `OnLButtonDown(UINT nFlags,Cpoint pt)` has two arguments, `nFlags` and `pt`. The first argument shows the status of the mouse, which

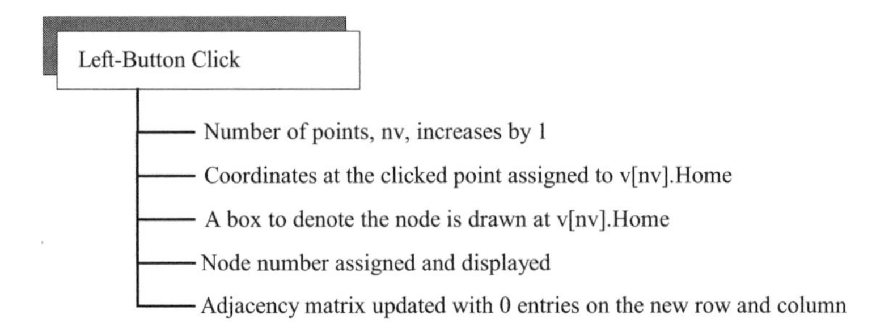

Figure 7.8 Contents of `OnLButtonDown()`.

is not used in this application. The second argument, `pt`, represents the mouse's clicked point, which is the home position of the current node. Using this argument, the nodes in the graph can only be displayed in the drawing area through the function `PtInRect(pt)`, using the following conditional test:

```
if (CRect(TopLeft,BottomRight).PtInRect(pt))
```

A node is drawn only if the above test is TRUE. Besides drawing the node, the total number of nodes `nv` is increased by one following this action. The adjacency matrix is also updated with the creation of a new row and column whose entries are all 0s.

The third event is the right-button click of the mouse for drawing the links in the graph. Basically, the link between any two nodes the graph is obtained by clicking the right button of the mouse consecutively on the two nodes. Each right-button click is detected by the message handler `WM_RBUTTONDOWN`, which immediately calls the function `OnRButtonDown()`, as follows:

```
void CCode7A::OnRButtonDown(UINT nFlags,CPoint pt)
{
	CClientDC dc(this);
	CString s;

	int i;
	CPen penGray(PS_SOLID,1,TextColor);
	dc.SelectObject(penGray); dc.SetBkColor(BgColor);
	for (i=1;i<=nv;i++)
	{
		if (v[i].Box.PtInRect(pt))
		{
			RButtonFlag++;
			if (RButtonFlag==1)
				Pt1=i;
			if (RButtonFlag==2)
			{
				Pt2=i; e[Pt1][Pt2]=1; nLink++;
			}
```

```
        }
        if (RButtonFlag==2)
        {
                dc.MoveTo(v[Pt1].Home); dc.LineTo(v[Pt2].Home);
                s.Format("%d",e[Pt1][Pt2]);
                dc.TextOut(MatText.x+TextGap*(Pt1-1),
                     MatText.y+TextGap+TextGap*(Pt2-1),s);
                dc.TextOut(MatText.x+TextGap*(Pt2-1),
                     MatText.y+TextGap+TextGap*(Pt1-1),s);
                s.Format("%d",++v[Pt1].degree);
                dc.TextOut(MatText.x+TextGap+TextGap*N,
                     MatText.y+TextGap+TextGap*(Pt1-1),s);
                s.Format("%d",++v[Pt2].degree);
                dc.TextOut(MatText.x+TextGap+TextGap*N,
                     MatText.y+TextGap+TextGap*(Pt2-1),s);
                RButtonFlag=0;
        }
    }
}
```

In the function `OnRButtonDown()`, a flag called `RButtonFlag` is introduced to differentiate the first and the second clicks of the mouse right button. Its value determines the status of the link according to a simple rule, as follows:

`RButtonFlag=0` denotes inactive link.

`RButtonFlag=1` denotes the tail node has been selected.

`RButtonFlag=2` denotes the head node has been selected.

`RButtonFlag` is activated only when the click point `pt` is inside one of the nodes in the graph, using the following conditional test:

```
if (v[i].Box.PtInRect(pt))
```

The first right-button click on a node assigns this node as the tail of the link. This action assigns the value `RButtonFlag=1` and `Pt1` as the first node. In the second click, `RButtonFlag` increases its value to 2 and assigns `Pt2` as the second node. This action also increases the total number of links in the graph, `nLinks`, by one. With this second click, the link between `Pt1` (the first node) and `Pt2` (the second node) is immediately established. The adjacency graph is also updated and the value 1 is assigned to its corresponding element. Finally, the `RButtonFlag` value is refreshed to 0. The whole operation is shown in Figure 7.9.

7.3 MINIMUM SPANNING TREE PROBLEM

A connected acyclic graph in which any pair of two nodes has a unique path is called a *tree.* A tree can also be defined as a graph in which every link is a bridge. It

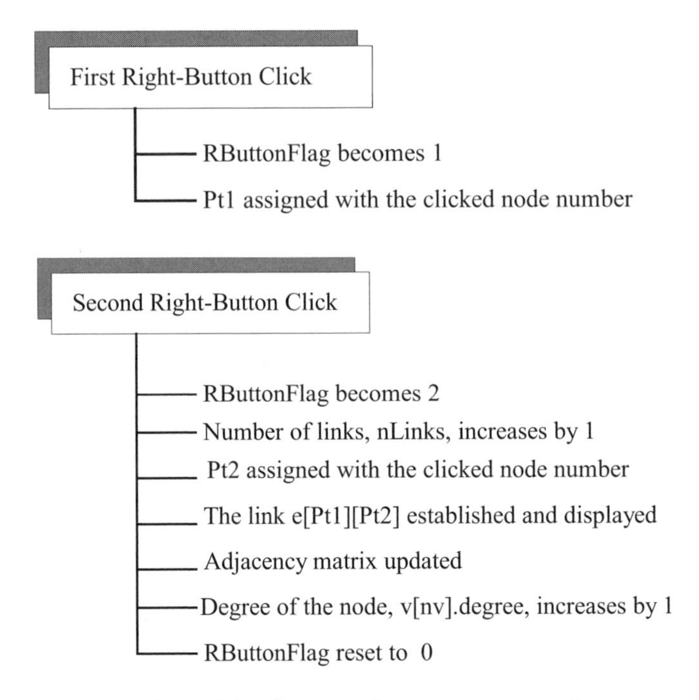

Figure 7.9 Contents of `OnRButtonDown()`.

is also easy to verify that a tree is always a bipartite graph. The end nodes of a tree are called *leaves*. A tree with a degree of two is called a *binary tree*. A set of disconnected trees is called a forest. A spanning tree of a graph G is defined as a tree subgraph of G in which all the nodes in G are included. A given graph can have many spanning trees formed from various combinations of the graph links. In a weighted graph, a spanning tree whose sum of the weights is minimum is called a *minimum spanning tree* (MST). Figure 7.10 shows a spanning tree for the graph in Figure 7.1.

The problem of finding the minimum spanning tree of a graph surfaces in many applications. For example, in radio communication, broadcasting involves transmitting a message from a node to all other nodes in a network. In this application, the message is to be transmitted to the nodes using the shortest possible path in terms of cost, such as the communication cost, time, and distance. In another application, the network manager of a cellular telephone network applies the concept of the minimum spanning tree in order to determine the shortest path for delivering a message from the source to its destination in the network.

The problem of finding the spanning trees of a graph originates from the fundamental properties of a tree. It is easy to verify that a tree with n nodes has $n - 1$ nodes, and that it has no cycle. From these properties, the minimum spanning tree problem reduces to the process of eliminating the edges of the graph one by one until it contains no cycle. This greedy process serves as the basic framework for con-

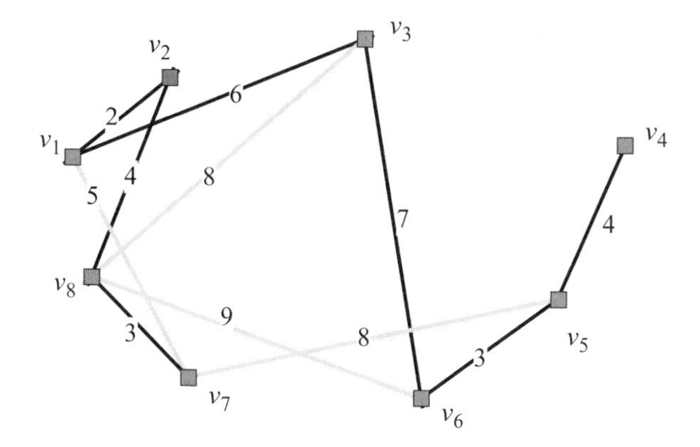

Figure 7.10 Minimum spanning tree of the graph in Fig. 7.1.

structing the minimum spanning tree of a graph, as demonstrated by several established algorithms. Two of the most popular algorithms for solving the minimum spanning tree problem are Kruskal's and Prim's algorithms (refer to Sedgewick [1, 2] for details). We begin our discussion with Kruskal's algorithm.

Kruskal's Algorithm

Kruskal's algorithm is easy to implement as it is based on a greedy approach. The algorithm starts with an empty subgraph called T of the graph G. The links of the graph are first sorted in increasing order based on their weights from the lowest to highest, and stored in a list called L. Iterations are performed to add the links from L to T starting from the highest order (smallest weight). At each iteration, a link from L is moved to T if the resultant subgraph does not contain a circuit; otherwise, the move is rejected. The process is repeated until T forms a spanning tree of G. This spanning tree is then the minimum spanning tree of the graph. The algorithm is summarized as follows:

```
Read the information on the weighted graph G
Create an empty graph, T
Create a list L of the sorted links in increasing order
do until T becomes a spanning tree of G
    Add the highest ordered link from L into T
    if T is a tree
        Accept the move
    else
        Reject the move by removing the link from T
    Remove the ordered link from L
```

Figure 7.11 shows the implementation of the Kruskal's algorithm in finding the minimum spanning tree of the graph in Figure 7.1.

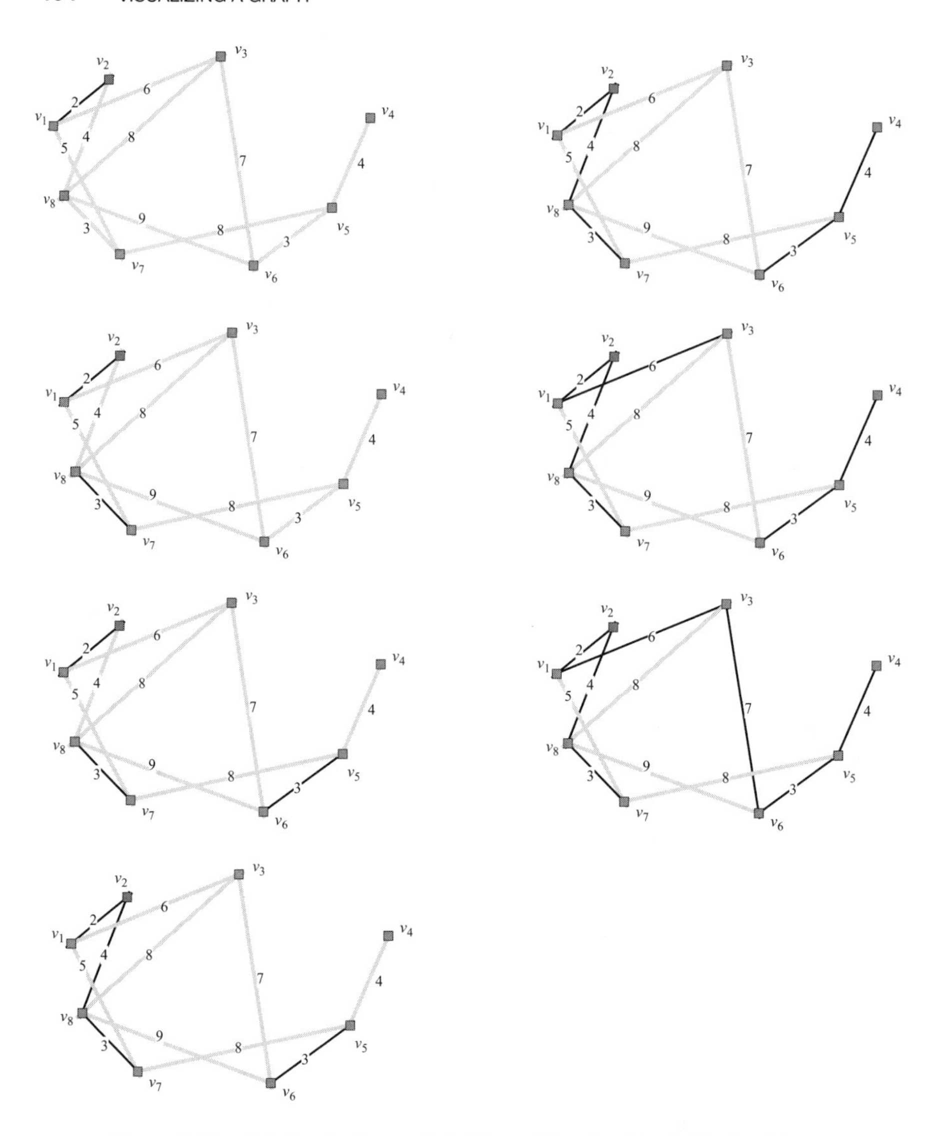

Figure 7.11 Solution to the graph in Figure 7.1 using Kruskal's algorithm.

Table 7.3 shows the steps for finding the minimum spanning tree of the graph using Kruskal's algorithm starting at v_1. The problem requires seven steps for converging to its solution using this algorithm.

Prim's Algorithm

Prim's algorithm is another common method for solving the minimum spanning tree problem. The algorithm begins with an empty graph T. The strategy in Prim's

Table 7.3 Solution using the Kruskal's algorithm for the graph in Figure 7.11

Iteration	Highest-ordered link in L	Cost	Decision	Resulting links in T
0				
1	e_{12}	2	accept	$\{e_{12}\}$
2	e_{78}	3	accept	$\{e_{12}, e_{78}\}$
3	e_{56}	3	accept	$\{e_{12}, e_{78}, e_{56}\}$
4	e_{28}	4	accept	$\{e_{12}, e_{78}, e_{56}, e_{28}\}$
5	e_{45}	4	accept	$\{e_{12}, e_{78}, e_{56}, e_{28}, e_{45}\}$
6	e_{13}	6	accept	$\{e_{12}, e_{78}, e_{56}, e_{28}, e_{45}, e_{13}\}$
7	e_{36}	7	accept	$\{e_{12}, e_{78}, e_{56}, e_{28}, e_{45}, e_{13}, e_{36}\}$
Total cost		29		

algorithm consists of building a tree in *T* one node at a time starting from any node in the graph. From the starting node, a link with the minimum weight to its neighboring node is added to *T*. Next, the shortest link emanating from the two nodes becomes the minimum link and is added into *T*. This added link must not form a circuit in *T*, otherwise the move is rejected. The step repeats from the second node to the third and, subsequently, until all the nodes have been included. Prim's algorithm is summarized as follows:

```
Read the information on the weighted graph G;
Create an empty graph, T;
Choose the first node, and add into T;
do until T becomes a spanning tree of G
    Choose the minimum link, originating from the node;
    if the new link does not have a circuit
        Add the link and its new node into T;
    else
        Reject the move;
```

Figure 7.12 shows the Prim's algorithm approach for finding the minimum spanning tree. The method is different from Kruskal's algorithm as the iteration starts from one node and continues with the rest of the iterations by forming a tree with the node as the root of the tree.

Table 7.4 shows the steps in finding the minimum spanning tree using Prim's algorithm. Like Kruskal's, seven steps are required in this problem.

Code7B: Visualizing the Minimum Spanning Tree

We discuss the development of a user-friendly interface for finding the minimum spanning tree of a graph using Prim's algorithm The visualization model is called Code7B. This model is an ideal extension of the Code7A retaining most of the data structure and the code in its development. Figure 7.13 shows an output pro-

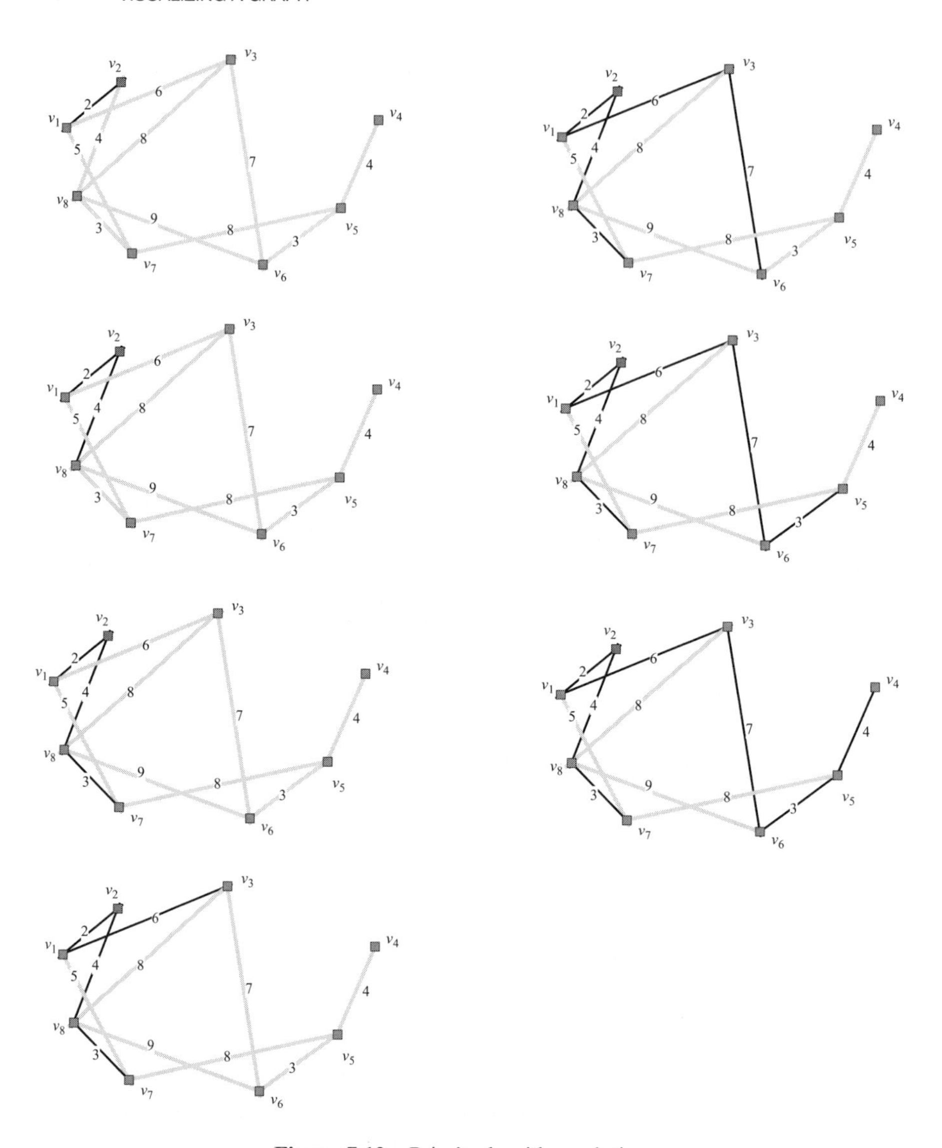

Figure 7.12 Prim's algorithm solution.

duced from the graph in Figure 7.1. A node in the graph is created through the left-button click of the mouse in the drawing area. A link in the graph is obtained by clicking the mouse's right button consecutively on two node boxes.

In this visualization model, a small dialog box appears on a link each time the link is created. This dialog box is an edit box for the user to enter a one-digit value that serves as the weight of the link. Once the connected graph has been established, the minimum spanning tree is shown by pressing the *Compute* `MST` button. The

Table 7.4 Prim's algorithm solution to the graph in Figure 7.12

Iteration	Available links	Minimum link	Cost	Resulting links in T
0				
1	e_{12}, e_{13}, e_{17}	e_{12}	2	$\{e_{12}\}$
2	e_{13}, e_{17}, e_{28}	e_{28}	4	$\{e_{12}, e_{28}\}$
3	$e_{13}, e_{17}, e_{87}, e_{86}, e_{83}$	e_{87}	3	$\{e_{12}, e_{28}, e_{87}\}$
4	$e_{13}, e_{17}, e_{86}, e_{83}, e_{75}$	e_{13}	6	$\{e_{12}, e_{28}, e_{87}, e_{13}\}$
5	$e_{17}, e_{86}, e_{83}, e_{75}, e_{36}$	e_{36}	7	$\{e_{12}, e_{28}, e_{87}, e_{13}, e_{36}\}$
6	$e_{17}, e_{86}, e_{83}, e_{75}, e_{65}$	e_{65}	3	$\{e_{12}, e_{28}, e_{87}, e_{13}, e_{36}, e_{65}\}$
7	$e_{17}, e_{86}, e_{83}, e_{75}, e_{54}$	e_{54}	4	$\{e_{12}, e_{28}, e_{87}, e_{13}, e_{36}, e_{65}, e_{54}\}$
Total cost			29	

spanning tree is marked as edges with bold lines in the graph and its cost is shown below the *Compute* MST button.

Figure 7.14 shows the organization of the Code7B project. The project consists of two files, Code7B.h and Code7B.cpp. Four events are mapped, namely, the initial display, left-button click of the mouse, right-button click of the mouse, and the *Compute* MST push button. These events are mapped and detected as WM_PAINT, WM_LBUTTONDOWN, WM_RBUTTONDOWN, and BN_CLICKED, respectively. In response to these events are the event handling functions OnPaint(), OnLButtonDown(), OnRButtonDown(), and OnClickCalc(), respectively. Since

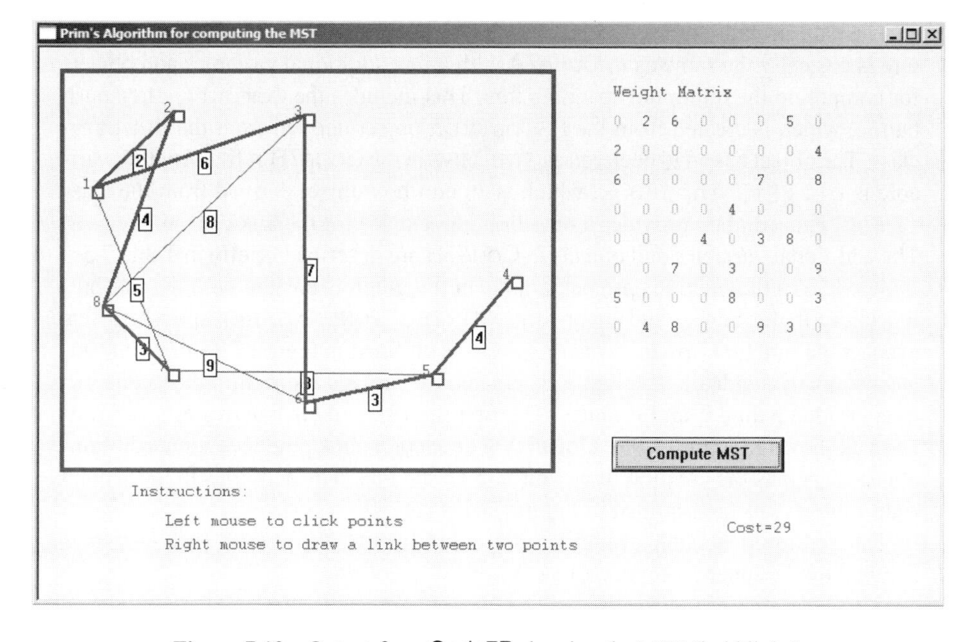

Figure 7.13 Output from Code7B showing the MST (bold links).

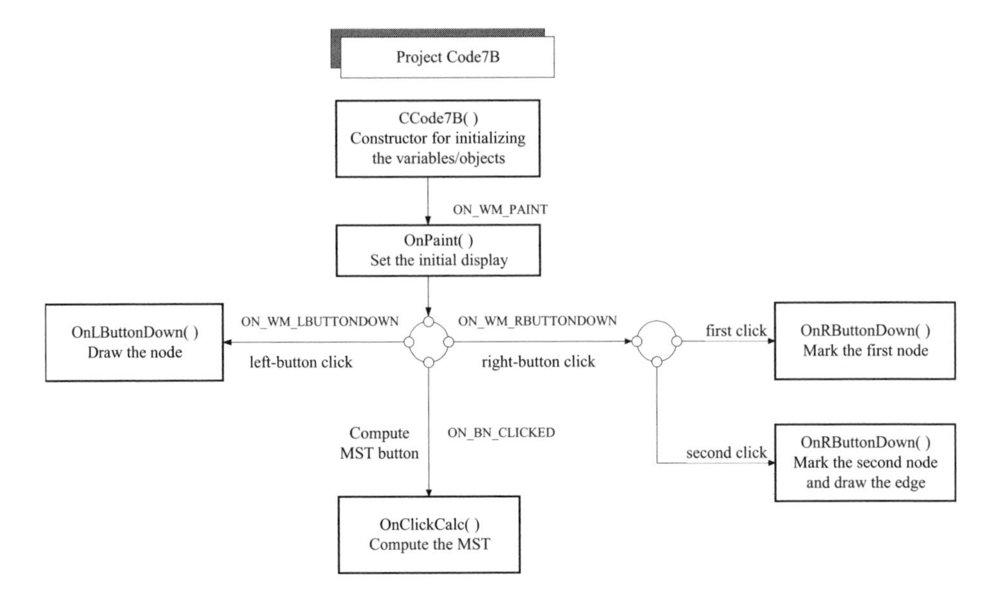

Figure 7.14 Outline of the Code7B project.

Code7B is an extension of Code7A, most of the code in these functions, with the exception of `OnClickCalc()`, are reused and changes are kept to the minimum.

The data structure of the Code7B project is described in Code7B.h. The structure is basically the same as in Code7A with a few additional variables and objects for computing the minimum spanning tree. This includes the `Compute MST` push button, which is created from the `MSTbutton` object derived from the `CButton` class. The object has an id declared as `IDC_MST` in the Code7B.h file. Another variable is `e[i][j].InputBox`, which is an edit box object derived from the class `CEdit`. This edit box provides a one-digit input space for the link e_{ij} from the user. The additional variables and objects in Code7B are described briefly in Table 7.5.

Just like in the previous project, a node in the graph is created with the left-button click of the mouse and mapped as `WM_LBUTTONDOWN`. In response to this event is the function `OnLButtonDown()`, as outlined in Figure 7.8 (minus the adjacency matrix update). The link e_{ij} is created by clicking the right button consecutively on the paired v_i and v_j nodes. The first right-button click activates a series of tasks as shown in Figure 7.9 in Code7A. The second click produces some additional tasks, such as the creation of an edit box on the link. The activities in the second right-button click are outlined in Figure 7.15.

The right-button click event is represented by `WM_RBUTTONDOWN` which is handled by the function `OnRButtonDown()`, as follows:

```
void CCode7A::OnRButtonDown(UINT nFlags,CPoint pt)
{
     CClientDC dc(this);
```

```
    int i,u,w;
    CPen penGray(PS_SOLID,1,TextColor);
    dc.SelectObject(penGray);
    for (i=1;i<=nv;i++)
        if (v[i].Box.PtInRect(pt))
        {
            RButtonFlag++;
            if (RButtonFlag==1)
                Pt1=i;
            if (RButtonFlag==2)
            {
                Pt2=i; nLink++;
                dc.MoveTo(v[Pt1].Home); dc.LineTo(v[Pt2].Home);
                u=(v[Pt1].Home.x+v[Pt2].Home.x)/2;
                w=(v[Pt1].Home.y+v[Pt2].Home.y)/2;
                e[Pt1][Pt2].InputBox.Create(WS_CHILD | WS_VISIBLE
                    | WS_BORDER,  CRect(CPoint(u,w),CSize(12,20)),
                    this, idc_WtInput++);
                e[Pt1][Pt2].Flag=1;
                RButtonFlag=0;
            }
        }
}
```

The above code represents a series of tasks that happen when the right button of the mouse is clicked. To differentiate between the first and second clicks, a variable called RButtonFlag is introduced. This flag describes the state of the link, as follows:

RButtonFlag=0 denotes inactive link.
RButtonFlag=1 denotes the tail node has been selected.
RButtonFlag=2 denotes the head node has been selected.

Table 7.5 Some of the objects and variables in Code7B

Variable/object	Class	Description
MSTbutton	CButton	Activation button for computing and displaying the MST of the graph
e[i][j].InputBox	CEdit	The dialog box as an input for the weight of the link, e_{ij}
idc_WtInput	int	The id of the dialog box e[i][j].InputBox
e[i][j].Wt	int	The integer value of e[i][j].InputBox
e[i][j].Flag	bool	A flag to see if the link e_{ij} exists (1) or not (0)
P[i],Q[j]	int	Beginning and ending nodes of an ordered link in the spanning tree
MSTcost	int	The cost of the spanning tree
MinLink	int	The link with the minimum weight originating from the end nodes of T that need to be added into the subgraph

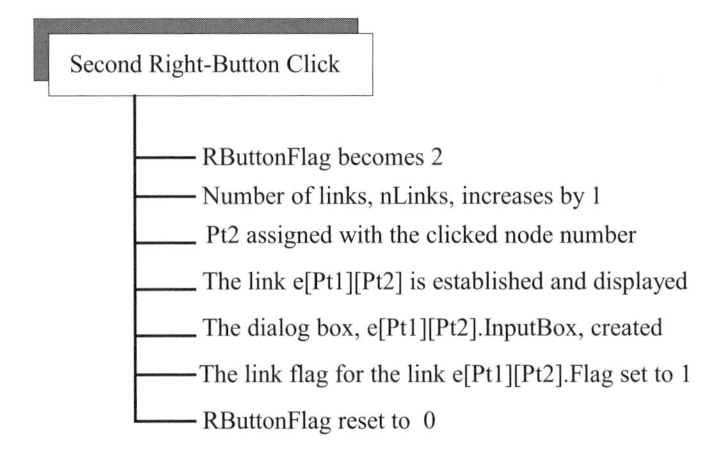

Figure 7.15 Activities activated by the second right-button click.

Initially, `RButtonFlag` is set to 0, which means no activity on the link drawing. The first click assigns `RButtonFlag` to the value of 1 and its node as `Pt1`. The second click assigns the value of 2 to `RButtonFlag` and `Pt2` to the node. The link is immediately established between `Pt1` and `Pt2`, and displayed on the graph. In addition, a small edit box is created on this link to allow input of its weight from the user. This edit box is represented as the `CEdit` object, `e[Pt1][Pt2].InputBox`.

To remember that this link has been established, another flag of type `bool` called `e[Pt1][Pt2].Flag` is introduced. Its value is indicated, as follows:

`e[Pt1][Pt2].Flag=0` denotes input on the link has been made.
`e[Pt1][Pt2].Flag=1` denotes input on the link has not been made.

Whenever a link is established, `e[Pt1][Pt2].Flag` takes the value of 1 (TRUE) to indicate that an input on its weight is yet to be completed by the user. The program checks for this flag on all links in the graph before finding the minimum spanning tree.

Figure 7.16 shows the activities that take place when the `Compute MST` push button is clicked. The event is detected as `BN_CLICKED` and it is handled by the function `OnClickCalc()`. The function is written as follows:

```
void CCode7A::OnClickCalc()
{
     CClientDC dc(this);
     CString s;
     int i,j;
     dc.SelectObject(FontCourier);
```

```
dc.SetTextColor(TextColor); dc.SetBkColor(BgColor);
dc.TextOut(MatText.x,MatText.y,"Weight Matrix");
for (i=1;i<=nv;i++)
{
     P[i]=((i==1)?1:0);
     Q[i]=((i==1)?0:i);
     for (j=1;j<=nv;j++)
     {
          if (e[i][j].Flag)
          {
               dc.SetTextColor(TextColor);
               e[i][j].InputBox.GetWindowText(s);
               e[i][j].Wt=e[j][i].Wt=atoi(s);
               s.Format("%d",e[i][j].Wt);
               dc.TextOut(MatText.x+TextGap*(i-1),
               MatText.y+TextGap+TextGap*(j-1),s);
               dc.TextOut(MatText.x+TextGap*(j-1),
               MatText.y+TextGap+TextGap*(i-1),s);
          }
          if (e[i][j].Wt==99)
          {
               dc.SetTextColor(RGB(180,180,180));
               s.Format("%d",0);
               dc.TextOut(MatText.x+TextGap*(i-1),
               MatText.y+TextGap+TextGap*(j-1),s);
               dc.TextOut(MatText.x+TextGap*(j-1),
               MatText.y+TextGap+TextGap*(i-1),s);
          }
     }
}
MSTcost=0;
```

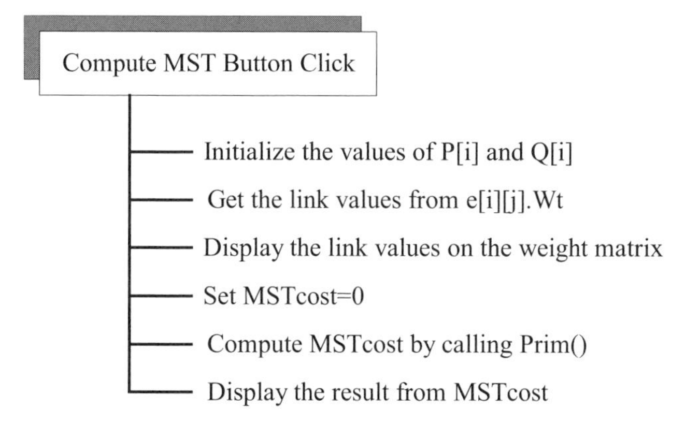
Compute MST Button Click
Initialize the values of P[i] and Q[i]
Get the link values from e[i][j].Wt
Display the link values on the weight matrix
Set MSTcost=0
Compute MSTcost by calling Prim()
Display the result from MSTcost

Figure 7.16 Outline of the activities in `OnClickCalc()`.

```
      for (i=1;i<=nv-1;i++)
            Prim();
      dc.SetTextColor(TextColor);
      s.Format("Cost=%d",MSTcost); dc.TextOut(600,400,s);
}
```

In the function, the arrays `P[]` and `Q[]` store the two endpoints of the links in the subgraph *T*. These two arrays are given some initial values to indicate their position as the tail or head of the links, which are needed for comparing the subsequent link values emanating from the nodes.

The minimum spanning tree cost is denoted by `MSTcost`, and this variable has an initial value of 0. Prior to evaluating the minimum spanning tree, the user has to complete the entries on the weights of the links in the graph. The program checks for this validity through the flag `e[i][j].Flag` using the following conditional test:

```
if (e[i][j].Flag)
```

The function `OnClickCalc()` computes the minimum spanning tree of the graph using Prim's algorithm. In constructing the spanning tree, `OnClick-Calc()` calls up the function `Prim()` each time a link needs to be added into the subgraph *T*. At each call, `Prim()` computes and identifies the link with a minimum value represented as the variable `MinWt`. This value is added to `MSTcost` at each call of the function. `MSTcost` eventually becomes the total cost for the minimum spanning tree after the formation of the tree is completed. The function is written as follows:

```
void CCode7B::Prim()
{
      CClientDC dc(this);
      CPen penMSTpath(PS_SOLID,3,RGB(100,100,100));
      int i,j,k,r,w,u,h;

      MinWt=99;
      for (i=1;i<=nv;i++)
      {
            if (P[i]==0)
                  break;
            for (j=1;j<=nv;j++)
                  if (Q[j]!=0)
                        for (k=1;k<=nv;k++)
                              if (Q[k]!=0)
                                    for (r=1;r<=nv;r++)
                                          if (Q[r]!=0)
                                                for (h=1;h<=nv;h++)
                                                      if (Q[h]!=0)
                                                            if
```

```
                    (e[P[i]][Q[j]].Wt<=e[P[i]][Q[k]].Wt
                    && e[P[i]][Q[k]].Wt<=e[P[i]][Q[r]].Wt
                    && e[P[i]][Q[r]].Wt<=e[P[i]][Q[h]].Wt
                    && e[P[i]][Q[j]].Wt<=MinWt)
                    {
                         MinWt=e[P[i]][Q[j]].Wt;
                         u=i; w=j;
                    }
     }
     MSTcost += MinWt;
     dc.SelectObject(penMSTpath);
     dc.MoveTo(v[P[u]].Home); dc.LineTo(v[Q[w]].Home);
     for (i=1;i<=nv;i++)
          if (P[i]==0)
          {
               P[i]=w;   break;
          }
     for (i=1;i<=nv;i++)
          if (Q[i]==w)
          {
               Q[i]=0;   break;
          }
}
```

7.4 SUMMARY AND CONCLUSION

This chapter describes two applications involving graph theory. The first application is about providing a user-friendly interface for constructing a graph. The interface allows the user to draw the nodes and the links of a graph by simply clicking the mouse at any place on the window. The second application discusses the problem of finding the minimum spanning tree of a graph. We apply Prim's algorithm to solve the problem by extending the work from the earlier application. Both applications are user friendly and this helps in making the problems more attractive.

Both Code7A and Code7B projects can be extended to support other applications in graph theory. As mentioned earlier, Code7A provides all the fundamental data structure and groundwork for tackling many problems in graph theory. This project can easily be modified for solving other problems such as the maximum clique, maximum independent set, and the vertex or edge coloring. It is also good to produce a visualization tool for determining if a graph is bipartite or if it isomorphic to another graph. For Code7B, a good extension is to produce a visual tool for solving the minimum spanning tree problem using Kruskal's algorithm. In this case, Kruskal's push button should be added to the visual interface on the window. Adding Kruskal's method will help in understanding several ways of solving the minimum spanning tree problem.

Another challenging application of Code7B is the problem of broadcasting a message in a network from a node. The term *broadcasting* means the message from the source node is to reach all the other nodes in the network. An interesting approach to this problem is to identify the minimum spanning tree with its root from the source node. The minimum spanning tree guarantees the path with the shortest route from the source node to the destination nodes, assuming no other factors such as the network startup and transmission costs. When transmitting a message, the main objective is to complete the job at the earliest time. In message transmission, a node can only send one message to another node at a time. In addition, the sending node cannot receive a message at the same time. It is also necessary for the message to follow a path that will guarantee a safe passage. In this case, problems such as data collision and congestion need to be avoided. The details are left for the reader to explore.

BIBLIOGRAPHY

1. R. Sedgewick, *Algorithms in C: Part 1-4,* 3rd ed., Addison-Wesley, 2001.
2. R. Sedgewick, *Algorithms in C++ Part 5: Graph Algorithms,* 3rd ed., Addison-Wesley, 2001.
3. J. L. Gross and J. Yellen, *Handbook of Graph Theory,* CRC Press, 2003.
4. J. Yellen and J. L. Gross, *Graph Theory and Its Applications,* CRC Press, 1998.
5. T. H. Cormen, C. E. Leiserson, R. L. Rivest, and C. Stein, *Introduction to Algorithms,* 2nd ed., MIT Press, 2001.

CODE LISTINGS

Code7A: Drawing a Graph

```
// Code7A.h
#include <afxwin.h>
#define N 10

class CCode7A : public CFrameWnd
{
private:
        CPoint TopLeft,BottomRight;
        CPoint InstText,MatText;
        CFont FontCourier;
        CSize BoxSize;
        int nv,nLink,RButtonFlag,Pt1,Pt2;
        int TextGap,TextColor,BgColor;
        int **e;
        typedef struct
        {
                CPoint Home;
                CRect Box;
                int degree;
        } NODE;
        NODE *v;
```

```
public:
	CCode7A();
	~CCode7A();
	void Prim();
	afx_msg void OnPaint();
	afx_msg void OnLButtonDown (UINT, CPoint);
	afx_msg void OnRButtonDown (UINT, CPoint);
	DECLARE_MESSAGE_MAP();
};

class CMyWinApp : public CWinApp
{
public:
  virtual BOOL InitInstance();
};

// Code7A.cpp: graph
#include "Code7A.h"

BOOL CMyWinApp::InitInstance()
{
	CCode7A* pFrame = new CCode7A;
	m_pMainWnd = pFrame;
	pFrame->ShowWindow(SW_SHOW);
	pFrame->UpdateWindow();
	return TRUE;
}

CMyWinApp  MyApplication;

BEGIN_MESSAGE_MAP(CCode7A,CFrameWnd)
	ON_WM_PAINT()
	ON_WM_LBUTTONDOWN()
	ON_WM_RBUTTONDOWN()
END_MESSAGE_MAP()

CCode7A::CCode7A()
{
	int i,j;
	v=new NODE [N+1];
	e=new int *[N+1];
	for (i=0;i<=N;i++)
		e[i]=new int [N+1];
	Create(NULL,"Drawing a graph",WS_OVERLAPPEDWINDOW,CRect(0,0,800,500));
	TopLeft=CPoint(20,20); BottomRight=CPoint(450,360);
	MatText=CPoint(500,30); InstText=CPoint(80,370);
	nv=0; nLink=0; RButtonFlag=0;
	for (i=1;i<=N;i++)
	{
		v[i].degree=0;
		for (j=1;j<=N;j++)
			e[i][j]=e[j][i]=99;
	}
	FontCourier.CreatePointFont(60,"Courier");
```

```
	BgColor=RGB(240,240,240); TextGap=20; BoxSize=CSize(10,10);
	TextColor=RGB(100,100,100);
}

CCode7A::~CCode7A()
{
	for (int i=0;i<=N;i++)
		delete e[i];
	delete v,e;
}

void CCode7A::OnPaint()
{
	CPaintDC dc(this);
	CRect rc;
	CBrush BgBrush(BgColor);
	GetClientRect(&rc);
	dc.FillRect(&rc,&BgBrush);

	CPen penDrawingBox(PS_SOLID,4,RGB(100,100,100));
	dc.SelectObject(penDrawingBox);
	dc.SelectStockObject(HOLLOW_BRUSH);
	rc=CRect(TopLeft,BottomRight);
	dc.Rectangle(rc);

	dc.SelectObject(FontCourier);
	dc.SetTextColor(TextColor); dc.SetBkColor(BgColor);
	dc.TextOut(InstText.x,InstText.y,"Instructions:");
	dc.TextOut(InstText.x+30,InstText.y+25,
		"Left mouse to click points");
	dc.TextOut(InstText.x+30,InstText.y+45,
		"Right mouse to draw a link between two points");
	dc.TextOut(MatText.x,MatText.y,"Adjacency Matrix");
	dc.TextOut(MatText.x+TextGap*N,MatText.y,"Degree");
}


void CCode7A::OnLButtonDown(UINT nFlags,CPoint pt)
{
	CClientDC dc(this);
	CString s;
	CPen penGray(PS_SOLID,2,TextColor);
	dc.SelectObject(penGray);

	dc.SelectObject(FontCourier);
	dc.SetTextColor(TextColor);
	dc.SetBkColor(BgColor);
	if (CRect(TopLeft,BottomRight).PtInRect(pt))
		if (nv<=N)
		{
			nv++;
			v[nv].Home=pt;
			v[nv].Box=CRect(CPoint(pt),CSize(BoxSize));
			dc.Rectangle(v[nv].Box);
```

```
                        s.Format("%d",nv);
                        dc.TextOut(v[nv].Home.x-10,v[nv].Home.y-10,s);
                        for (int i=1;i<=nv;i++)
                        {
                                dc.TextOut(MatText.x+TextGap*(i-1),
                                        MatText.y+TextGap+TextGap*(nv-1),"0");
                                dc.TextOut(MatText.x+TextGap*(nv-1),
                                        MatText.y+TextGap+TextGap*(i-1),"0");
                                s.Format("%d",v[nv].degree);
                                dc.TextOut(MatText.x+TextGap+TextGap*N,
                                        MatText.y+TextGap+TextGap*(nv-1),s);
                        }
                }
}

void CCode7A::OnRButtonDown(UINT nFlags,CPoint pt)
{
        CClientDC dc(this);
        CString s;

        int i;
        CPen penGray(PS_SOLID,1,TextColor);
        dc.SelectObject(penGray); dc.SetBkColor(BgColor);
        for (i=1;i<=nv;i++)
        {
                if (v[i].Box.PtInRect(pt))
                {
                        RButtonFlag++;
                        if (RButtonFlag==1)
                                Pt1=i;
                        if (RButtonFlag==2)
                        {
                                Pt2=i; e[Pt1][Pt2]=1; nLink++;
                        }
                }
                if (RButtonFlag==2)
                {
                        dc.MoveTo(v[Pt1].Home); dc.LineTo(v[Pt2].Home);
                        s.Format("%d",e[Pt1][Pt2]);
                        dc.TextOut(MatText.x+TextGap*(Pt1-1),
                                MatText.y+TextGap+TextGap*(Pt2-1),s);
                        dc.TextOut(MatText.x+TextGap*(Pt2-1),
                                MatText.y+TextGap+TextGap*(Pt1-1),s);
                        s.Format("%d",++v[Pt1].degree);
                        dc.TextOut(MatText.x+TextGap+TextGap*N,
                                MatText.y+TextGap+TextGap*(Pt1-1),s);
                        s.Format("%d",++v[Pt2].degree);
                        dc.TextOut(MatText.x+TextGap+TextGap*N,
                                MatText.y+TextGap+TextGap*(Pt2-1),s);
                        RButtonFlag=0;
                }
        }
}
```

Code7B: Minimum Spanning Tree

```
// Code7B.h
#include <afxwin.h>
#define N 10
#define IDC_MST 500

class CCode7B : public CFrameWnd
{
private:
	CPoint TopLeft,BottomRight;
	CPoint InstText,MatText;
	CFont FontCourier;
	CSize BoxSize;
	CRect MSTbox;
	CButton MSTbutton;

	int nv,nLink,RButtonFlag,Pt1,Pt2,idc_WtInput;
	int *P,*Q,MSTcost,MinWt;
	int TextGap,TextColor,BgColor;
	typedef struct
	{
		int Wt;
		bool Flag;
		CEdit InputBox;
	} LINK;
	LINK **e;
	typedef struct
	{
		CPoint Home;
		CRect Box;
	} NODE;
	NODE *v;
public:
	CCode7B();
	~CCode7B();
	void Prim();
	afx_msg void OnClickCalc();
	afx_msg void OnPaint();
	afx_msg void OnLButtonDown (UINT, CPoint);
	afx_msg void OnRButtonDown (UINT, CPoint);
	DECLARE_MESSAGE_MAP();
};

class CMyWinApp : public CWinApp
{
public:
  virtual BOOL InitInstance();
};


// Code6B.cpp: computing the minimum spanning tree
#include "Code7B.h"
```

```
BOOL CMyWinApp::InitInstance()
{
       CCode7B* pFrame = new CCode7B;
       m_pMainWnd = pFrame;
       pFrame->ShowWindow(SW_SHOW);
       pFrame->UpdateWindow();
       return TRUE;
}

CMyWinApp  MyApplication;

BEGIN_MESSAGE_MAP(CCode7B,CFrameWnd)
       ON_WM_PAINT()
       ON_WM_LBUTTONDOWN()
       ON_WM_RBUTTONDOWN()
       ON_BN_CLICKED(IDC_MST,OnClickCalc)
END_MESSAGE_MAP()

CCode7B::CCode7B()
{
       int i,j;
       P=new int [N+1];
       Q=new int [N+1];
       v=new NODE [N+1];
       e=new LINK *[N+1];
       for (i=0;i<=N;i++)
              e[i]=new LINK [N+1];
       Create(NULL,"Prim's Algorithm for computing the MST",
              WS_OVERLAPPEDWINDOW,CRect(0,0,800,500));
       TopLeft=CPoint(20,20); BottomRight=CPoint(450,360);
       MSTbutton.Create("Compute MST",WS_CHILD | WS_VISIBLE | BS_DEFPUSH BUTTON,
               CRect(500,330,650,360),this,IDC_MST);
       MatText=CPoint(500,30); InstText=CPoint(80,370);
       nv=0; nLink=0; RButtonFlag=0; idc_WtInput=1001;
       for (i=1;i<=N;i++)
              for (j=1;j<=N;j++)
              {
                     e[i][j].Wt=e[j][i].Wt=99;
                     e[i][j].Flag=e[j][i].Flag=0;
              }
       FontCourier.CreatePointFont(60,"Courier");
       BgColor=RGB(240,240,240); TextGap=25; BoxSize=CSize(10,10);
       TextColor=RGB(100,100,100);
}

CCode7B::~CCode7B()
{
       for (int i=0;i<=N;i++)
              delete e[i];
       delete v,e,P,Q;
}

void CCode7B::OnPaint()
{
```

```
        CPaintDC dc(this);
        CRect rc;
        CBrush BgBrush(BgColor);
        GetClientRect(&rc);
        dc.FillRect(&rc,&BgBrush);

        CPen penDrawingBox(PS_SOLID,4,RGB(100,100,100));
        dc.SelectObject(penDrawingBox);
        dc.SelectStockObject(HOLLOW_BRUSH);
        rc=CRect(TopLeft,BottomRight);
        dc.Rectangle(rc);

        dc.SelectObject(FontCourier);
        dc.SetTextColor(TextColor);
        dc.SetBkColor(BgColor);
        dc.TextOut(InstText.x,InstText.y,"Instructions:");
        dc.TextOut(InstText.x+30,InstText.y+25,"Left mouse to click points");
        dc.TextOut(InstText.x+30,InstText.y+45,
               "Right mouse to draw a link between two points");
}

void CCode7B::OnLButtonDown(UINT nFlags,CPoint pt)
{
        CClientDC dc(this);
        CString s;
        CPen penGray(PS_SOLID,2,TextColor);
        dc.SelectObject(penGray);

        dc.SelectObject(FontCourier);
        dc.SetTextColor(TextColor);
        dc.SetBkColor(BgColor);
        if (CRect(TopLeft,BottomRight).PtInRect(pt))
               if (nv<N)
               {
                      nv++;
                      v[nv].Home=pt;
                      v[nv].Box=CRect(CPoint(pt),CSize(BoxSize));
                      dc.Rectangle(v[nv].Box);
                      s.Format("%d",nv);
                      dc.TextOut(v[nv].Home.x-10,v[nv].Home.y-10,s);
               }
}

void CCode7B::OnRButtonDown(UINT nFlags,CPoint pt)
{
        CClientDC dc(this);

        int i,u,w;
        CPen penGray(PS_SOLID,1,TextColor);
        dc.SelectObject(penGray);
        for (i=1;i<=nv;i++)
               if (v[i].Box.PtInRect(pt))
               {
                      RButtonFlag++;
```

```
                    if (RButtonFlag==1)
                          Pt1=i;
                    if (RButtonFlag==2)
                    {
                          Pt2=i; nLink++;
                          dc.MoveTo(v[Pt1].Home); dc.LineTo(v[Pt2].Home);
                          u=(v[Pt1].Home.x+v[Pt2].Home.x)/2;
                          w=(v[Pt1].Home.y+v[Pt2].Home.y)/2;
                          e[Pt1][Pt2].InputBox.Create(WS_CHILD | WS_VISIBLE
                             | WS_BORDER,CRect(CPoint(u,w),CSize(12,20)),
                             this, idc_WtInput++);
                          e[Pt1][Pt2].Flag=1;
                          RButtonFlag=0;
                    }
              }
}

void CCode7B::OnClickCalc()
{
      CClientDC dc(this);
      CString s;
      int i,j;

      dc.SelectObject(FontCourier);
      dc.SetTextColor(TextColor); dc.SetBkColor(BgColor);
      dc.TextOut(MatText.x,MatText.y,"Weight Matrix");
      for (i=1;i<=nv;i++)
      {
            P[i]=((i==1)?1:0);
            Q[i]=((i==1)?0:i);
            for (j=1;j<=nv;j++)
            {
                  if (e[i][j].Flag)
                  {
                        dc.SetTextColor(TextColor);
                        e[i][j].InputBox.GetWindowText(s);
                        e[i][j].Wt=e[j][i].Wt=atoi(s);
                        s.Format("%d",e[i][j].Wt);
                        dc.TextOut(MatText.x+TextGap*(i-1),
                              MatText.y+TextGap+TextGap*(j-1),s);
                        dc.TextOut(MatText.x+TextGap*(j-1),
                              MatText.y+TextGap+TextGap*(i-1),s);
                  }
                  if (e[i][j].Wt==99)
                  {
                        dc.SetTextColor(RGB(180,180,180));
                        s.Format("%d",0);
                        dc.TextOut(MatText.x+TextGap*(i-1),
                              MatText.y+TextGap+TextGap*(j-1),s);
                        dc.TextOut(MatText.x+TextGap*(j-1),
                              MatText.y+TextGap+TextGap*(i-1),s);
                  }
            }
      }
```

```
        MSTcost=0;
        for (i=1;i<=nv-1;i++)
                Prim();
        dc.SetTextColor(TextColor);
        s.Format("Cost=%d",MSTcost); dc.TextOut(600,400,s);
}

void CCode7B::Prim()
{
        CClientDC dc(this);
        CPen penMSTpath(PS_SOLID,3,RGB(100,100,100));
        int i,j,k,r,w,u,h;

        MinWt=99;
        for (i=1;i<=nv;i++)
        {
                if (P[i]==0)
                        break;
                for (j=1;j<=nv;j++)
                        if (Q[j]!=0)
                                for (k=1;k<=nv;k++)
                                        if (Q[k]!=0)
                                                for (r=1;r<=nv;r++)
                                                        if (Q[r]!=0)
                                                                for (h=1;h<=nv;h++)
                                                                        if (Q[h]!=0)

                        if (e[P[i]][Q[j]].Wt<=e[P[i]][Q[k]].Wt

                        && e[P[i]][Q[k]].Wt<=e[P[i]][Q[r]].Wt

                        && e[P[i]][Q[r]].Wt<=e[P[i]][Q[h]].Wt

                        && e[P[i]][Q[j]].Wt<=MinWt)
                        {
                                MinWt=e[P[i]][Q[j]].Wt;
                                u=i; w=j;
                        }
        }
        MSTcost += MinWt;
        dc.SelectObject(penMSTpath);
        dc.MoveTo(v[P[u]].Home); dc.LineTo(v[Q[w]].Home);
        for (i=1;i<=nv;i++)
                if (P[i]==0)
                {
                        P[i]=w;         break;
                }
        for (i=1;i<=nv;i++)
                if (Q[i]==w)
                {
                        Q[i]=0;         break;
                }
}
```

CHAPTER 8

GRAPH APPLICATIONS

8.1 GRAPH–NETWORK RELATIONSHIP

An understanding of graphs contributes to the design of several types of network models. A *network* consists of several elements linked physically or logically that work cooperatively in a group as a system. A physical network has nodes and links that are visible. For example, in a local area network, the elements are the computers whereas the links are the cables that connect the computers to each other. A logical network is formed to represent a solution to a particular problem. In business, a network model is built as a logical process arising from logistics control, production planning, financial planning, and capacity analysis. Some examples include determining the routes for delivering parcels, assigning airline crews to aircraft, and optimizing the shipping sources to destinations.

A network model consists of the transformation of a problem into the form of a network so that the relationship between the participating elements can be formulated. A network is best modeled as a graph as the latter has properties and solutions that can be shared and applied to the former. In general, a graph G described in the last chapter, with nodes v_i, and edges e_{ij}, is a fundamental form of a network. A node in the graph may represent an event or element, whereas a link between a pair of nodes is the flow path showing the relationship or precedence between the nodes.

A simple network may be represented as a small graph consisting of a few nodes and links. A large network, like the human body, may consist of thousands, millions, or even billions of nodes and countless number of links. The network in the human body is so large that it is necessary to break it into several smaller networks according to their functions. The blood circulation system is a network that includes the heart, veins, and arteries, whereas the hearing system forms another network that includes the cochlea and ear drum. These smaller networks work cooperatively and are interrelated to each other in order for the body to function properly.

Network design is a complicated problem involving many variables, depending on the type of application. Several issues and problems arise in the development of a network. Fortunately, many problems in network design involve problems in graph theory as well, such as the maximum flow, shortest path, minimum spanning tree, maximum clique, and graph coloring. Hence, graph properties and solutions to these problems contribute greatly to solving many problems involving the design of networks.

In this chapter, we discuss one of the main problems in graph theory, namely, the shortest-path problem, which has a wide application in the design of networks. One interesting topic in our model is the Floyd–Warshall algorithm, which solves the all-pairs shortest-path problem of a graph. We also discuss one application of the shortest-path problem in a mesh network.

8.2 SHORTEST-PATH PROBLEM

The shortest path problem is stated as follows: *Given a weighted graph, find the minimal cost linking a pair of nodes in the graph.* The cost referred to in this problem can be weights, energy, distance, time, and so on. These items are the performance measures associated with the given problem. A path may consist of one or more links that connect a source node and the destination node.

The main objective in the shortest-path problem is to find a path in which the sum of the weights is minimum. The normal dynamic programming solution to the problem consists of the natural decomposition of the problem into several stages. Each stage in the problem has its own state in which a decision in one stage recursively transforms the current state into a state in the next stage.

Two cases of the shortest-path problems are normally discussed. In the single-pair case, the problem is confined to finding a single path connecting two nodes only. In the all-pairs case, the problem requires finding the shortest paths between all pairs of nodes in the graph. Figure 8.1 illustrates the second case but only two paths are shown. The first path connects v_1 and v_5 and the path is $v_1 \rightarrow v_7 \rightarrow v_5$ having cost = 13. The second path connects v_2 and v_4, having the path $v_2 \rightarrow v_8 \rightarrow v_6 \rightarrow v_5 \rightarrow v_4$ with cost = 20.

Shortest-path problems have many applications in real life. For example, consider a geographical region consisting of several cities (nodes) and a set of roads (edges) linking these cities. A fuel supplier finds it useful to have information on the shortest paths between the cities so that fuel deliveries to these cities can be made in the most economical way using the shortest paths. In image segmentation, the shortest path problem has its application in establishing a connection between any two pixels of an image. In one typical problem, a path that crosses black pixels in the image may result in a high cost. Therefore, the objective here is to find a path that minimizes crossing the black pixels. This problem evolves into finding the shortest paths between pairs of pixels in the image.

The shortest-path problem can also make significant contributions to the design of a printed circuit board (PCB). There are two main problems in PCB design:

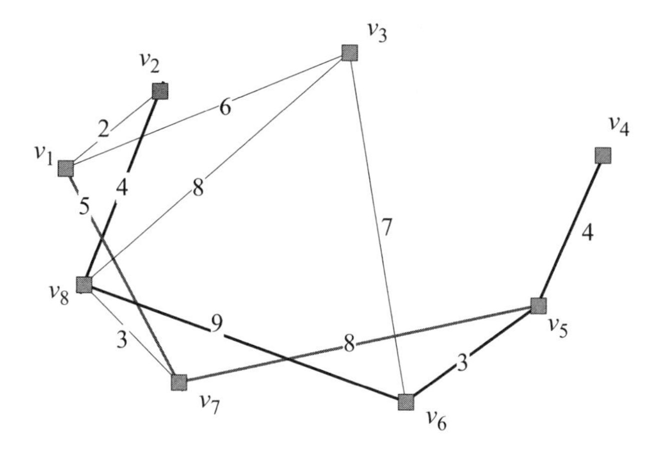

Figure 8.1 Two shortest paths. Path 1: $v_1 \rightarrow v_7 \rightarrow v_5$. Path 2: $v_2 \rightarrow v_8 \rightarrow v_6 \rightarrow v_5 \rightarrow v_4$.

placement and routing. Placement involves a systematic installation of the electronic components onto the small area of the PCB. Routing involves wiring between the electronic components. A typical routing problem in PCB design requires finding the shortest path for connecting two or more components. The wiring is to be done according to the shape of a planar graph, and crossings between the wires are not allowed. The main objective in routing is to produce a realization that minimizes the overall congestion in the network.

We discuss briefly two common methods for solving the shortest-path problem. The first method is Dijkstra's algorithm, which is suitable for finding the single-source shortest path between a pair of nodes in a graph. The second method is the Floyd–Warshall algorithm, which computes the all-pairs shortest-path problem. The difference between the two methods lies in their approaches: Dijkstra's method involves only one pair of nodes at a time, whereas the Floyd–Warshall method considers all the possible combinations of the nodes.

Dijkstra's Algorithm

Shortest-path computation involving a single pair of nodes is commonly solved using Dijkstra's algorithm. This algorithm is a greedy method of solving the problem and it functions almost similarly to Prim's algorithm for finding the minimum spanning tree. In finding the minimum spanning tree, Prim's algorithm starts from one node and extends outward within the graph until all the vertices have been reached. Dijkstra's algorithm proceeds in stages in which the shortest link among several others in the graph is included in the path at one time. The procedure repeats with the next-shortest link, and so on, until the path from the source to its destination is completed. The algorithm has a complexity of $O(N^2)$ where N is the number of nodes in the graph.

Dijkstra's algorithm begins by assigning any node in the graph a permanent label with the value of 0, and all vertices a temporary label with the value of 0. The algorithm then proceeds to select the least-cost link connecting a node with a permanent label to a node with a temporary label. The second node'a label is then updated from temporary to permanent, and its value is then determined by the addition of the cost of the link with the first node's value. The next step is to find the next least-cost link by extending it to a node with a temporary label from either the first node or second node. The third node's label is next, and the distance from the first node is determined in a similar manner. This process is repeated until the labels of all vertices in the graph become permanent.

The Floyd–Warshall Algorithm

The Floyd–Warshall algorithm computes the all-pairs shortest-path problem with a complexity of $O(N^3)$, where N is the number of nodes in the graph. The weight of a link between the nodes v_i and v_j is denoted w_{ij}. In order to compute the shortest path, w_{ij} between v_i and v_j, for $i = 1, 2, \ldots, N$ and $j = 1, 2, \ldots, N$, must be given. If the link between v_i and v_j does not exist, then the weight is denoted by ∞. This weight is assigned to $w_{ij}^{\min}$ as the initial minimal distance between v_i and v_j. The Floyd–Warshall algorithm assumes the links in the graph to have nonnegative weights and nonnegative cycles.

The Floyd–Warshall algorithm starts by finding all the minimal distances between the pairs of nodes without passing through any intermediate nodes. These values are recorded in the minimal distance table. The rule is then relaxed by allowing one node to be the only intermediate node, starting with v_0. We define a_{ijk} as the shortest distance from v_i to v_j using only the nodes $v_0, v_0, \ldots, v_k$ as the intermediate nodes. With a_k as the intermediate node, the minimal distance is a_{ijk}. The minimal distances between the pairs of nodes are determined by comparing the previous values with the present ones. New updates are recorded in the minimal distance table. This is followed by v_1, v_2, and so on until the last node, v_N, is reached. The process repeats with two nodes and so on, using the earlier results for comparison. At each step, a comparison is made with the values in the table and any new minimal distance values are recorded in the table. The algorithm follows.

Input: Weight w_{ij} between v_i and v_j, for $i = 1, 2, \ldots, N$ and $j = 1, 2, \ldots, N$.
Output: Shortest path $w_{ij}^{\min}$ between v_i and v_j.
Steps:

```
for i = 1 to N
    for j = 1 to N
        Let w_ij^min = w_ij;
    endfor
endfor
for i = 1 to N
    for j = 1 to N
        for k = 1 to N
            if w_ij^min ≠ ∞ or w_ik^min ≠ ∞ or w_jk^min ≠ ∞
                if w_ji^min + w_ik^min < w_jk^min
```

```
                    Let n = 1;
                    Compute w_ji^min = w_ji^min + w_ik^min;
                    for m = 1 to N
                        Let a_jkm = ∞;
                    endfor
                    for m = 1 to N
                        if a_jim ≠ ∞
                            Compute a_jkn ← a_jkm;
                            Set n ← n + 1;
                        endif
                    endfor
                    Set a_jkn = i;
                    Set n ← n + 1;
                    for m = 1 to N
                        if a_ikm ≠ ∞
                            Compute a_jkn ← a_ikm;
                            Set n ← n + 1;
                        endif
                    endfor
                endif
            endif
        endfor
    endfor
endfor
```

Table 8.1 describes the symbols used in the Floyd–Warshall algorithm. In this algorithm, w_{ij}^{min} is the minimal distance between v_i and v_j. The variable a_{ijk} represents the path for the minimal distance between v_i and v_j, with v_k as an intermediate node. The path between v_i and v_j may have no intermediate node, v_k, or it may have more than one node, depending on its distance.

Code8A: Shortest-Path Visualization

We discuss the development of a visualization model for the shortest-path problem. The project is called Code8A, having the files Code8A.h and Code8A.cpp. The model adopts the Floyd–Warshall algorithm for computing the shortest paths be-

Table 8.1 Symbols used in the Floyd–Warshall algorithm

Symbol	Description
w_{ij}	Weight of a link between v_i and v_j
w_{ij}^{min}	Shortest (minimal) distance between v_i and v_j
∞	Value assigned to denote that the link between v_i and v_j does not exist
a_{ijk}	Path for the minimal distance between v_i and v_j, with v_k as the intermediate node

tween all pairs of nodes in the graph. Figure 8.2 shows the output of Code8A. It consists of a graph in the drawing area in the top half of the window, whereas the path is shown in the text area in the bottom half. There are 20 nodes in the graph, shown as shaded rectangles. The number of nodes is represented by the macro constant `N`, declared from `#define`. A different number of nodes can be obtained by changing this macro value.

In Figure 8.2, the nodes in the graph are distributed at the randomly determined coordinates in the drawing area. The links between the nodes, `e[i][j]`, are drawn as thin lines. The weight `e[i][j].Wt` in each link is also determined randomly. We introduce a rule called the *nearest neighbor rule* which assigns a link to a pair of nodes if the distance between the two nodes is less than or equal to a threshold value, `LinkRange`. In this application, `LinkRange` is set to 200 units. From this simple rule, the nodes that are within the range distance of `LinkRange` are directly linked to each other in the graph.

The nodes in the graph appear to be well distributed with their positions determined randomly in the window. The graph is not congested as the links are drawn only if the nodes are close to each other. The graph modeled in this example has many applications in wireless communication networks, such as in a type called an *ad hoc* network. An ad hoc network consists of battery-powered nodes where each

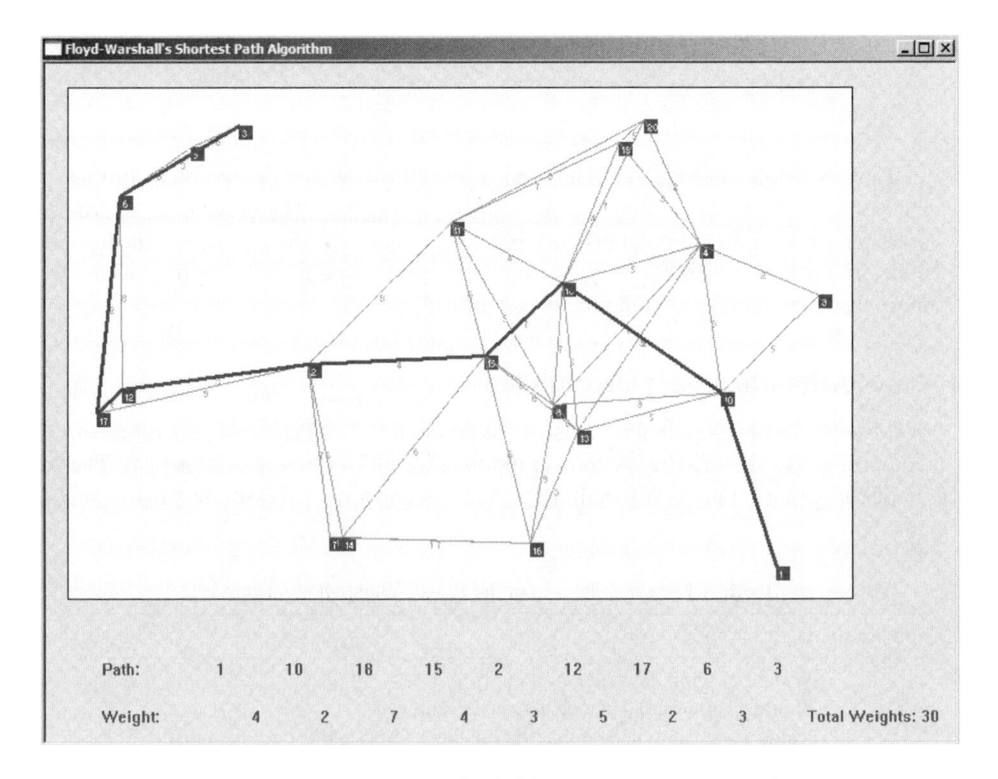

Figure 8.2 Output from Code8A showing the shortest path.

node has both a transmitter and a receiver. The network has no basic infrastructure such as a base station. Therefore, the nodes in the network organize themselves by sending signals to each other within their limited transmission range.

The shortest path between a pair of nodes in the graph is obtained by clicking the nodes consecutively using the left button of the mouse. The path is shown as a thick line from the source node to its destination. The program also allows up to eight shortest paths to be drawn by repeating the same rule each time. The paths are drawn as thick lines using eight randomly determined colors, to differentiate one from another. Each time the shortest path line is drawn, the visited nodes are also listed in the text area. This facility provides some extension capability to the program. For example, the reader can easily add the tracing feature in the shortest path for detecting path collision and evaluating the cost incurred.

Code8A involves only one application class called `CCode8A`. The variables and objects in the project in the class are declared in Code8A.h, and they are described in Table 8.2. Node i, or v_i, is represented by `v[i]`. A node is formed from a structure called `NODE` with links with its members called `Home` and `rc`. The structure `NODE` is declared as follows:

```
typedef struct
{
      CPoint Home;
      CRect rc;
} NODE;
NODE *v;
```

Table 8.2 Graph elements of Code8A

Variable/object	Type	Description
`v[i]`	`NODE`	Node v_i
`v[i].Home`	`CPoint`	Top-left coordinates of v_i
`v[i].rc`	`CRect`	Rectangular box representation of v_i
`e[i][j]`	`LINK`	The link e_{ij} between the nodes v_i and v_j
`e[i][j].Wt`	`int`	Weight of e_{ij}
`e[i][j].sd`	`int`	Shortest (minimal) distance between v_i and v_j
`e[i][j].via[k]`	`int`	kth intermediate node of the link e_{ij}
`Source`	`int`	Source node of a path
`Dest`	`int`	Destination node of a path.
`LineFlag`	`int`	A flag to signal the click point is the first (1) or second (2) node in the path
`alpha[i][j][k]`	`int`	The path in the Floyd–Warshall algorithm in which v_i and v_j are the pairing nodes, and v_k is the intermediate node

A link e_{ij} between two nodes v_i and v_i is `e[i][j]`. It is declared from the structure `LINK`, as follows:

```
typedef struct
{
        int Wt,sd;
        int via[10];
} LINK;
LINK **e;
```

In the above structure, the weight of a link is `Wt`. A link between a pair of nodes may or may not exist. If a link does not exist, the path between the two nodes in a connected graph must pass through one or more intermediate nodes, denoted as `via[]`.

In computing the shortest path, several symbols are used, as described in Table 8.2. The shortest path between a pair of nodes is represented as `sd`. The source and destination nodes in a path are represented as `Source` and `Dest`, respectively. The path a_{ijk} between the nodes v_i and v_j, with v_k as the intermediate node, is denoted as `alpha[i][j][k]`. Another variable is `LineFlag` which is assigned with a default value of 0. A click on a node changes this value to 1, to signal that the node is a source in the path. A click at another node changes this value to 2, which makes the node the destination in the path.

Several other variables involving the styles of the text and graphic display are listed in Table 8.3. The `CPoint` objects `TopLeft` and `BottomRight` form the top-left and bottom-right corners of the drawing area, respectively. The text is presented using two grayscales, `Color1` and `Color2`, with `BgColor` as the standard background color. The fonts used in this application are represented by the `CFont` objects, `fontCourier` and `fontArial`.

Figure 8.3 shows the organization of Code8A.cpp. There are seven member functions in Code8A, as listed in Table 8.4. Two of these functions are event handlers, namely, `OnPaint()` and `OnLButtonDown()`.

The constructor `Code8A()` allocates memory for the class and initializes several variables, objects, and arrays in the program. The memory for the arrays is allocated dynamically, as follows:

```
// allocate memory
v=new NODE [N+1];
e=new LINK *[N+1];
alpha=new int **[N+1];
for (i=0;i<=N;i++)
{
        e[i]=new LINK [N+1];
        alpha[i]=new int *[N+1];
        for (j=0;j<=N;j++)
             alpha[i][j]=new int [N+1];
}
```

Random numbers are integer numbers generated from the clock cycles in the computer. The coordinates of the nodes are assigned with random numbers so that

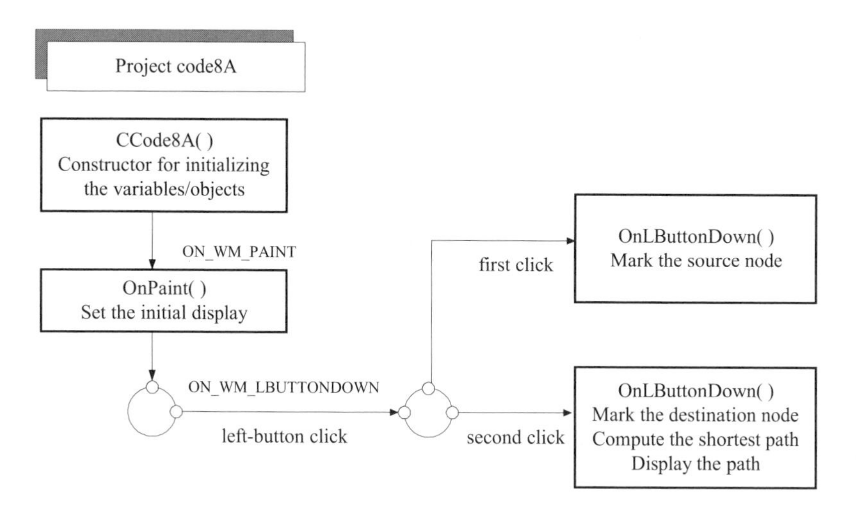

Figure 8.3 Organization of Code8A.

Table 8.3 Display variables/objects in Code8A

Variable/object	Class	Description
`TopLeft`	`CPoint`	Top-left coordinates of the drawing area
`BottomRight`	`CPoint`	Bottom-right coordinates of the drawing area
`Color1,Color2`	`int`	Color scheme for text
`BgColor`	`int`	Color scheme for the background
`fontCourier`	`CFont`	Courier font
`fontArial`	`CFont`	Arial font

Table 8.4 Functions in Code8A

Function	Description
`CCode8A()`	The constructor that initializes the variables and arrays. The function also computes the all-pairs shortest paths by calling up the function `ComputePath()`
`~CCode8A()`	The destructor
`DisplayGraph()`	Draws the graph randomly in the main window
`ComputePath()`	Computes the shortest path using the Floyd–Warshall algorithm
`DrawPath()`	Draws the shortest path between two nodes in the graph
`OnPaint()`	Sets up the initial display and draws the graph by calling up the function `DisplayGraph()`
`OnLButtonDown()`	Marks the source and destination nodes in the graph

their position inside the drawing area is determined randomly. These random values are created through the C++ function `rand()` after executing the following two lines of code:

```
time_t seed=time(NULL);
srand((unsigned)seed);
```

The function `rand()` produces integer random numbers from –32,768 to 32,768. The above code converts these numbers to positive integers only. A random number within a given range is obtained by applying the remainder operator % to `rand()`. The following examples show some of these operations:

Example	Description
`rand()`	An integer from -32,768 to 32,768.
`rand()%10`	An integer from 0 to 9.
`1+rand()%10`	An integer from 1 to 10.
`(double)1/(1+rand()%10)`	A real number from 0.1 to 1.

The following routine in the constructor assigns random coordinates to the top-left corner of the node boxes:

```
for (i=1;i<=N;i++)
    v[i].Home=CPoint(TopLeft.x+rand()%(BottomRight.x-30),
                TopLeft.y+rand()%(BottomRight.y-30));
```

The weights in the links are also assigned random values. A random integer value between 0 to 9 is assigned as the weight of a link if the link between two nodes has been established. If the link is absent, then a remote number, 99, is assigned as its weight, which represents ∞ in the Floyd–Warshall algorithm. The following code realizes this idea:

```
for (i=1;i<=N;i++)
{
    e[i][i].Wt=0;
    for (j=i;j<=N;j++)
        if (sqrt(pow(v[i].Home.x-v[j].Home.x,2)
            +pow(v[i].Home.y-v[j].Home.y,2))<=LinkRange)
            e[i][j].Wt=e[j][i].Wt=1+rand()%9;
        else
            e[i][j].Wt=e[j][i].Wt=99;
}
```

In assigning the weights, we consider the *nearest neighbor* rule. In this rule, a link between a pair of nodes is said to exist if their straight-line distance is less than or equal to a threshold value, `LinkRange`. The straight-line distance between the points (a_1, b_1) and (a_2, b_2) can be implemented by checking the Euclidean distance formula, given as follows:

Euclidean distance between (a_1, b_1) and $(a_2, b_2) = \sqrt{(a_1^2 - b_1^2)^2 + (a_2^2 - b_2^2)^2}$

In the constructor, we assign the weights of the links as the initial shortest distances between the nodes, `e[i][j].sd=e[i][j].Wt`. The shortest distance between v_i and v_j may involve one or more intermediate nodes v_k computed from `alpha[i][j][k]`. This is done as follows:

```
for (i=1;i<=N;i++)
     for (j=1;j<=N;j++)
     {
          e[i][j].sd=e[i][j].Wt;
          for (k=1;k<=N;k++)
               alpha[i][j][k]=99;
     }
```

Text is displayed using grayscale colors defined by the variables `Color1` and `Color2`, and a background color `BgColor`. Two fonts are used for displaying text: Courier and Arial. The variables are initialized as follows:

```
// set fonts and colors
Color1=RGB(100,100,100);
Color2=RGB(170,170,170);
BgColor=RGB(230,230,230);
CFont fontArial;
::ZeroMemory (&lfArial,sizeof(lfArial));
lfArial.lfHeight=60;
fontArial.CreatePointFontIndirect (&lfArial);
::ZeroMemory(&lfCourier,sizeof(lfCourier));
lfCourier.lfHeight=60;
fontCourier.CreatePointFontIndirect(&lfCourier);
```

The last task in the constructor is to compute the shortest paths between all pairs of nodes in the graph. The all-pairs shortest-path problem requires the a priori calculation of all paths simultaneously. This is necessary as the graph is static where its shape will not change throughout the running of the program. To realize this objective, the Floyd–Warshall algorithm is implemented in a function called `ComputePath()`. This function is called up from the constructor as it has all the required input for computing the shortest path, namely, the information about the graph. `ComputeGraph()` is written as follows:

```
void CCode8A::ComputePath()
{
     int i,j,k,m,n;
     for (i=1;i<=N;i++)
          for (j=1;j<=N;j++)
               for (k=1;k<=N;k++)
                  if (e[j][i].sd!=99 || e[i][k].sd!=99 || e[j][k].sd!=99)
```

```
                    if (e[j][i].sd+e[i][k].sd<e[j][k].sd)
                    {
                         n=1;
                         e[j][k].sd=e[j][i].sd+e[i][k].sd;
                         for (m=1;m<=N;m++)
                              alpha[j][k][m]=99;
                         for (m=1; m<=N; m++)
                              if (alpha[j][i][m]!=99)
                                   alpha[j][k][n++]=alpha[j][i][m];
                         alpha[j][k][n++]=i;
                         for (m=1; m<=N; m++)
                              if (alpha[i][k][m]!=99)
                                   alpha[j][k][n++]=alpha[i][k][m];
                    }
}
```

The shortest-path value (or the minimal distance) between v_i and v_j is stored as the array `e[i][j].sd`, which `alpha[i][j][k]` shows to be the path between these two nodes.

Two events are associated with this application. First is the initial display, which is detected as `WM_PAINT` and handled by `OnPaint()`. The initial display consists of a graph that is generated randomly in a rectangular box, shown in the following code fragments:

```
// clear the window
CBrush BgBrush(BgColor);
GetClientRect(&rc);
dc.FillRect(&rc,&BgBrush);

// set the initial display
rc=CRect(TopLeft,BottomRight);
dc.Rectangle(&rc);
DisplayGraph();
```

`OnPaint()` calls up the function `DisplayGraph()` to display the graph in the drawing area of the main window. `DisplayGraph()` displays the nodes and labels them according to the following code fragments:

```
CClientDC dc(this);
CString s;
CPen penGray(PS_SOLID,1,RGB(150,150,150));

dc.SetBkMode(TRANSPARENT);
dc.SelectObject(&fontCourier);
dc.SetTextColor(RGB(255,255,255));
dc.SelectObject(penGray);
for (int i=1;i<=N;i++)        // draw the nodes
```

```
{
	v[i].rc=CRect(v[i].Home,v[i].Home+CPoint(12,12));
	dc.FillSolidRect(&v[i].rc,Color1);
	s.Format("%d",i);
	dc.TextOut(v[i].Home.x+2,v[i].Home.y+2,s);
}
```

Once the nodes have been drawn, the next step is to draw the links between the nodes based on the nearest-neighbor rule. A link is present if the distance between any pair of nodes is less than the threshold value, `LinkRange`. The following code fragments show how this is done:

```
for (i=1;i<=N;i++)  // draw the links
	for (int j=i;j<=N;j++)
		if (e[i][j].Wt!=99)
		{
			dc.MoveTo(v[i].Home);
			dc.LineTo(v[j].Home);
			s.Format("%d",e[i][j].Wt);
			if (i!=j)
				dc.TextOut((v[i].Home.x+v[j].Home.x)/2,
					(v[i].Home.y+v[j].Home.y)/2,s);
		}
```

The nodes are drawn when the user clicks the mouse's left button in the drawing area. This event is detected as `WM_LBUTTONDOWN` and handled by the function `OnLButtonDown()`, as follows:

```
void CCode8A::OnLButtonDown(UINT nFlags,CPoint pt)
{
	for (int i=1;i<=N;i++)
		if (v[i].rc.PtInRect(pt))
		{
			LineFlag++;
			if (LineFlag==1)
				Source=i;
			if (LineFlag==2)
			{
				Dest=i;
				DisplayGraph(); DrawPath();
				LineFlag=0;
			}
		}
}
```

A global variable called `LineFlag` plays an important role in the program as it denotes the status of the left-button click on the nodes. Initially, `LineFlag` is assigned with 0 to denote no activity concerning the left-button click of the mouse. The first click on the node of the graph marks the first node (source). The subsequent click on the second node (destination) marks this node, and the shortest path is immediately computed and displayed on the graph. When the first node is clicked, `LineFlag` is updated to 1 to denote that the node is the source. The second click on a node causes `LineFlag` to increase to 2, and this signals that the node is the destination. The second click also calls the functions `DisplayGraph()` and `DrawPath()` to redraw the graph and draw the shortest path between the two nodes, respectively. The operation terminates by resetting the value of `LineFlag` to 0.

The function `DrawPath()` draws the shortest path from the source node to the destination node in the graph. The source and destination nodes are denoted as `Source` and `Dest`, respectively, whereas the path through an intermediate node `r` is `e[Source][Dest].via[r]`. The process begins by erasing the text area for showing the shortest path, as follows:

```
CClientDC dc(this);
CString s;
CRect rc;
CPen penDark(PS_SOLID,3,Color1);
CBrush BgBrush(BgColor);
int k,p,q,u,w,r;
rc=CRect(150,500,800,600);      // erase the text area
dc.FillRect(&rc,&BgBrush);
dc.SetTextColor(Color1);
dc.SetBkColor(BgColor);
```

In the text area, the path and its weights are shown according to their node numbers in the graph. The path is shown in the graph as a thick line using the MFC functions `MoveTo()` and `LineTo()`. The construction starts at the source node by setting `e[Source][Dest].via[r]=Source`, as follows:

```
p=150; q=120;                     // display the source
r=1; e[Source][Dest].via[r]=Source;
dc.TextOut(p-100,500,"Path:");
dc.TextOut(p-100,540,"Weight:");
s.Format("%d",e[Source][Dest].via[r]);
dc.TextOut(p,500,s);
dc.SelectObject(&penDark);
w=Source; dc.MoveTo(v[w].Home);
```

From the source node, a path to the destination node must pass through the intermediate nodes or vias, `e[Source][Dest].via[r]`. This is shown in `DrawPath()`, as follows:

```
for (k=1; k<=N; k++)            // display the vias
{
       u=alpha[Source][Dest][k];
       if (u!=99)
       {
              r++; e[Source][Dest].via[r]=u;
              dc.LineTo(v[u].Home);
              p += 60;
              s.Format("%d",e[Source][Dest].via[r]);
              dc.TextOut(p,500,s);
              q += 60;
              s.Format("%d",e[u][w].Wt);
              dc.TextOut(q,540,s);
              w=u;
       }
}
```

And, finally, the path reaches the destination node by setting `e[Source][Dest].via[r]=Dest` with its shortest path, `e[Source][Dest].sd`, shown as follows:

```
if (e[Dest][w].sd!=99)         // display the destination
       dc.LineTo(v[Dest].Home);
e[Source][Dest].via[r]=Dest;
s.Format("%d",Dest);
p += 60; q += 60;
dc.TextOut(p,500,s);
s.Format("%d",e[Dest][w].sd);
dc.TextOut(q,540,s);
s.Format("Total Weights: %d", e[Source][Dest].sd);
dc.TextOut(q+60,540,s);
```

8.3 MESH NETWORK APPLICATION

A mesh network is a parallel computing network having processors (nodes) arranged in a rectangular array. This topology and processor arrangement provides several advantages for many problems that are in the shape of two-dimensional arrays. The mesh network provides a platform for mapping the problems directly to the rectangular regions, which can then be solved in parallel in the network. For example, in image processing, an image is represented as pixels arranged in a rectangle. For a sharp and colorful image, massive storage in the form of a large array is required in its representation as bits of data. This array occupies a huge amount of memory in the computer. An array operation in an image, such as the computation of its eigenvalues, involves a massive calculation and updating of the arrays. These operations consume a lot of memory, which has the effect of slowing down the computer. An ideal solution is to distribute the operations to the processors of a parallel computer network. In a mesh parallel computing network, for example, the job is represented as one or more arrays that can be mapped directly on the processors. In this case, the domain is decomposed into smaller modules, which helps in solving the problem concurrently. Therefore, the whole job can be completed more effectively in a cooperative manner much less time.

There are several types of mesh networks. The standard mesh network, as shown in Figure 8.4 (left), has nodes, each having two or three links. The intermediate node in the network has four links through its north, south, east, and west ports. The node at the corner has two links, whereas the node at the noncorner boundary has three links. The standard mesh network is suitable for solving a problem in which the graph representing this problem is rectangular in shape.

A *toroidal mesh* network is another form of the mesh network in which all the nodes have exactly four links each. This feature provides the symmetrical properties that improve the communication capability of the network. Figure 8.4 (right) shows a toroidal mesh network having 16 processors, arranged in a 4×4 grid. A toroidal mesh provides a better communication facility than the standard mesh with these extra links in the boundary processors.

Computing the shortest paths between all pairs of processors is a necessity in a mesh network to allow active data movement in the network. A path is defined as the route from the source node to its destination node. Communication between processors is a component of a general problem known as *task scheduling*. In task scheduling, the problem is to find a feasible schedule for mapping a set of tasks onto a set of processors. The main objective in task scheduling is to design a schedule that will minimize the overall completion time. The communication problem between processors crops up frequently as data from one processor is needed in another processor before a task in the latter can start.

In general, communication between the processors depends on factors such as the initial rate of the source processor startup, transmission rate on the link, conges-

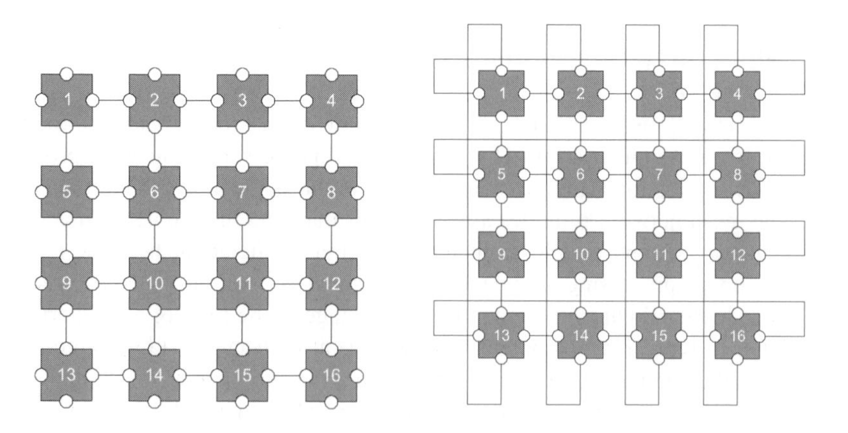

Figure 8.4 Two types of square mesh networks: standard (left) and toroidal (right).

tion rate, and the dependency graph. A dependency graph that represents a particular problem may consist of directed links that show the partial orders of the directed graph. In a directed link, a task at the tail node must be completed before the task at the head node can start. As communication between processors may incur some significant delays in the execution of the tasks, finding the shortest path between them definitely contributes to reducing the cost.

Code8B: Shortest Path on a Mesh Network

In this section, we apply the shortest-path problem discussed in the last section to the mesh parallel computing network. The visualization model is called Code8B and the code is strongly based on the previous Code8A. The same technique of computing the all-pairs shortest paths using the Floyd–Warshall algorithm in Code8A is applied to Code8B. This is necessary as Code8A applies to a general computing network and all changes pertaining to a particular network should be kept to the minimum in order to allow the portability of the code.

Our visualization model, Code8B, simulates the standard mesh network on a 5 × 10 topology having 50 processors, as shown in Figure 8.5. The processors are numbered starting from the left in the top row. The figure shows three different paths in three shades of gray. The shortest path between two nodes in the mesh network is obtained and displayed by clicking the mouse's left button on the two nodes. The model allows up to eight different paths to be displayed on the mesh.

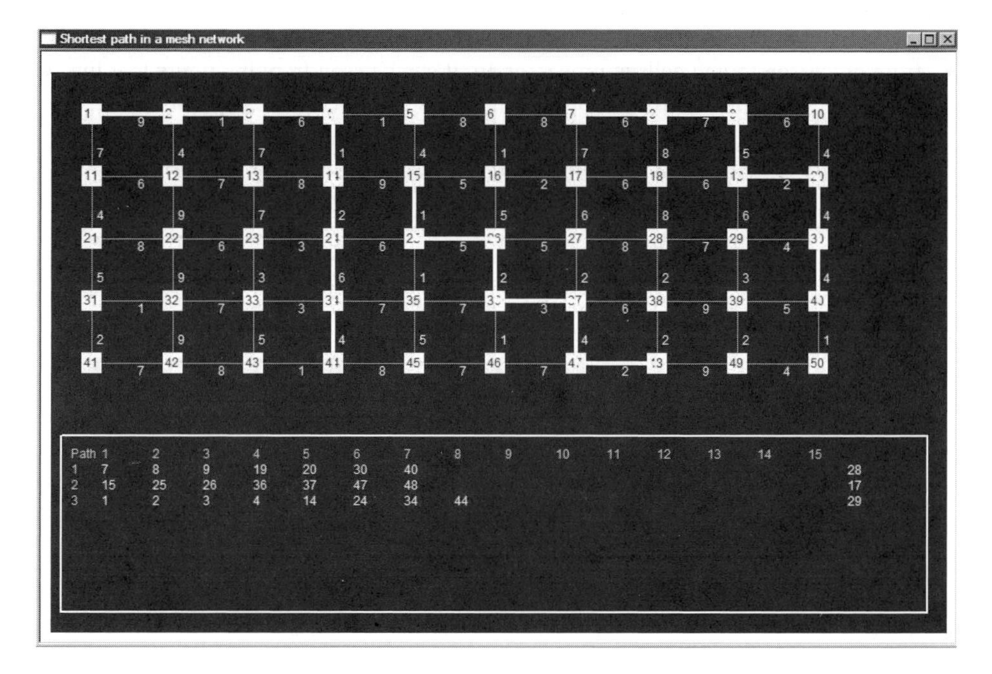

Figure 8.5 Output from Code8B using the 5 × 10 mesh network.

These paths are shown as lines of different using the randomly generated colors. The paths and the links in each path are also listed in the text area of the window.

The project consists of the files Code8B.h and Code8B.cpp. Most of the variables and objects in Code8A.h, as described in Table 8.1, are retained in Code8B.h. The application class in this application is `CCode8B`. The additional variables and objects are listed in Table 8.5.

A new structure called `PATH` is introduced to describe the paths selected by the user. This structure has members as defined in Table 8.5. The first selected path is labeled `Path[1]`, the second as `Path[2]`, and so on. Another structure is `PATHLINK`, which describes the links in each path. A link has a beginning node (`b`) and ending node (`e`). The structure `PATHLINK` is nested inside `PATH` to enable the members in the latter to have access to the members in the former. The two structures are declared as follows:

```
typedef struct
{
      int b,e;
} PATHLINK;
typedef struct
{
      int Distance,nPLink,st;
      PATHLINK PLink[N+1];
} PATH;
```

Figure 8.6 shows two paths in the mesh network, `Path[1]` and `Path[2]`, and their linked members from the structures `PATH` and `PATHLINK`. `Path[1]` starts from processor 15 and ends at processor 49. It can be seen from the figure that link 2 in path 1 begins at processor 25 and ends at processor 26, represented as `Path[1].Plink[2].b` and `Path[1].Plink[2].e`, respectively.

Code8B also includes some new constants, namely, `Nx` and `Ny`, which represent the number of the rows and columns of the processors in the mesh network, re-

Table 8.5 Additional variables and objects in Code8B

Variables/objects	Type	Description
`Path[i]`	`PATH`	Path *i*
`Path[i].PLink[j]`	`PATHLINK`	The *j*th link in path *i*
`Path[i].PLink[j].b`	`int`	Beginning node of the *j*th link in the *i*th path
`Path[i].Plink[j].e`	`int`	Ending node of the *j*th link in the *i*th path
`Path[i].Distance`	`int`	Total weights in path *i*
`Path[i].nPLink`	`int`	Total number of links in path *i*
`Path[i].st`	`int`	Start time of path *i*
`nPath`	`int`	Total number of paths selected
`nLink`	`int`	Total number of links

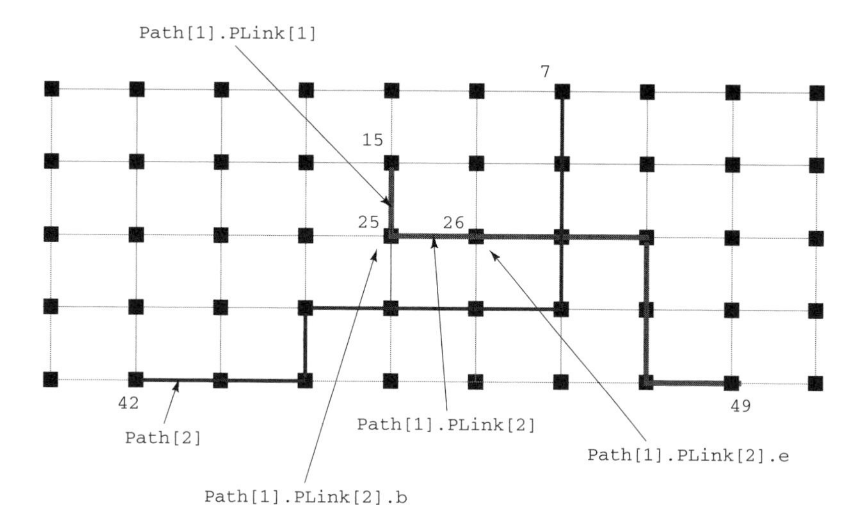

Figure 8.6 The mesh network showing the relationship between the members.

spectively. Another constant is `MaxPath` which is the maximum number of paths allowed in the network. The values of the constants can be changed using the `#define` directive in Code8B.h. The communication model in the mesh network modeled here is in its simplest form. Factors like the initial startup cost, the transmission rate, and the physical media of the links are not considered in evaluating the communication cost involving shortest paths.

Basically, the constructor `CCode8B()` has the same code as its corresponding function in Code8A.cpp. However, there are some minor changes involving the allocation of memory for the arrays in the structures, as follows:

```
// allocate memory
v=new NODE [N+1];
e=new LINK *[N+1];
alpha=new int **[N+1];
Path=new PATH [N+1];
for (i=0;i<=N;i++)
{
      e[i]=new LINK [N+1];
      alpha[i]=new int *[N+1];
      for (j=0;j<=N;j++)
            alpha[i][j]=new int [N+1];
}
```

In a standard mesh network, the links between the processors are established according to the rectangular requirement of the network. Each intermediate processor has four links in its north, south, east, and west ports. Each processor at the corners has two links only, whereas each intermediate boundary processor has three links. The following routine assigns the links according to these requirements:

```
time_t seed=time(NULL);    srand((unsigned)seed);
nLink=0; nPath=0; LineFlag=0;
for (i=1;i<=N;i++)
{
        e[i][i].Wt=99;
        for (j=i;j<=N;j++)
        {
                e[i][j].Wt=e[j][i].Wt=99;
                if (((j-i==1&&j-i!=Nx-1)||(j-i==Nx&&j-i!=Nx*(Ny-1)))
                        && !(j-i==1 && i%Nx==0) )
                {
                        e[i][j].Wt=e[j][i].Wt=1+rand()%9;
                        nLink++;
                }
        }
}
```

The drawing area consists of an area defined by the `CRect` object `DrawArea`. Similarly, the object `TextArea` is derived from the same class and it defines the text area in the window. The processors in the network are arranged in a rectangular array with their home coordinates defined in the constructor, as follows:

```
DrawArea=CRect(10,20,900,350);
k=0;
for (j=1;j<=Ny;j++)
        for (i=1;i<=Nx;i++)
        {
                k++;
                v[k].Home.x=40+DrawArea.left+80*(i-1);
                v[k].Home.y=40+DrawArea.top+60*(j-1);
        }
```

Random numbers from 0 to 9 are used as the weights of the links in the network. A similar approach to Code8A using the function `rand()` is applied to determine these weights.

The main window displays the mesh network using the function `OnPaint()`. The following routine in the function `OnPaint()` draws the links and labels their weights:

```
for (i=1;i<=N;i++)
        for (j=i;j<=N;j++)
                if (e[i][j].Wt!=99)
                {
                        dc.MoveTo(v[i].Home);
                        dc.LineTo(v[j].Home);
```

```
                s.Format("%d",e[i][j].Wt);
                dc.TextOut(5+(v[i].Home.x+v[j].Home.x)/2,(
                      v[i].Home.y+v[j].Home.y)/2,s);
          }
```

The processors are drawn as yellow rectangles, as follows:

```
dc.SetTextColor(RGB(0,0,0));
for (i=1;i<=N;i++)
{
      v[i].rc=CRect(v[i].Home.x-10,v[i].Home.y-10,
            v[i].Home.x+10,v[i].Home.y+10);
      dc.FillSolidRect(&v[i].rc,RGB(250,250,0));
      s.Format("%d",i);
      dc.TextOut(v[i].Home.x-8,v[i].Home.y-8,s);
}
```

The function `DrawPath()` draws the selected path on the graph and displays the node numbers in the text area. This function is called up by the function `OnLButtonDown()` when the flag value `LineFlag=2` is reached. `DrawPath()` gets the results of the all-pairs shortest paths from the function `ComputePath()`.

A path from the source to its destination is determined from the last value returned in `alpha[Source][Dest][k]`. Each path consists of one or more links, represented as `PLink[r]`. The source and destination nodes are rewritten as `Path[nPath].PLink[r].b` and `Path[nPath].PLink[r].e`, respectively. The intermediate nodes in the path are determined from `PLink[r]` in this array. The links in each path are determined using the following routine:

```
int r=Path[nPath].st=1;
Path[nPath].PLink[r].b=Source;
for (k=1; k<=N; k++)
{
     u=alpha[Source][Dest][k];
     if (u!=99)
     {
          Path[nPath].PLink[r].e=Path[nPath].PLink[r+1].b=u;
          r++;
     }
}
Path[nPath].nPLink=r;
Path[nPath].PLink[r].e=Dest;
```

Each path is drawn on the graph using a different color. The color for a path is generated randomly using the MFC function `rand()` by setting the red, blue, and green components of the function `RGB()`. The same color is used to display the node number of the path in the text area. The following code in `DrawPath()` implements this idea:

```
time_t seed=time(NULL);
srand((unsigned)seed);
PathColor=RGB(rand()%256,rand()%256,rand()%256);
dc.SelectObject(&fontTimes);
dc.SetTextColor(RGB(200,200,200));
dc.SetBkMode(TRANSPARENT);
s.Format("%d",nPath);
dc.TextOut(20+TextArea.left,20+TextArea.top+15*nPath,s);
CPen PathPen(PS_SOLID,3,PathColor);
dc.SelectObject(&PathPen);
```

The highlight of Code8B is to display the shortest path of the selected pair of nodes in the graph. This is done as follows:

```
u=Path[nPath].PLink[1].b;
dc.MoveTo(v[u].Home);
dc.SetTextColor(PathColor);
for (r=1;r<=Path[nPath].nPLink;r++)
{
     w=Path[nPath].PLink[r].e;
     dc.LineTo(v[w].Home);
     s.Format("%d",Path[nPath].PLink[r].b);
     dc.TextOut(50+TextArea.left+TextGap*(r-1+Path[nPath].st-1),
          20+TextArea.top+15*nPath,s);
}
s.Format("%d",Path[nPath].PLink[r-1].e);
dc.TextOut(50+TextArea.left+TextGap*(r-1+Path[nPath].st-1),
     20+TextArea.top+15*nPath,s);
s.Format("%d",Path[nPath].Distance);
dc.TextOut(TextArea.right-100,20+TextArea.top+15*nPath,s);
```

8.4 SUMMARY AND CONCLUSION

In this chapter, two models involving the applications of graph theory were discussed. The first model describes the shortest-path problem. We continue the work from the previous chapter on the friendly graph interface by adding the shortest-path problem. We apply the Floyd–Warshall algorithm for finding the all-pairs shortest paths in the graph. The shortest path is computed and displayed in the window by simply clicking any two nodes of the graph. The second model continues the work of the first by applying it to a mesh network. Here, each node is a processor, and the Floyd–Warshall algorithm helps in computing the communication cost for transferring data from one processor to another.

The models developed in this chapter can further be extended to several other network applications. The project Code8A provides the basic framework for any graph-related applications. As mentioned earlier, one good application is in the design of ad hoc wireless networks. In this application, the nodes need to organize themselves to form a network without help from external sources. Therefore, it is necessary to supply many other attributes to the nodes. A node in the network should have information about its neighbors such as its connectivity to its immediate neighbors and its message transmission history. To achieve these abilities, a node must have its own table in the form a simple database about its neighbors. This information requires an upgrade of its structure, NODE, to include other information as well. Also, the way a message flows using the shortest path in Code8A will definitely contribute to the design of a routing protocol in an ad hoc network.

Code8B also contributes to the design of the routing strategy for a mesh network. The project is a fundamental work on communication but it has a great potential for the development of an efficient communication model on this network. An extension to the work includes designing a protocol to allow effective communication between the processors by taking into consideration a number of factors and constraints affecting the network's performance. These include the initial setup cost, the transmission rate, the contention rate, and the physical structure of the network.

BIBLIOGRAPHY

1. R. Sedgewick, *Algorithms in C: Part 1-4,* 3rd ed., Addison-Wesley, 2001.
2. R. Sedgewick, *Algorithms in C++ Part 5: Graph Algorithms,* 3rd ed., Addison-Wesley, 2001.
3. J. L. Gross and J. Yellen, *Handbook of Graph Theory,* CRC Press, 2003.

CODE LISTINGS

Code8A: Shortest Path in a Graph

```
#include <afxwin.h>
#include <math.h>
#define N 20
#define LinkRange 200

class CCode8A : public CFrameWnd
{
private:
      int Source,Dest;
      int Color1,Color2,BgColor;
      int ***alpha,LineFlag;
      typedef struct
      {
            int Wt,sd;
            int via[10];
      } LINK;
      LINK **e;
      typedef struct
```

```
      {
            CPoint Home;
            CRect rc;
      } NODE;
      NODE *v;
      CPoint TopLeft,BottomRight;
      CFont fontArial,fontCourier;
      LOGFONT lfCourier,lfArial;
public:
      CCode8A();
      ~CCode8A();
      void DisplayGraph();
      void ComputePath();
      void DrawPath();
      afx_msg void OnPaint();
      afx_msg void OnLButtonDown (UINT, CPoint);
      DECLARE_MESSAGE_MAP()
};

class CMyWinApp : public CWinApp
{
public:
      virtual BOOL InitInstance();
};

#include "code8A.h"

CMyWinApp  MyApplication;

BOOL CMyWinApp::InitInstance()
{
    CCode8A* pFrame = new CCode8A;
    m_pMainWnd = pFrame;
    pFrame->ShowWindow(SW_SHOW);
    pFrame->UpdateWindow();
    return TRUE;
}

BEGIN_MESSAGE_MAP(CCode8A, CFrameWnd)
    ON_WM_PAINT()
      ON_WM_LBUTTONDOWN()
END_MESSAGE_MAP()

CCode8A::CCode8A()
{
      int i,j,k;
      Create(NULL, "Floyd-Warshall's Shortest Path Algorithm",
          WS_OVERLAPPEDWINDOW,CRect(0,0,800,600));

      // allocate memory
      v=new NODE [N+1];
      e=new LINK *[N+1];
      alpha=new int **[N+1];
      for (i=0;i<=N;i++)
```

```
	{
		e[i]=new LINK [N+1];
		alpha[i]=new int *[N+1];
		for (j=0;j<=N;j++)
			alpha[i][j]=new int [N+1];
	}

	// initialize variables
	LineFlag=0;
	TopLeft=CPoint(20,20); BottomRight=CPoint(700,450);
	time_t seed=time(NULL); srand((unsigned)seed);
	for (i=1;i<=N;i++)
		v[i].Home=CPoint(TopLeft.x+rand()%(BottomRight.x-30),
				TopLeft.y+rand()%(BottomRight.y-30));
	for (i=1;i<=N;i++)
	{
		e[i][i].Wt=0;
		for (j=i;j<=N;j++)
			if (sqrt(pow(v[i].Home.x-v[j].Home.x,2)
				+pow(v[i].Home.y-v[j].Home.y,2))<=LinkRange)
				e[i][j].Wt=e[j][i].Wt=1+rand()%9;
			else
				e[i][j].Wt=e[j][i].Wt=99;
	}
	for (i=1;i<=N;i++)
	  for (j=1;j<=N;j++)
		{
		 e[i][j].sd=e[i][j].Wt;
			for (k=1;k<=N;k++)
				alpha[i][j][k]=99;
		}

	// set fonts and colors
	Color1=RGB(100,100,100);
	Color2=RGB(170,170,170);
	BgColor=RGB(230,230,230);
	CFont fontArial;
	::ZeroMemory (&lfArial,sizeof(lfArial));
	lfArial.lfHeight=60;
	fontArial.CreatePointFontIndirect (&lfArial);
	::ZeroMemory(&lfCourier,sizeof(lfCourier));
	lfCourier.lfHeight=60;
	fontCourier.CreatePointFontIndirect(&lfCourier);

	// compute the shortest paths
	ComputePath();
}

CCode8A::~CCode8A()  // destroy the arrays
{
	int i,j;
	for (i=0;i<=N;i++)
	{
		e[i]=new LINK [N+1];
```

```
            for (j=0;j<=N;j++)
                delete alpha[i][j];
            delete e[i],alpha[i];
        }
        delete v,e,alpha;
}

void CCode8A::OnPaint()
{
        CPaintDC dc(this);
        CString s;
        CRect rc;

        // clear the window
        CBrush BgBrush(BgColor);
        GetClientRect(&rc);
        dc.FillRect(&rc,&BgBrush);

        // set the initial display
        rc=CRect(TopLeft,BottomRight);
        dc.Rectangle(&rc);
        DisplayGraph();
}

void CCode8A::DisplayGraph()
{
        CClientDC dc(this);
        CString s;
        CPen penGray(PS_SOLID,1,RGB(150,150,150));

        dc.SetBkMode(TRANSPARENT);
        dc.SelectObject(&fontCourier);
        dc.SetTextColor(RGB(255,255,255));
        dc.SelectObject(penGray);
        for (int i=1;i<=N;i++)   // draw the nodes
        {
            v[i].rc=CRect(v[i].Home,v[i].Home+CPoint(12,12));
            dc.FillSolidRect(&v[i].rc,Color1);
            s.Format("%d",i);
            dc.TextOut(v[i].Home.x+2,v[i].Home.y+2,s);
        }
        dc.SetTextColor(Color2);
        for (i=1;i<=N;i++)        // draw the links
            for (int j=i;j<=N;j++)
                if (e[i][j].Wt!=99)
                {
                    dc.MoveTo(v[i].Home);
                    dc.LineTo(v[j].Home);
                    s.Format("%d",e[i][j].Wt);
                    if (i!=j)
                        dc.TextOut((v[i].Home.x+v[j].Home.x)/2,
                            (v[i].Home.y+v[j].Home.y)/2,s);
                }
}
```

```
void CCode8A::ComputePath()
{
    int i,j,k,m,n;
    for (i=1;i<=N;i++)
        for (j=1;j<=N;j++)
            for (k=1;k<=N;k++)
                if (e[j][i].sd!=99 || e[i][k].sd!=99 || e[j][k].sd!=99)
                    if (e[j][i].sd+e[i][k].sd<e[j][k].sd)
                    {
                        n=1;
                        e[j][k].sd=e[j][i].sd+e[i][k].sd;
                        for (m=1;m<=N;m++)
                            alpha[j][k][m]=99;
                        for (m=1; m<=N; m++)
                            if (alpha[j][i][m]!=99)
                                alpha[j][k][n++]=alpha[j][i][m];
                        alpha[j][k][n++]=i;
                        for (m=1; m<=N; m++)
                            if (alpha[i][k][m]!=99)
                                alpha[j][k][n++]=alpha[i][k][m];
                    }
}

void CCode8A::DrawPath()
{
      CClientDC dc(this);
      CString s;
      CRect rc;
      CPen penDark(PS_SOLID,3,Color1);
      CBrush BgBrush(BgColor);
      int k,p,q,u,w,r;

      rc=CRect(150,500,800,600);                // erase the text area
      dc.FillRect(&rc,&BgBrush);
      dc.SetTextColor(Color1);
      dc.SetBkColor(BgColor);
      p=150; q=120;                             // display the source
      r=1; e[Source][Dest].via[r]=Source;
      dc.TextOut(p-100,500,"Path:");
      dc.TextOut(p-100,540,"Weight:");
      s.Format("%d",e[Source][Dest].via[r]);
      dc.TextOut(p,500,s);

      dc.SelectObject(&penDark);
      w=Source; dc.MoveTo(v[w].Home);
      for (k=1; k<=N; k++)                      // display the vias
      {
           u=alpha[Source][Dest][k];
           if (u!=99)
           {
                r++; e[Source][Dest].via[r]=u;
                dc.LineTo(v[u].Home);
                p += 60;
                s.Format("%d",e[Source][Dest].via[r]);
```

```
                dc.TextOut(p,500,s);
                q += 60;
                s.Format("%d",e[u][w].Wt);
                dc.TextOut(q,540,s);
                w=u;
            }
        }
        if (e[Dest][w].sd!=99)          // display the destination
            dc.LineTo(v[Dest].Home);
        e[Source][Dest].via[r]=Dest;
        s.Format("%d",Dest);
        p += 60; q += 60;
        dc.TextOut(p,500,s);
        s.Format("%d",e[Dest][w].sd);
        dc.TextOut(q,540,s);
        s.Format("Total Weights: %d",e[Source][Dest].sd);
        dc.TextOut(q+60,540,s);
}

void CCode8A::OnLButtonDown(UINT nFlags,CPoint pt)
{
        for (int i=1;i<=N;i++)
            if (v[i].rc.PtInRect(pt))
            {
                LineFlag++;
                if (LineFlag==1)
                    Source=i;
                if (LineFlag==2)
                {
                    Dest=i;
                    DisplayGraph(); DrawPath();
                    LineFlag=0;
                }
            }
}
```

Code8B: Shortest Paths in a Mesh Network

```
// code8B.h
#include <afxwin.h>
#include <math.h>
#define Nx 10
#define Ny 5
#define N Nx*Ny
#define LinkRange 200
#define MaxPath 8

class CCode8B : public CFrameWnd
{
private:
        int Source,Dest;
        int nPath,nLink,TextGap;
        int ***alpha,LineFlag;
```

```
        typedef struct
        {
            CPoint Home;
            CRect rc;
        } NODE;
        NODE *v;
        typedef struct
        {
            int Wt,sd;
        } LINK;
        LINK **e;
        typedef struct
        {
            int b,e;
        } PATHLINK;
        typedef struct
        {
            int Distance,nPLink,st;
            PATHLINK PLink[N+1];
        } PATH;
        PATH *Path;
        CPoint TextHome;
        CRect DrawArea,TextArea;
        CFont fontCourier,fontTimes;
        LOGFONT lfCourier,lfTimes;
public:
        CCode8B();
        ~CCode8B();
        void ComputePath(),DrawPath();
        afx_msg void OnPaint();
        afx_msg void OnLButtonDown(UINT, CPoint);
        DECLARE_MESSAGE_MAP()
};

class CMyWinApp : public CWinApp
{
public:
        virtual BOOL InitInstance();
};

// code8B.cpp: Shortest path in a mesh network
#include "code8B.h"

CMyWinApp  MyApplication;

BOOL CMyWinApp::InitInstance()
{
    CCode8B* pFrame = new CCode8B;
    m_pMainWnd = pFrame;
    pFrame->ShowWindow(SW_SHOW);
    pFrame->UpdateWindow();
    return TRUE;
}
```

```
BEGIN_MESSAGE_MAP(CCode8B, CFrameWnd)
    ON_WM_PAINT()
    ON_WM_LBUTTONDOWN()
END_MESSAGE_MAP()

CCode8B::CCode8B()
{
      int i,j,k;
      Create(NULL, "Shortest path in a mesh network",
           WS_OVERLAPPEDWINDOW,CRect(0,0,920,600));

      // allocate memory
      v=new NODE [N+1];
      e=new LINK *[N+1];
      alpha=new int **[N+1];
      Path=new PATH [N+1];
      for (i=0;i<=N;i++)
      {
           e[i]=new LINK [N+1];
           alpha[i]=new int *[N+1];
           for (j=0;j<=N;j++)
                alpha[i][j]=new int [N+1];
      }

      // initialize variables
      time_t seed=time(NULL);
      srand((unsigned)seed);
      nLink=0; nPath=0; LineFlag=0;
      for (i=1;i<=N;i++)
      {
           e[i][i].Wt=99;
           for (j=i;j<=N;j++)
           {
                e[i][j].Wt=e[j][i].Wt=99;
                if (((j-i==1 && j-i!=Nx-1) || (j-i==Nx && j-i!=Nx*(Ny-1)))
                     && !(j-i==1 && i%Nx==0) )
                {
                     e[i][j].Wt=e[j][i].Wt=1+rand()%9;
                     nLink++;
                }
           }
      }
      for (i=1;i<=N;i++)
        for (j=1;j<=N;j++)
           {
                     e[i][j].sd=e[i][j].Wt;
                for (k=1;k<=N;k++)
                     alpha[i][j][k]=99;
           }

      // set fonts, colors, text and drawing area
      TextGap=50; TextHome=CPoint(70,520);
      DrawArea=CRect(10,20,900,350);
      TextArea=CRect(10,360,900,560);
```

```
	k=0;
	for (j=1;j<=Ny;j++)
		for (i=1;i<=Nx;i++)
		{
			k++;
			v[k].Home.x=40+DrawArea.left+80*(i-1);
			v[k].Home.y=40+DrawArea.top+60*(j-1);
		}
	ZeroMemory(&lfCourier,sizeof(lfCourier));
	lfCourier.lfHeight=60;
	fontCourier.CreatePointFontIndirect(&lfCourier);
	ZeroMemory(&lfTimes,sizeof(lfTimes));
	lfTimes.lfHeight=90;
	fontTimes.CreatePointFontIndirect(&lfTimes);

	// compute the shortest paths
	ComputePath();
}

CCode8B::~CCode8B()
{
	for (int i=0;i<=N;i++)
	{
		for (int j=0;j<=N;j++)
			delete alpha[i][j];
		delete e[i],alpha[i];
	}
	delete v,e,alpha,Path;
}

void CCode8B::OnPaint()
{
	CPaintDC dc(this);
	int i,j,k;
	CString s;
	CPen LinePen(PS_SOLID,1,RGB(150,150,150));
	CPen TextBoxPen(PS_SOLID,2,RGB(0,50,255));
	CBrush BgBrush(RGB(0,0,0));
	CRect rc=CRect(DrawArea.left,DrawArea.top,
			TextArea.right,TextArea.bottom);
	dc.FillRect(&rc,&BgBrush);
	rc=CRect(TextArea.left+10,TextArea.top+10,
			TextArea.right-10,TextArea.bottom-10);
	dc.SelectObject(TextBoxPen);
	dc.MoveTo(TextArea.left+10,TextArea.top+10);
	dc.LineTo(TextArea.right-20,TextArea.top+10);
	dc.LineTo(TextArea.right-20,TextArea.bottom-20);
	dc.LineTo(TextArea.left+10,TextArea.bottom-20);
	dc.LineTo(TextArea.left+10,TextArea.top+10);

	dc.SelectObject(LinePen); dc.SelectObject(&fontTimes);
	dc.SetBkMode(TRANSPARENT); dc.SetTextColor(RGB(0,0,255));
	for (i=1;i<=N;i++)
		for (j=i;j<=N;j++)
```

```
                if (e[i][j].Wt!=99)
                {
                     dc.MoveTo(v[i].Home); dc.LineTo(v[j].Home);
                     s.Format("%d",e[i][j].Wt);
                     dc.TextOut(5+(v[i].Home.x+v[j].Home.x)/2,(
                          v[i].Home.y+v[j].Home.y)/2,s);
                }
      dc.SetTextColor(RGB(0,0,0));
      for (i=1;i<=N;i++)
      {
           v[i].rc=CRect(v[i].Home.x-10,v[i].Home.y-10,
                v[i].Home.x+10,v[i].Home.y+10);
           dc.FillSolidRect(&v[i].rc,RGB(250,250,0));
           s.Format("%d",i);
           dc.TextOut(v[i].Home.x-8,v[i].Home.y-8,s);
      }
      dc.SelectObject(&fontTimes);
      dc.SetTextColor(RGB(200,200,200));
      dc.SetBkMode(TRANSPARENT);
      dc.TextOut(20+TextArea.left,TextArea.top+20,"Path");
      for (k=1;k<=15;k++)
      {
           s.Format("%d",k);
           dc.TextOut(50+TextArea.left+TextGap*(k-1),20+TextArea.top,s);
      }
}

void CCode8B::OnLButtonDown(UINT nFlags,CPoint pt)
{
      int i;
      if (nPath<=MaxPath-1)
           for (i=1;i<=N;i++)
                if (v[i].rc.PtInRect(pt))
                {
                     LineFlag++;
                     if (LineFlag==1)
                          Source=i;
                     if (LineFlag==2)
                     {
                          Dest=i;
                          LineFlag=0;
                          if (e[Source][Dest].sd!=99)
                          {
                               nPath++;
                               Path[nPath].Distance=e[Source][Dest].sd;
                               DrawPath();
                          }
                     }
                }
}

void CCode8B::ComputePath()
{
    int i,j,k,m,n;
```

```
    for (i=1;i<=N;i++)
        for (j=1;j<=N;j++)
            for (k=1;k<=N;k++)
               if (e[j][i].sd!=99 || e[i][k].sd!=99 || e[j][k].sd!=99)
                    if (e[j][i].sd+e[i][k].sd<e[j][k].sd)
                    {
                     n=1;
                          e[j][k].sd=e[j][i].sd+e[i][k].sd;
                     for (m=1;m<=N;m++)
                          alpha[j][k][m]=99;
                     for (m=1; m<=N; m++)
                          if (alpha[j][i][m]!=99)
                                         alpha[j][k][n++]=alpha[j][i][m];
                              alpha[j][k][n++]=i;
                              for (m=1; m<=N; m++)
                                     if (alpha[i][k][m]!=99)
                                        alpha[j][k][n++]=alpha[i][k][m];
                          }
}

void CCode8B::DrawPath()
{
      CClientDC dc(this);
      CString s;

      int u,w,k,PathColor;
      int r=Path[nPath].st=1;
      Path[nPath].PLink[r].b=Source;
      for (k=1; k<=N; k++)
      {
           u=alpha[Source][Dest][k];
           if (u!=99)
           {
                Path[nPath].PLink[r].e=Path[nPath].PLink[r+1].b=u;
                r++;
           }
      }
      Path[nPath].nPLink=r;
      Path[nPath].PLink[r].e=Dest;

      time_t seed=time(NULL); srand((unsigned)seed);
      PathColor=RGB(rand()%256,rand()%256,rand()%256);
      dc.SelectObject(&fontTimes);
      dc.SetTextColor(RGB(200,200,200));
      dc.SetBkMode(TRANSPARENT);
      s.Format("%d",nPath);
      dc.TextOut(20+TextArea.left,20+TextArea.top+15*nPath,s);
      CPen PathPen(PS_SOLID,3,PathColor);
      dc.SelectObject(&PathPen);
      u=Path[nPath].PLink[1].b;
      dc.MoveTo(v[u].Home);
      dc.SetTextColor(PathColor);
      for (r=1;r<=Path[nPath].nPLink;r++)
      {
```

```
            w=Path[nPath].PLink[r].e;
            dc.LineTo(v[w].Home);
            s.Format("%d",Path[nPath].PLink[r].b);
            dc.TextOut(50+TextArea.left+TextGap*(r-1+Path[nPath].st-1),
                 20+TextArea.top+15*nPath,s);
        }
        s.Format("%d",Path[nPath].PLink[r-1].e);
        dc.TextOut(50+TextArea.left+TextGap*(r-1+Path[nPath].st-1),
             20+TextArea.top+15*nPath,s);
        s.Format("%d",Path[nPath].Distance);
        dc.TextOut(TextArea.right-100,20+TextArea.top+15*nPath,s);
}
```

CHAPTER 9

MULTIPROCESSOR SCHEDULING PROBLEM

9.1 PARALLEL COMPUTING SYSTEMS

Most computers at home and in the office today are single-processor computers, or *sequential computers,* that are fast enough to do most jobs. A sequential computer has only one processor and enough random access memory capacity for fulfilling all the requirements at home and in a small office. However, for a number-crunching type of job such as in high-level graphics and numerical applications, a sequential computer may not be able to perform these operations at the desired speed. For one thing, the computer is slow when it comes to handling operations involving floating point numbers, large arrays, and repeating loops. These operations require a computer with a fast central processing unit (CPU) and large random access memory (RAM) to process and store the arrays with high-precision accuracy.

It is generally known that it would take something like one year for one person to build a house alone. However, if the job is shared among four people the length of time for completing the house will definitely be reduced, perhaps only two months. One good solution for tackling the speed issue in numerically intensive applications is to use several processors in parallel and distributed computer systems. A parallel and distributed computing system has more than one CPU and large random access memory, making it suitable for applications in number-crunching operations.

A *parallel computer system* is a computer that has more than one processing element (PE) enclosed within its system. A processing element has both a central processing unit and some limited local memory (LM). A processing element is commonly referred as a processor, and they mean the same thing here. Several processors may be connected in a network to share a large global memory within a system, as illustrated in Figure 9.1. This shared memory network model has the ad-

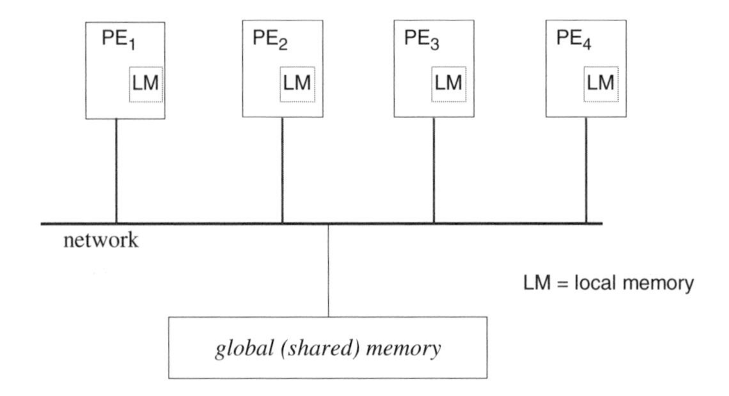

Figure 9.1 A shared-memory parallel computer network.

vantage of providing fast memory transfer between the processors, as the memory is centralized in one location only.

Another model is the distributed memory computer system, as shown in Figure 9.2. In this system, each processing unit has its own memory. Data can be transferred from one PE to another through its message-passing mechanism, by moving it from a memory unit in the first PE to the memory unit of another PE. Communication in this case is slower but the system has the advantage of having processors with a greater autonomy in their operations.

A *distributed computer system* is a network of autonomous computers covering a substantial working area, such as a university campus. Each computer in this system is capable of processing its own jobs independently of other processors in the network. The computers in the network share a parallel computing environment such as the parallel virtual machine (PVM) and message-passing interface (MPI), which provides facilities for collaborative processing of jobs. PVM provides a coherent and portable parallel computing environment across several heterogeneous computers for message passing and processing. The system was developed by Oak Ridge National Laboratory, USA, in early 1990s. MPI is another system developed by more than 175 individuals from 40 organizations in the USA in the early 1990s with the purpose of providing an efficient and portable computing platform for message passing.

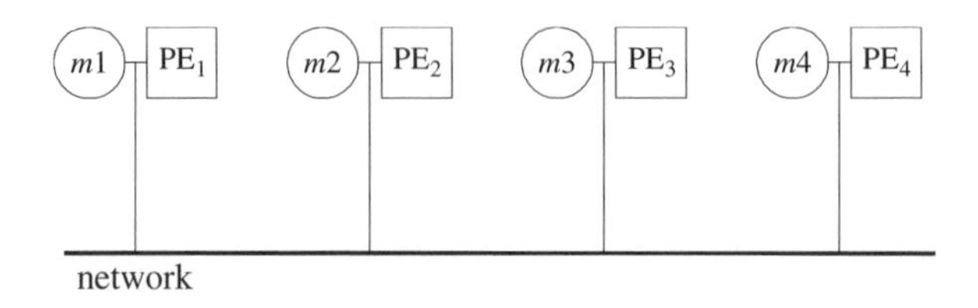

Figure 9.2 A distributed memory computer network.

9.2 TASK SCHEDULING PROBLEM

Task scheduling is a combinatorial optimization problem that has many applications in the design of feasible schedules for a parallel computer system. A good task scheduler utilizes the full capability of the computer system to produce maximum results in the form of fast execution of programs and efficient handling of jobs. Among other things, task scheduling contributes to compiler design, operating system management, development of drivers, and file management. The areas of applications involving task scheduling have also expanded over the years to include several new forms of distributed computing environments involving mobile and grid computing.

Task scheduling is basically a software issue. The problem is important in a parallel and distributed computer system as it contributes to the overall performance index of the system. It is important for the system to explore its full capability by using its available resources for achieving its maximum performance. There is no point in having very sophisticated hardware in the form of a parallel machine if the software that controls its operation does not explore its full capability.

The task scheduling problem is described as follows: *Given a set of tasks from a program and a set of processors in a parallel computing system, how can these tasks be mapped to the processors in such a way that the whole operation will complete at the earliest time possible?*

A *task* v_i is defined as a module or a unit of job in a computer program consisting of as few as one line and as many as a few hundred lines of code. The complete set of tasks having relationships between them in the form of directed edges is called a *task graph*. Figure 9.3 shows a task graph in which the tasks are represented as nodes in white boxes and the directed links between the nodes denote the relationships between them. A directed edge between two tasks in a task graph shows the partial or-

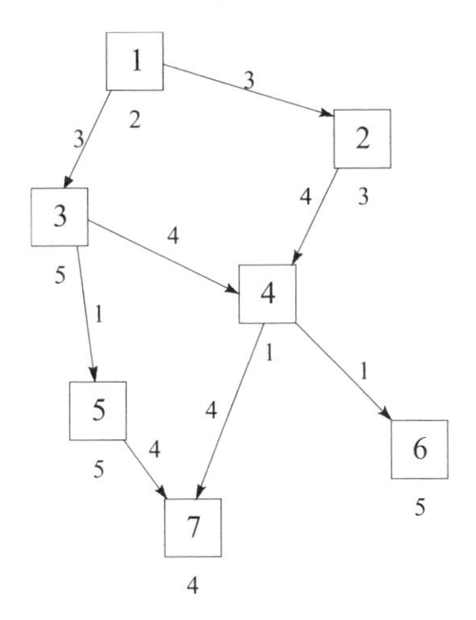

Figure 9.3 Task graph having seven nodes and eight directed links.

der relationships between them. A partial order between two tasks is said to exist if a task depends on its predecessor task(s) for some shared data or information. In Figure 9.3, the partial order $v_5 < v_3$ of the task graph is shown as a directed edge between the two nodes. This partial order shows a precedence relationship in which v_3 is the predecessor of v_5. The relationship implies that v_5 will not be able to start executing unless v_3 has been completed. A task may have more than one predecessor task, as in the case of v_4 and v_7 in the figure. In this case, the two tasks will have to wait for the completion of all their predecessor tasks before they can start executing.

Task scheduling involves the mapping of a set of tasks from a task graph to the processors with the main objective of minimizing the overall completion time. Figure 9.4 shows a task graph having seven tasks and its mapping onto a set of four processors to produce a scheduling output in the form of Gantt charts. The main objective in task scheduling is to produce an optimum schedule with minimum overall completion time. This minimum completion time is also called the *schedule length*

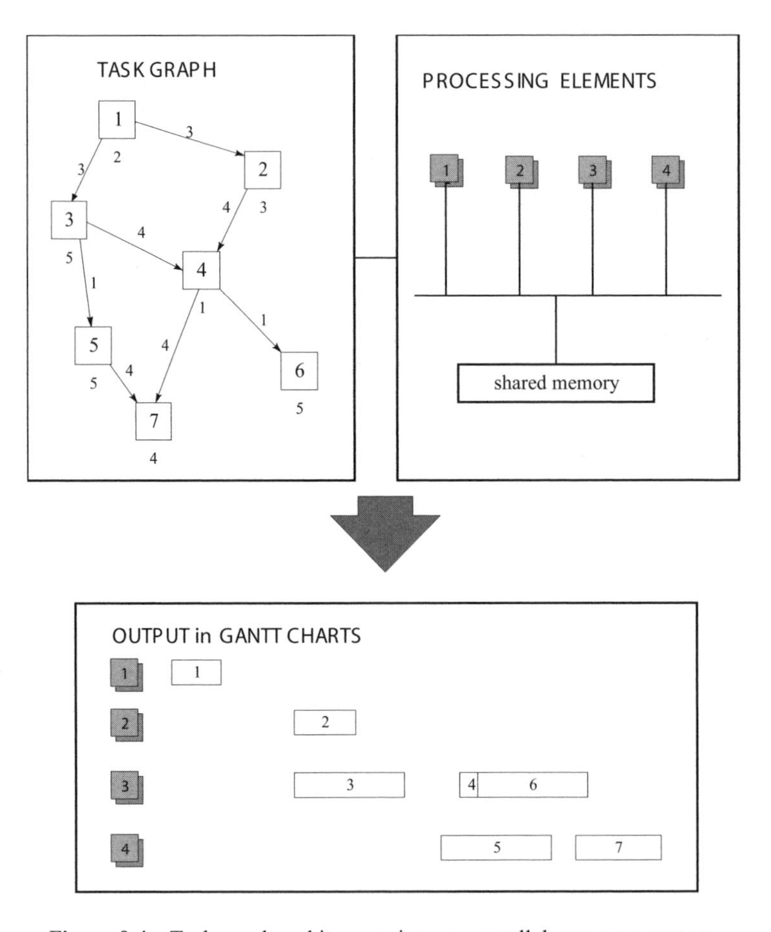

Figure 9.4 Task graph and its mapping on a parallel computer system.

or the makespan of the task graph. Another common objective is to distribute the tasks evenly among the processors; this is called *load balancing.* This objective is necessary to make sure that all the processors are fully utilized, and to avoid the case of uneven job distribution, which causes some processors to be too busy while some others are idle.

It is also an objective in task scheduling to minimize the laxity rate in the processors. A laxity in a processor is defined as the lax time between the completion of a task and the start of another task in the processor. A high laxity rate reflects inefficiency in the scheduling system of the network, and this is intolerable as the running time for the system is expensive.

For a real-time system, it is important for a schedule to be designed so as to meet the deadline of executions. A *soft* deadline in the schedule indicates that compliance to it is not very critical to the overall functioning of the system. The system may still function if there is some delay in the deadline although this may cause some discomfort or inconvenience. In a system with a *hard* deadline, compliance to the deadline is very critical to system survival. The system may lose some critical data or even crash if the deadline is not met. For example, an aircraft due to land at an airport has a schedule with a tolerable soft deadline and intolerable hard deadline. The aircraft has prepared itself to land before the soft deadline time of t_1 but due to some unforeseen circumstances at the airport, it is unable to do so. It can tolerate this delay by circling the airport a few times while waiting for clearance for landing. The ultimate deadline, or hard deadline, for landing is then set at t_2. A failure to land on or before t_2 will result in a disaster as the aircraft may run out of fuel. In reality, this problem has happened before; a South American aircraft crashed near an airport in New York, causing many deaths.

An optimum schedule in the mapping of tasks to processors helps to produce a healthy computing system in which the capability of the resources in the system is fully explored. Some established methods for achieving this goal include methods based on list scheduling, clustering, mathematical programming, queueing theory, graph theoretic, and enumerated search. There are also approaches using the latest tools in neural networks, simulated annealing, genetic algorithms, and other evolutionary algorithms.

There are two types of task scheduling. In *static scheduling,* all the information about the tasks is known a priori, that is, before they are scheduled. The information includes the length of the task, its precedence relationship with other tasks, and the related communication costs. Very often, the task graph is known prior to scheduling, and this provides enough time to decide on the mapping. Static task scheduling is often associated with offline scheduling in which its implementation is not critical and can be delayed.

On the other hand, if the state of the tasks is only known on the fly, that is, as the execution is in progress, then this is a case of *dynamic scheduling.* Dynamic scheduling is more difficult to apply as it involves some extra overhead in determining the state of the arriving tasks and the processors at every time slot prior to scheduling the tasks. It is not always possible to get an optimum result in the form of minimum schedule length for this reason. Therefore, the main objective in dynamic scheduling is not the minimum schedule length. Instead, the more important objective is to produce a feasible schedule that meets the deadline of execution.

Dynamic scheduling is often caused by the nondeterminism factor on the states of the tasks and the PEs prior to their execution. Nondeterminism in a program originates from factors such as uncertainties in the number of cycles (such as loops), the *and/or* branches, and the variable task and arc sizes. The scheduler has very little a priori knowledge about these task characteristics and the system state estimation is obtained on the fly as the execution is in progress. This is an important step before a decision is made on how the tasks are to be distributed. Dynamic scheduling is often associated with real-time scheduling, which involves periodic tasks and tasks with critical deadlines.

Task Scheduling Concepts

There are several rules to follow in scheduling tasks in a multiprocessor system. The general rule to follow is, only one task can be accommodated by a processor in a single time slot. This means a processor can do only one thing at a time, and no multitasking capability on a single processor is assumed. Other rules related to task precedence are:

1. In a partial order $v_j < v_i$, the earliest time v_j can start executing is at the completion time of v_i. If v_j is scheduled in the same processor as v_i, then the completion time of v_i is the starting time of v_j, provided the time slot is available.
2. Otherwise, if v_j is scheduled in a different processor, then the communication cost between them must be added.
3. If more than one predecessor exists, then the earliest time the task can start is the maximum of the sum of the execution time of the predecessor tasks and their communication costs. This is known as the synchronization process performed on the processors to make sure that the precedence rules on the tasks are not broken.
4. If a time slot is occupied, then the scheduled task on this slot will have to wait until the currently executing task completes its execution.

The task scheduling problem is illustrated in Figure 9.5. The figure shows four tasks—v_1, v_2, v_3, and v_4—marked as clear boxes. Two processors are used in the simulation. Each task has the length written under its box. The task's precedence relationship with other tasks is shown as directed edges, together with the associated cost at each edge. The tasks can be scheduled in many ways, using several existing methods. One solution is to place v_1, v_3, and v_4 into PE_1, and v_2 into PE_2, in the schedule. The task v_1 has a length of 4 units to complete its execution at $t = 4$ in PE_1. As v_2 has v_1 as its predecessor, the former has to wait for the latter to complete its execution before it can start. Since v_2 is mapped on a different PE from v_1, there is a communication cost to add and this causes some delay in its start time for execution.

The example in Figure 9.5 produces a schedule length of 18 units, as shown by the Gantt charts in the figure. v_1 becomes the first task to be executed while the last

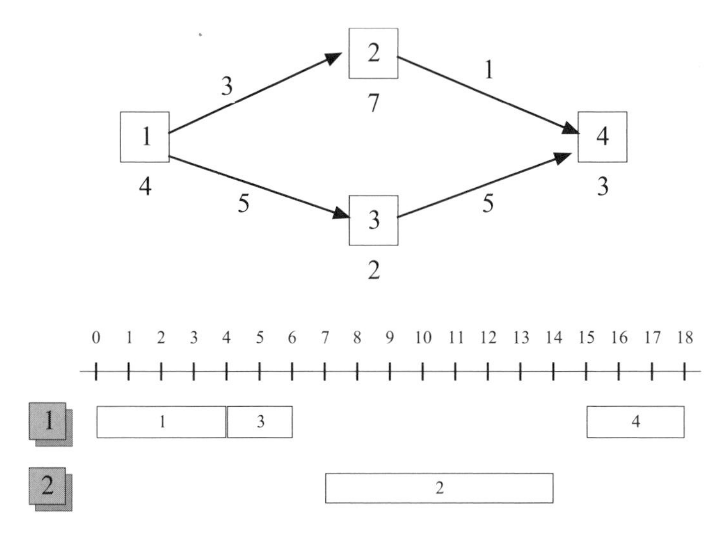

Figure 9.5 Scheduling of four tasks on two processors.

one is v_4. The schedule may not be optimum, depending on the method used. A different schedule with a higher or lower schedule length results when a different mapping method is applied. Therefore, the combination here is an important factor in determining the optimum solution and that is what a good scheduler tries to achieve.

Path Maximum Magnitude Scheduling Model

We propose a scheduling model called the *path maximum magnitude* (PMM), in which the priorities of the tasks in the task graph for mapping are based on their path maximum magnitude (*pmm*) in the task graph. A task can have several paths leading to the root of its graph. A *rooted path* of a task is defined as a path from the task to the root of the graph. The *magnitude* of a rooted path is the sum of the length of all the tasks and the communication costs in this path, excluding the task's length. The *path maximum magnitude* of a task, denoted by v_i^{pmm}, is then the maximum of the magnitude of these paths.

To illustrate the above terminologies, we refer to the task graph in Figure 9.3. In this graph, v_4 has two rooted paths: $v_1 \rightarrow v_2 \rightarrow v_4$ and $v_1 \rightarrow v_3 \rightarrow v_4$. The path magnitude as follows:

Path $v_1 \rightarrow v_2 \rightarrow v_4$: path magnitude $= 4 + 3 + 3 + 2 = 12$
Path $v_1 \rightarrow v_3 \rightarrow v_4$: path magnitude $= 4 + 5 + 3 + 2 = 14$

This gives the path maximum magnitude v_4^{pmm}, as follows:

$$v_4^{pmm} = \text{Path maximum magnitude of } v_4 = \max(12, 14) = 14$$

The priority list for scheduling the tasks onto the processors is based on their path maximum magnitude. Tasks are sorted according to their pmm value, in ascending order from lowest to highest. In the priority list, the assignment of tasks to the processors is made starting with the highest-ranking tasks.

Table 9.1 shows the output from PMM in a network of four processors. The tasks are first sorted into their priority list according to their pmm values. From the list, the tasks are mapped to the available processors according to their sorted order: v_1, v_3, v_2, v_5, v_4, v_6, and v_7. This way of mapping produces a schedule that has a completion time at $t = 18$. The Gantt charts in this example are shown in Figure 9.4.

9.3 TASK SCHEDULING VISUALIZATION MODEL

The model PMM for scheduling the tasks is illustrated visually in the project Code9. Figure 9.6 shows the output from its simulation of a task graph with 18 nodes on a four-processor system. The display consists of a drawing area in the upper half of the window and an output area in the lower half. The drawing area provides input from the user for drawing the task graph. A node in the task graph is drawn by clicking the left button of the mouse in this area. The click also assigns a random value as the length of the node. This random value is generated from the clock cycle facility inside the computer. A directed edge is obtained by clicking a pair of nodes consecutively. With the clicks, the first clicked node becomes the predecessor task of the second node.

A random value is also assigned to the edge to represent the communication cost from the first node to the second. The program also provides the Task Scheduler push-button window. A click on this push button activates the task scheduler to produce the scheduling results in the form of Gantt charts. In addition, a list view scroll window is shown in the drawing area for displaying the information about the tasks created by the user. This information is produced and displayed in the list view window when the push button *Task Scheduler* is clicked.

The project Code9 consists of the files Code9.h and Code9.cpp. The header file Code9.h defines the constants as listed in Table 9.2.

Table 9.1 Scheduling results for the task graph in Figure 9.3

Task	Length	pmm	priority	PE	ast	act	pre
v_1	2	0	1	1	0	2	0
v_2	3	5	3	3	5	8	1
v_3	5	5	2	2	5	10	1
v_4	1	14	5	2	11	12	2
v_5	5	11	4	4	11	16	1
v_6	5	16	6	2	12	17	1
v_7	4	20	7	4	16	20	2

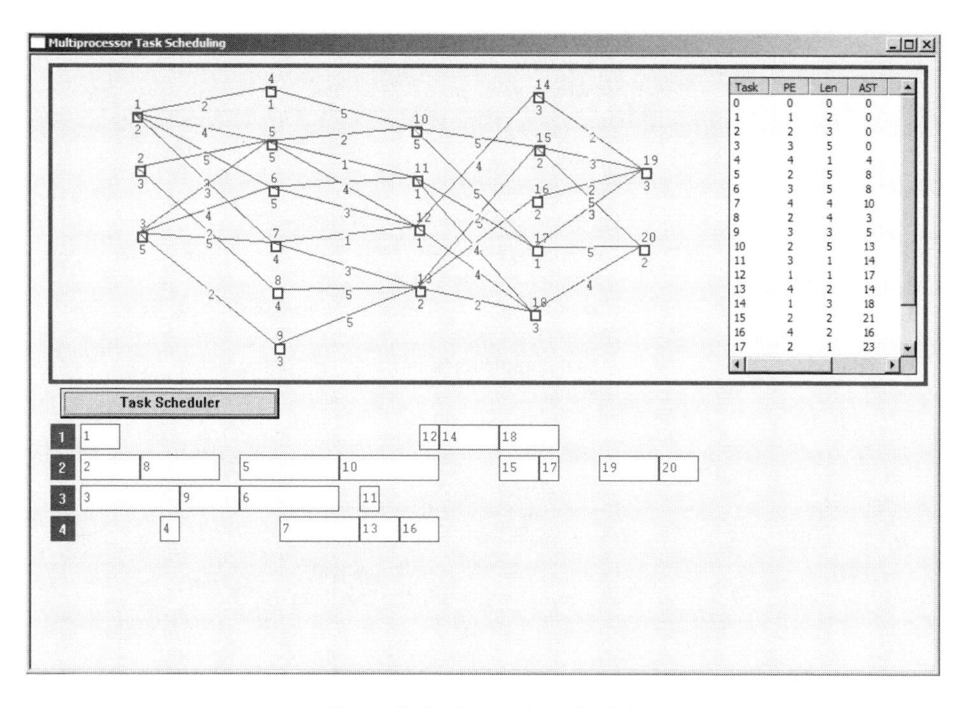

Figure 9.6 Output from Code9.

The two main players in this application are the task, v_i, and its mapping on the processing element, PE_k. They are referred as the ith task in the graph or `v[i]`, and the kth processing element or `PE[k]`, respectively, in the project. The information about v_i is represented by the structure `NODE`, whereas PE_k is represented by `PROC`. Both structures are defined in Code9.h.

The structure `NODE` in Code9.h, as shown in Table 9.3, describes the links between a task, v_i, and its elements. The *length* (size) of v_i, denoted by `v[i].len`, is defined as the elapsed time for the execution of the task sequentially on a processor. This length is also referred to as the task *execution time* or the task *worst-case computation time.* The partial order $v_i < v_j$ creates a precedence relationship between the two tasks. The variable `v[i].pre[0]` has been reserved to denote the number of

Table 9.2 Constants in Code9

Constants	Description
`N`	Maximum number of tasks allowed in the graph
`nPE`	Number of processors
`IDC_SCHEDULE`	Id for the push button *Task Scheduler*
`IDC_TASKINFO`	Id for the list view window
`nFIELDS`	Number of fields in the list view table

Table 9.3 Task information in Code9

Element	Description
`v[i]`	Task i, or v_i
`v[i].len`	Length of v_i
`v[i].aPE`	The assigned PE of v_i
`v[i].pmm`	Path maximum magnitude of v_i, or v_i^{pmm}
`v[i].pre[0]`	Number of predecessor tasks of v_i
`v[i].pre[j]`	jth predecessor task of v_i, for $j \geq 1$
`v[i].preCom[j]`	Communication cost between v_i and its jth predecessor task
`v[i].ast`	Actual start time of v_i
`v[i].act`	Actual completion time of v_i
`v[i].lrt`	Low ready time of v_i
`v[i].hrt`	High ready time of v_i
`v[i].sort`	Sorted v_i according to its pmm values in order from lowest to highest
`v[i].Home`	Coordinates of v_i in the drawing area
`v[i].GHome`	Coordinates of v_i in the Gantt charts
`v[i].Box`	Box representing v_i in the drawing area
`v[i].GBox`	Box representing v_i in the Gantt charts

predecessor tasks of v_i. The subsequent index j in `v[i].pre[j]` denotes the jth predecessor of v_i. For example, in Figure 9.3, `v[4].pre[0]`=2, `v[4].pre[1]` =2 (which is v_2), and `v[4].pre[2]`=3 (or v_3). The communication costs as a result of these precedence relationships are given by `v[4].preCom[1]`=4 and `v[i].preCom[2]`=4, respectively.

Our strategy for scheduling the tasks is to give higher priorities to the tasks with low path maximum magnitude values. The path maximum magnitude of a task is defined as the maximum of the sum of all the lengths of its predecessor tasks and the communication costs in the path from the task to its root. Therefore, all the tasks in the graph are sorted according to their pmm values before they are mapped onto the processors. The sorted tasks in the list are represented as `v[i].sort`.

The attributes of a processing element, PE_k, are described in Table 9.4. Each processing element has a list of the successfully assigned tasks, denoted as `PE[k].aTS[i]`. It can be easily verified from Figure 9.4 that `PE[4].aTS[1]` `=5` and `PE[4].aTS[2]=7`. At each stage during the execution, a processor has an indicator to determine its ready time for accepting a new task assignment. This ready time, denoted by `PE[k].prt`, depends very much on the rules for task assignments discussed earlier, and it is necessary in order to maintain a proper synchronization in the system.

The last attribute of a processing element is the processor execution length, `PE[k].pel`, measured as the total length from the start of the first task in the processor to the end of the last task. This variable can effectively be determined at the last stage of the execution, in the case of static scheduling. For the dynamic

Table 9.4 Information about a processing element

Element	Description
PE[k]	Processing element k, or PE_k
PE[k].aTS[i]	ith assigned task in PE_k
PE[k].prt	Processor ready time, or the earliest time PE_k is ready to accept a new task assignment
PE[k].pel	Processor execution length, which is the length measured from the starting time of the first task to the completion time of the last task in PE_k
PE[k].Home	Coordinates of PE_k in the Gantt charts area

case, `PE[k].pel` can be determined as the execution is progressing to give an updated value at every time slot.

The main function of the constructor `CCode9()` is to allocate memory for the class `CCode9`. The initial setup consists of the main window and two child windows, which are created and displayed in the constructor. The main window consists of the drawing area in the upper half and the Gantt charts area in the lower half. The push button *Task Scheduler* and the list view facilities for displaying information about the tasks are provided as child windows. The following code fragments create the three windows:

```
Create(NULL,"Multiprocessor Task Scheduling",
     WS_OVERLAPPEDWINDOW,CRect(0,0,920,630));
TSbutton.Create("Task Scheduler",WS_CHILD | WS_VISIBLE
     | BS_DEFPUSH BUTTON, CRect(30,325,250,355),this,IDC_SCHEDULE);
TaskInfoView.Create(WS_VISIBLE | WS_CHILD | WS_BORDER | LVS_REPORT
    | LVS_NOSORTHEADER,CRect(BottomRight.x-200,TopLeft.y+10,
     BottomRight.x-10,BottomRight.y-10), this, IDC_TASKINFO);
```

Table 9.5 Functions in Code9

Function	Description
CCode9()	The constructor
~CCode9()	The destructor
OnPaint()	Provides the intial display on the main window
OnLButtonDown()	Draws the nodes of the graph and labels them
OnRButtonDown()	Draws the links between nodes in the graph
OnClickCalc()	Activates the push button *Task Scheduler*
PreScheduler()	Prepares the tasks for the initial assignments
Scheduler()	Computes the high ready time and low ready time, and assigns the task to the processor
TaskInfo()	Displays the task information in the list view window
PMM()	Computes the path maximum magnitude of the tasks in the task graph

The list view window is used to display the information about all the tasks created by the user. The window displays the information about the assigned tasks in the graph, namely, their task number, assigned PE (aPE), length (len), actual start time (ast), completion time (act), the number of predecessor tasks (pre), the path maximum magnitude (pmm), and the sorted priority list (sort). The columns and their titles are created in the constructor as follows:

```
char* column[nFIELDS+1]
     ={"Task","PE","Len","AST","CT","Pre","pmm","Order"};
int columnWidth[nFIELDS+1]={40,40,40,40,40,40,40,40};
LV_COLUMN lvColumn;
lvColumn.mask = LVCF_WIDTH | LVCF_TEXT | LVCF_FMT | LVCF_SUBITEM;
lvColumn.fmt = LVCFMT_CENTER;
lvColumn.cx = 85;
for (int i=0;i<=nFIELDS;i++)
{
     lvColumn.iSubItem = 0;
     lvColumn.pszText = column[i];
     TaskInfoView.InsertColumn(i,&lvColumn);
     TaskInfoView.SetColumnWidth(i,columnWidth[i]);
}
```

The memory for the arrays v[] and PE[] are allocated dynamically using the function new. The two arrays are linked to the structures NODE and PROC, respectively:

```
v=new NODE [N+1];
PE=new PROC [nPE+1];
```

The last part of the constructor is concerned with initializing the global variables. These variables include the number of input nodes (nv), the right-button flag (RButtonFlag), the information about the processors, and the information for displaying text. The following code performs these operations:

```
nv=0; RButtonFlag=0;
FontCourier.CreatePointFont(60,"Courier");
BgColor=RGB(240,240,240);
TextGap=25; BoxSize=CSize(10,10);
TextColor=RGB(100,100,100);
for (int k=1;k<=nPE;k++)
{
     PE[k].aTS[0]=0;
     PE[k].prt=0;
     PE[k].pel=0;
     PE[k].Home=CPoint(20,360+(k-1)*30);
}
```

Figure 9.7 summarizes all the operations in Code9. The project is associated with four events in simulating the task scheduling problem. The first is the initial display, detected by the macro `WM_PAINT` and handled by the function `OnPaint()`. The next two are the left-button and right-button clicks of the mouse. The left-button click is detected by `WM_LBUTTONDOWN` and handled by the function `OnLButtonDown()`. The right-button click is detected by `WM_RBUTTONDOWN` and handled by `OnRButtonDown()`. The other event is the push-button click on the *Task Scheduler* button. The event is detected by `BN_CLICKED` and handled by the function `OnClickCalc()`.

The function `OnPaint()` provides the initial display of the output. The function starts by erasing the screen and divides the window into two areas: the drawing area (input) and the Gantt charts area (output). The function is written as follows:

```
void CCode9::OnPaint()
{
      CPaintDC dc(this);
      CRect rc;
      CString s;
      CPen penBlue(PS_SOLID,5,RGB(0,0,255));
      CBrush BgBrush(BgColor);
      GetClientRect(&rc);
      dc.FillRect(&rc,&BgBrush);

      CPen penDrawingBox(PS_SOLID,4,RGB(100,100,100));
      dc.SelectObject(penDrawingBox);
      dc.SelectStockObject(HOLLOW_BRUSH);
      rc=CRect(TopLeft,BottomRight);
      dc.Rectangle(rc);
      dc.SelectObject(&penBlue);
      dc.SetTextColor(RGB(255,255,255));
      for (int k=1;k<=nPE;k++)
      {
            rc=CRect(PE[k].Home,CSize(25,25));
            dc.FillSolidRect(&rc,RGB(100,100,100));
            s.Format("%d",k); dc.TextOut(PE[k].Home.x+8,PE[k].Home.y+5,s);
      }
}
```

The mouse's left-button click is an event handled by the function `OnLButtonClick()`. The operations performed following this event are summarized in Figure 9.8. Basically, the event is handled almost in the same manner as the ones in Chapters 6 and 7, and the same code from those chapters is used here. A click in the drawing area produces a small box that represents a node in the task graph. The click also causes the number of nodes `nv` to increase its value by one and assigns the home coordinates of the node at the point on the screen where the node is clicked (`pt`).

The function `OnLButtonDown()` also assigns some initial values to the newly created node, such as the number of predecessor tasks (`pre[0]`), actual starting time for execution (`ast`), its completion time (`act`), and the execution status

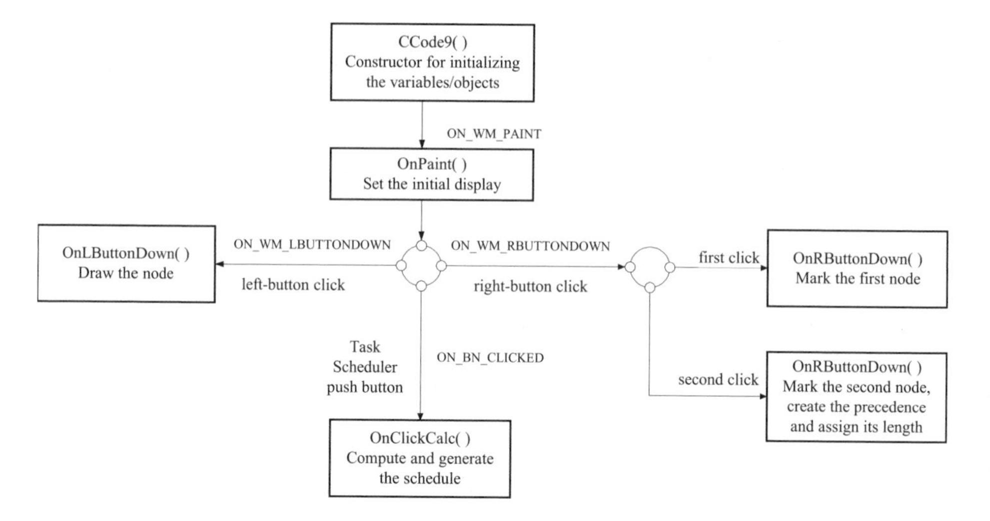

Figure 9.7 Summary of the operations in Code9.

(sta). A small random number is generated and assigned as the length (len) of this node. The following code in OnLButtonDown() performs these operations:

```
void CCode9::OnLButtonDown(UINT nFlags,CPoint pt)
{
	CClientDC dc(this);
	CString s;
	CPen penGray(PS_SOLID,2,TextColor);
	dc.SelectObject(penGray);

	dc.SelectObject(FontCourier);
	dc.SetTextColor(TextColor); dc.SetBkColor(BgColor);
	if (CRect(TopLeft,BottomRight).PtInRect(pt))
		if (nv<=N)
```

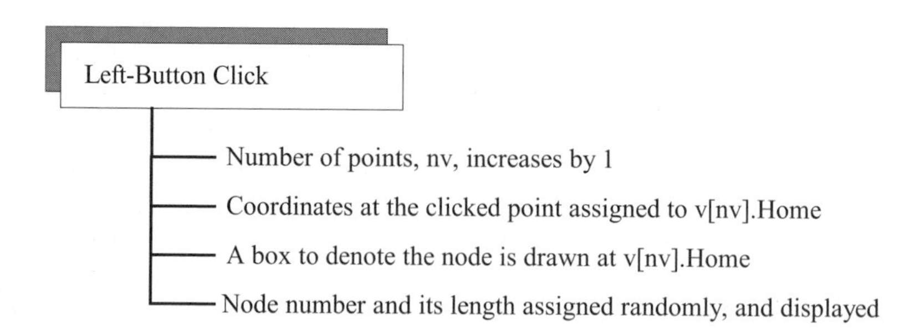

Figure 9.8 Summary of operations from the left-button click event.

```
        {
                nv++;
                v[nv].Home=pt;
                v[nv].pre[0]=0;
                v[nv].ast=0;
                v[nv].act=0;
                v[nv].sta=0;// set Node status to inactive
                v[nv].Box=CRect(CPoint(pt),CSize(BoxSize));
                dc.Rectangle(v[nv].Box);
                s.Format("%d",nv);
                dc.TextOut(v[nv].Home.x,v[nv].Home.y-15,s);
                v[nv].len=1+rand()%5;
                s.Format("%d",v[nv].len);
                dc.TextOut(v[nv].Home.x,v[nv].Home.y+10,s);
        }
}
```

The mouse's right-button click is an event detected by `WM_RBUTTONDOWN`. The event updates the task graph through the function `OnRButtonDown()`. Figure 9.9 summarizes the operations performed in this function. The main operation here is to create the links between the nodes in the graph to represent the precedence relationship of the tasks. The edge is created and displayed on the graph by consecutively clicking any two nodes. In this case, the first node is assigned as the variable `Pt1`

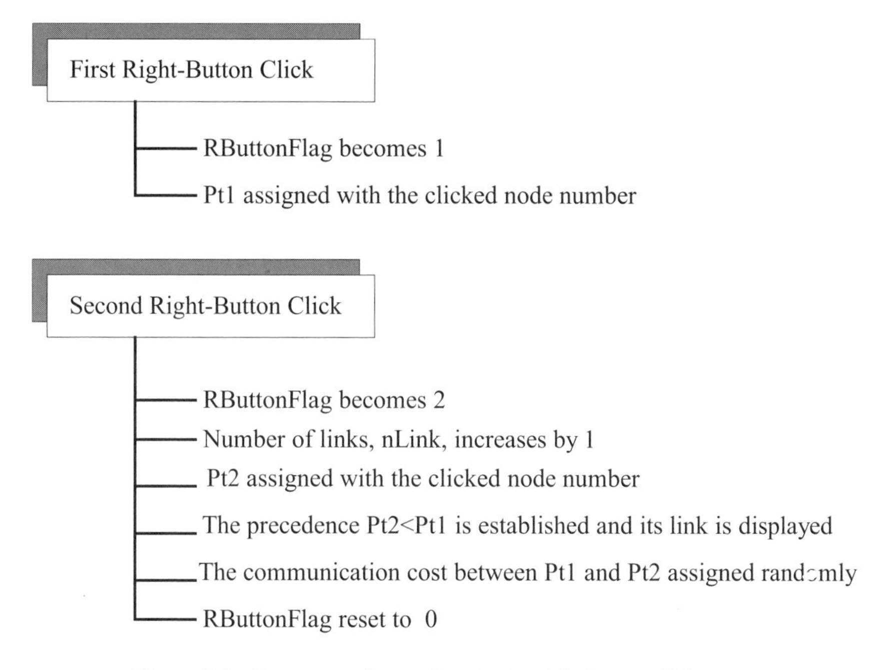

Figure 9.9 Summary of operations in the right-button click event.

and the second node is Pt2. The precedence relationship is Pt1 is the predecessor of Pt2. The function also assigns a random integer number as the communication cost for this precedence relationship.

The function OnRButtonDown() is written as follows:

```
void CCode9::OnRButtonDown(UINT nFlags,CPoint pt)
{
      CClientDC dc(this);
      CString s;

      int i,u,w,r;
      CPen penGray(PS_SOLID,1,TextColor);
      dc.SelectObject(penGray);
      time_t seed=time(NULL);
      srand((unsigned)seed);
      dc.SelectObject(FontCourier);
      dc.SetTextColor(TextColor);
      dc.SetBkColor(BgColor);
      for (i=1;i<=nv;i++)
            if (v[i].Box.PtInRect(pt))
            {
                  RButtonFlag++;
                  if (RButtonFlag==1)
                        Pt1=i;
                  if (RButtonFlag==2)
                  {
                        Pt2=i;
                        RButtonFlag=0;
                        r=++v[Pt2].pre[0];
                        v[Pt2].pre[r]=Pt1;
                        v[Pt2].preCom[r]=1+rand()%5;
                        dc.MoveTo(v[Pt1].Home);
                        dc.LineTo(v[Pt2].Home);
                        u=(v[Pt1].Home.x+v[Pt2].Home.x)/2;
                        w=(v[Pt1].Home.y+v[Pt2].Home.y)/2;
                        s.Format("%d",v[Pt2].preCom[r]);
                        dc.TextOut(u,w,s);
                  }
            }
}
```

In the function OnRButtonDown(), the first click on a node increases the value of the flag RButtonFlag by 1. This assigns the clicked node number to the variable Pt1. The second click increases the flag value to 2 and assigns the node number to the variable Pt2. A precedence relationship is now established, with

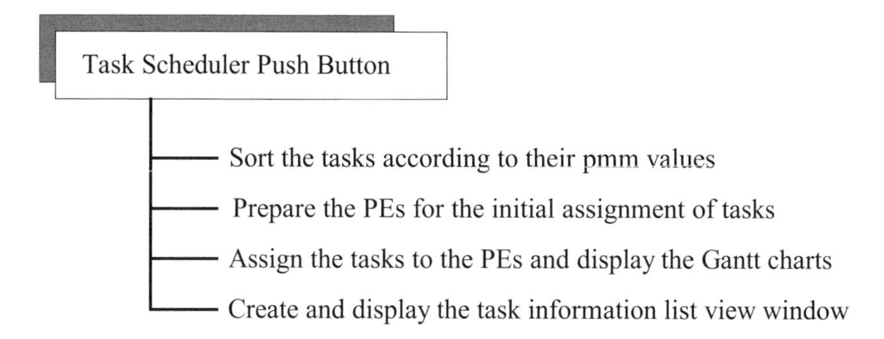

Figure 9.10 Summary of operations from the push button *Task Scheduler* event.

`Pt1` becoming the predecessor task of `Pt2`. The number of predecessor tasks `v[Pt2].pre[0]` increases its value by one where `v[Pt2].pre[r]` stores the node number, `Pt1`. Here, r=`v[Pt2].pre[0]` is a temporary variable that represents the number of precedence tasks for `Pt2`.

A random number between 1 and 5 is then generated and assigned to `v[Pt2].preCom[r]`. This number represents the communication cost from `v[Pt1]` to `v[Pt2]`, as a result of their precedence relationship. Once the relationship has been established, a line is drawn from `v[Pt1]` to `v[Pt2]`, with the communication cost shown on the link. Finally, the flag `RButtonFlag` is refreshed to 0 for the next pair of nodes.

Once the graph has been formed, it is ready for mapping onto the processors. This is achieved by clicking the mouse's left button on the push button *Task Scheduler*. The event is detected as `BN_CLICKED`, which activates the function `OnClickCalc()`. Figure 9.10 summarizes the operations in `OnClickCalc()`, which represent the push button *Task Scheduler* event.

The push-button event is divided into four main operations, represented by the client functions `PMM()`, `PreScheduler()`, `Scheduler()`, and `TaskInfo()`. These functions compute the pmm values of the tasks and the processor high and ready times, and assign the tasks to the processors according to the PMM model.

The function `PMM()` reads the information about the tasks and computes their pmm values. The pmm values are stored in the priority list as the variable `v[i].sort`, and these values are used to determine the order in which the tasks are mapped. The following routine performs this idea:

```
void CCode9::PMM()  // Sort the tasks according to levels
{
    int i,j,k,r,tmp;
    for (i=1;i<=nv;i++)
    {
          v[i].sort=i;
          if (v[i].pre[0]==0)
             v[i].pmm=0;
```

```
            else
            {
               r=v[i].pre[1];
               tmp=v[r].len+v[i].preCom[1]+v[r].pmm;
               for (j=1;j<=v[i].pre[0];j++)
               {
                    r=v[i].pre[j];
                    tmp=(tmp<v[r].len+v[i].preCom[j]+v[r].pmm)?
                         v[r].len+v[i].preCom[j]+v[r].pmm:tmp;
               }
               v[i].pmm=tmp;
            }
            for (k=1;k<=i-1;k++)
            {
                 r=v[k].sort;
                 if (v[i].pmm<v[r].pmm)
                 {
                    for (j=i;j>=k+1;j–)
                         v[j].sort=v[j-1].sort;
                    v[k].sort=i;
                    break;
                 }
            }
      }
}
```

Once the priority list has been determined, the next step is to warm up the processors by assigning each one of them with a task, each using the function `PreScheduler()`. This function assigns the processor using a temporary local variable called `AsPE`. This is necessary to make sure all the processors in the network are utilized by assigning them with at least one initial task.

The step begins by assigning the highest priority task in the list to PE_1, the second highest to PE_2, and so on until all the processors have one task each. The sorted tasks are represented as `u=v[i].sort`. For the start, the highest priority tasks are the tasks with pmm value equal 0, or the tasks that have no predecessors (`v[i].pre[0]=0`). The program checks for these tasks and they are immediately assigned to the processors if their number is less than or equal to the number of processors. If the number of tasks with no predecessors is higher than the number of available processors, then a selection is made whereby the priorities are given to those with higher lengths. All tasks that have been assigned are removed from the sorted list by setting their status flag to `v[i].sta=1`. The rest of the unscheduled tasks are put in the waiting list to be mapped in the function `Scheduler()`. The following code shows this process:

```
for (i=1;i<=nv;i++)
{
      u=v[i].sort;
```

```
        if (v[i].pre[0]==0)
        {
              if (i<=nPE)
              {
                  AsPE=i;
                  v[u].aPE=AsPE;
                  v[u].ast=0;
                  v[u].act=v[u].ast+v[u].len;
                  v[u].sta=1;
                  PE[AsPE].aTS[0]++;
                  PE[AsPE].aTS[1]=u;
                  PE[AsPE].prt=v[u].act;
                  PE[AsPE].pel += v[u].len;
              }
              else
              {
                  tmp=PE[1].prt;
                  AsPE=1;
                  for (k=1;k<=nPE;k++)
                        if (tmp>PE[k].prt)
                        {
                              tmp=PE[k].prt;
                              AsPE=k;
                        }
                  v[u].aPE=AsPE;
                  v[u].ast=tmp;
                  v[u].act=v[u].ast+v[u].len;
                  v[u].sta=1;
                  r=++PE[AsPE].aTS[0];
                  PE[AsPE].aTS[r]=u;
                  PE[AsPE].prt=v[u].act;
                  PE[AsPE].pel += v[u].len;
              }
        }
}
```

This is followed by updating their mapping information, including the actual start time for execution (`ast`) and its completion time (`act`). The processor also updates its information by evaluating its execution length (`pel`) and the assigned task information. Prior to assigning a task, the scheduler checks the ready time (`prt`) of the processors to make sure that none of them will receive more than one task. The following routine in `PreScheduler()` performs these operations:

```
v[u].aPE=AsPE;
v[u].ast=tmp;
```

```
v[u].act=v[u].ast+v[u].len;
v[u].sta=1;
r=++PE[AsPE].aTS[0];
PE[AsPE].aTS[r]=u;
PE[AsPE].prt=v[u].act;
PE[AsPE].pel += v[u].len;
```

If the number of tasks having precedence is higher than the number of processors, then a different strategy is adopted. All the tasks with no predecessors are assigned first and the remaining places are allocated to the tasks with the highest priorities. The same step as above is then performed on the assigned tasks and their corresponding PEs, for updates on their current state. The following routine in `PreScheduler()` performs these operations:

```
for (j=1;j<=nv;j++)
{
        if (i<=nPE)
        {
            AsPE=i;
            v[u].aPE=AsPE;
            r=v[u].pre[1];
            tmp=v[r].act+v[u].preCom[1];
            for (j=1;j<=v[u].pre[0];j++)
            {
                r=v[u].pre[j];
                if (tmp<v[r].act+v[u].preCom[j])
                    tmp=v[r].act+v[u].preCom[j];
            }
            v[u].ast=tmp;
            v[u].act=v[u].ast+v[u].len;
            v[u].sta=1;
            PE[AsPE].aTS[0]++;
            PE[AsPE].aTS[1]=u;
            PE[AsPE].prt=v[u].act;
            PE[AsPE].pel += v[u].len;
        }
}
```

Next is the function `Scheduler()`, which is the actual scheduler in this application. This function continues the work started by `PreScheduler()` by looking at the list of sorted tasks. Tasks that have been assigned in `PreScheduler()` are removed from the list, and this is achieved by setting their status flag to `v[i].sta=1`. Therefore, the remaining tasks in the priority list are recognized through the flag value `v[i].sta=0`.

A task v_i is said to be *ready* to be assigned to a processor if it has received all the required data from the predecessor tasks. The *ready time* for v_i is defined as the earliest time that v_i can be assigned to any available processor. We further define the ready time for v_i as either high ready time or low ready time. *High ready time,* `v[i].hrt`, of a ready task v_i is the highest value of the sum of the predecessor task and its communication to v_i. The *low ready time,* `v[i].lrt`, is the next highest value. The processor with `v[i].hrt` may skip some or all of the communication cost from its latest predecessor to the incoming task to enable it to start executing the incoming task at the time t=`v[i].lrt`. For other processors, the earliest time they can execute the task v_i is at t=`v[i].hrt`.

A task highest in the priority list for assignment to a processor is called a *candidate.* The candidate is to be mapped based on some kind of competition between the processors. The strategy here is to choose a processor that has the required readiness status, and one that will contribute in making the schedule length as short as possible. To achieve this goal, the candidate's high ready time and low ready time are computed to determine the dominance of the candidate's predecessor tasks.

Figure 9.11 illustrates an instance of time showing the status of four processors, each having a task. The tasks have been sorted according to their pmm values. They are named v_6, v_7, v_8, and v_9, with their mapping schedule marked as clear rectangles, as shown in the figure. In this example, the candidate is v_{10}, which is not shown in the figure as it is not assigned yet. It is also assumed that v_6, v_7, and v_9 are the predecessor tasks of the candidate, with their communication costs shown as shaded rectangles.

It can be seen from the figure that high ready time (`hrt`) occurs at $t = 25$ in v_7 and the next highest, or the low ready time (`lrt`), at $t = 23$ in v_6. Therefore, PE_4, where v_7 has been assigned, becomes the top choice for assigning the candidate v_{10}. The temporary variable `HiDom` is assigned as the processor number for this assign-

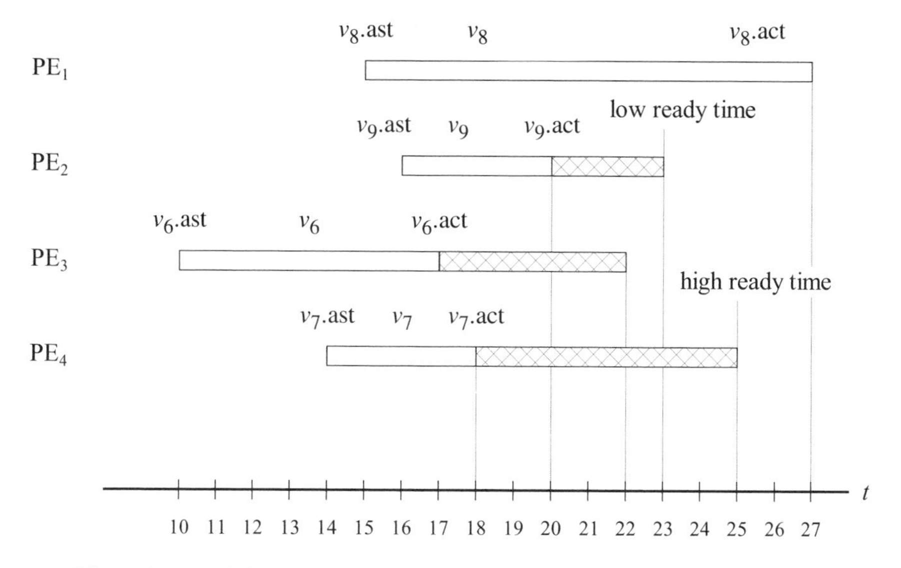

Figure 9.11 High and low ready time for choosing a processor as the candidate.

ment, while another temporary variable LoDom is assigned to PE_2 for having the low ready time task. Finally, the scheduler checks for the availability of HiDom, and it becomes the assigned PE if it is available. The following routine implements this idea:

```
for (j=1;j<=nv;j++)
{
     u=v[j].sort;
     if (!v[u].sta)
     {
          // calculate hrt
          HiDomIndex=1;
          HiDom=v[u].pre[HiDomIndex];
          v[u].hrt=v[HiDom].act+v[u].preCom[HiDomIndex];
          for (i=1;i<=v[u].pre[0];i++)
          {
               r=v[u].pre[i];
               if (v[u].hrt<v[r].act+v[u].preCom[i])
               {
                    v[u].hrt=v[r].act+v[u].preCom[i];
                    HiDom=r;HiDomIndex=i;
               }
          }
          // calculate lrt
          if (v[u].pre[0]==1)
               v[u].lrt=v[u].hrt;
          else
          {
               LoDomIndex=((HiDomIndex==1)?2:1);
               LoDom=v[u].pre[LoDomIndex];
               v[u].lrt=v[LoDom].act+v[u].preCom[LoDomIndex];
               for (i=1;i<=v[u].pre[0];i++)
                    if (i!=HiDomIndex)
                    {
                      r=v[u].pre[i];
                      if (v[u].lrt<v[r].act+v[u].preCom[i])
                           v[u].lrt=v[r].act+v[u].preCom[i];
                    }
          }
          HiDomPE=v[HiDom].aPE;
          if (PE[HiDomPE].prt<=v[u].hrt)
          {
               AsPE=HiDomPE;
               v[u].aPE=AsPE;
               if (v[u].pre[0]==1)
                    v[u].ast=PE[HiDomPE].prt;
               else
                    v[u].ast=((PE[HiDomPE].prt>=v[u].lrt)?
                              PE[HiDomPE].prt:v[u].lrt);
```

```
                v[u].act=v[u].ast+v[u].len;
                v[u].sta=1;
                PE[AsPE].prt=v[u].act;
                r=++PE[AsPE].aTS[0];
                PE[AsPE].aTS[r]=u;
                PE[AsPE].pel+=v[u].len;
            }
            else
            {
                AsPE=((HiDomPE==1)?2:1);
                tmp=PE[AsPE].pel;
                for (k=1;k<=nPE;k++)
                    if (k!=HiDomPE)
                        if (tmp>PE[k].pel)
                        {
                            AsPE=k;
                            tmp=PE[k].pel;
                        }
                v[u].aPE=AsPE;
                v[u].ast=((PE[AsPE].prt>=v[u].hrt)?
                            PE[AsPE].prt:v[u].hrt);
                v[u].act=v[u].ast+v[u].len;
                v[u].sta=1;
                PE[AsPE].prt=v[u].act;
                r=++PE[AsPE].aTS[0];
                PE[AsPE].aTS[r]=u;
                PE[AsPE].pel += v[u].len;
            }
        }
}
```

The last function called up by `Scheduler()` is `TaskInfo()`. This function displays all the necessary information about the assigned tasks in the list view window, as follows:

```
void CCode9:: TaskInfo()
{
    CString s;
    TaskInfoView.DeleteAllItems();
    LV_ITEM lvItem;
    lvItem.mask = LVIF_TEXT | LVIF_STATE;
    lvItem.state = 0;
    lvItem.stateMask = 0;
    for (int i=0;i<=nv;i++)
    {
        lvItem.iItem=i;
        lvItem.iSubItem=0;
        lvItem.pszText="";
        TaskInfoView.InsertItem(&lvItem);
```

```
            if (i+1<=nv && i+1<=N)
            {
                  s.Format("%d",i+1); TaskInfoView.SetItemText(i,0,s);
                  s.Format("%d",v[i+1].aPE); TaskInfoView.SetItemText(i,1,s);
                  s.Format("%d",v[i+1].len); TaskInfoView.SetItemText(i,2,s);
                  s.Format("%d",v[i+1].ast); TaskInfoView.SetItemText(i,3,s);
                  s.Format("%d",v[i+1].act); TaskInfoView.SetItemText(i,4,s);
                  s.Format("%d",v[i+1].pre[0]); TaskInfoView.SetItemText(i,5,s);
                  s.Format("%d",v[i+1].pmm); TaskInfoView.SetItemText(i,6,s);
                  s.Format("%d",v[i+1].sort); TaskInfoView.SetItemText(i,7,s);
            }
      }
}
```

9.4 SUMMARY AND CONCLUSION

Task scheduling is a big issue in a parallel and distributed computing system. The simulation and visualization model presented here only covers some basic elements of the problem. Among other things, this chapter presents the development of a simulation model for scheduling tasks in a dependency graph using a model called the path maximum magnitude (PMM). The simulation model provides a visual interface for drawing a dependency graph, with the magnitude of the tasks and the links determined randomly by the computer. The scheduling method using PMM produces results in the form of Gantt charts, which clearly show the start and completion time of each task in its assigned processor. The scheduling technique includes the precedence relationship between the tasks and their communication costs in the mapping.

The PMM method can easily be used as a prototype model for other scheduling methods. The model can be extended to include other parameters in task scheduling, such as factors affecting machine performance. These include the initial startup cost, the transmission rate, the contention rate, and the cabling mechanism. In a real parallel and distributed system, these parameters severely affect the communication cost of transferring information from one point in the network to another. Due to different requirements, some applications may not agree with the PMM method discussed here. In this case, other scheduling techniques may be used instead of PMM.

Other than parallel and distributed systems, the techniques discussed here can also be applied to problems involving machine scheduling, transportation, mobile computing, and grid computing. Our simulation model can be applied to cases in which there are similarities in the way the tasks are mapped in these problems.

BIBLIOGRAPHY

1. H. El-Rewini, T. G. Lewis, and H. H. Ali, *Task Scheduling in Parallel and Distributed Systems,* Prentice-Hall, 1994.

2. S. Salleh and A. Y. Zomaya, *Scheduling in Parallel Computer Systems: Fuzzy and Annealing Techniques,* Kluwer Acamedic Publishing, 1999.
3. R. Rotithor, Taxonomy of Dynamic Task Scheduling Schemes in Distributed Computing Systems, *IEE Proceedings on Computer and Digital Techniques,* vol. 141, no. 1, 1994.
4. M. J. Quinn, *Parallel Computing: Theory and Practice,* 2nd ed., McGraw-Hill, 1994.
5. T. Yang and A. Gerasoulis, DSC: Scheduling Parallel Tasks on an Unbounded Number of Processors, *IEEE Trans. Parallel and Distributed Systems,* vol. 5, no. 9, 1994.
6. A. Y. Zomaya, Parallel and Distributed Computing: The Scene, the Props, the Players, in *Parallel and Distributed Computing Handbook,* A. Y. Zomaya, ed., McGraw-Hill., 1996.

CODE LISTINGS

Code9: Task Scheduling Using Four Processors

```
// Code9.h
#include <afxwin.h>
#include <afxcmn.h>
#define N 20                          // Max #nodes
#define nPE 4                         // #PE
#define IDC_SCHEDULE 500
#define IDC_TASKINFO 501
#define nFIELDS 7

class CCode9 : public CFrameWnd
{
private:
      CPoint TopLeft,BottomRight;
      CFont FontCourier;
      CSize BoxSize;
      CButton TSbutton;
      CListCtrl TaskInfoView;

      int nv,RButtonFlag,Pt1,Pt2;
      int TextGap,TextColor,BgColor;
      typedef struct
      {
            int len;             // length
            int aPE;             // assigned Processor
            int pmm;             // path max. magnitude
            int pre[5];          // pred task, pre[0]=#pred tasks
            int preCom[5];       // comm cost of pred tasks
            int ast,act;         // ast=actual start time, act=completion time
            int hrt,lrt;         // high ready time, low ready time
            int sort;            // sorted nodes according to colevels
            bool sta;            // status
            CPoint Home,GHome;   // node coordinates in text & graphic areas
            CRect Box,GBox;      // node representation as a box
      } NODE;
      NODE *v;
      typedef struct
```

```
	{
		int aTS[N+1];		// task# (aTs[0]=#tasks in Pr)
		int prt,pel;		// proc ready time, execution length
		CPoint Home;
	} PROC;
	PROC *PE;
public:
	CCode9();
	~CCode9()			{}
	void PreScheduler(),Scheduler();
	void TaskInfo(),PMM();
	afx_msg void OnClickCalc();
	afx_msg void OnPaint();
	afx_msg void OnLButtonDown (UINT, CPoint);
	afx_msg void OnRButtonDown (UINT, CPoint);
	DECLARE_MESSAGE_MAP();
};

class CMyWinApp : public CWinApp
{
public:
  virtual BOOL InitInstance();
};

// Code9.cpp: task scheduling
#include "Code9.h"

BOOL CMyWinApp::InitInstance()
{
	CCode9* pFrame = new CCode9;
	m_pMainWnd = pFrame;
	pFrame->ShowWindow(SW_SHOW);
	pFrame->UpdateWindow();
	return TRUE;
}

CMyWinApp  MyApplication;

BEGIN_MESSAGE_MAP(CCode9,CFrameWnd)
	ON_WM_PAINT()
	ON_WM_LBUTTONDOWN()
	ON_WM_RBUTTONDOWN()
	ON_BN_CLICKED(IDC_SCHEDULE,OnClickCalc)
END_MESSAGE_MAP()

CCode9::CCode9()
{
	TopLeft=CPoint(20,10); BottomRight=CPoint(900,320);
	Create(NULL,"Multiprocessor Task Scheduling",
		WS_OVERLAPPEDWINDOW,CRect(0,0,920,630));
	TSbutton.Create("Task Scheduler",WS_CHILD | WS_VISIBLE
		| BS_DEFPUSH BUTTON, CRect(30,325,250,355),this,IDC_SCHEDULE);
	TaskInfoView.Create(WS_VISIBLE | WS_CHILD | WS_BORDER | LVS_REPORT
	   | LVS_NOSORTHEADER,CRect(BottomRight.x-200,TopLeft.y+10,
```

```
            BottomRight.x-10,BottomRight.y-10), this, IDC_TASKINFO);
        char* column[nFIELDS+1]
            ={"Task","PE","Len","AST","CT","Pre","pmm","Order"};
        int columnWidth[nFIELDS+1]={40,40,40,40,40,40,40,40};
        LV_COLUMN lvColumn;
        lvColumn.mask = LVCF_WIDTH | LVCF_TEXT | LVCF_FMT | LVCF_SUBITEM;
        lvColumn.fmt = LVCFMT_CENTER;
        lvColumn.cx = 85;
        for (int i=0;i<=nFIELDS;i++)
        {
            lvColumn.iSubItem = 0;
            lvColumn.pszText = column[i];
            TaskInfoView.InsertColumn(i,&lvColumn);
            TaskInfoView.SetColumnWidth(i,columnWidth[i]);
        }
        v=new NODE [N+1];
        PE=new PROC [nPE+1];

        nv=0; RButtonFlag=0;
        FontCourier.CreatePointFont(60,"Courier");
        BgColor=RGB(240,240,240);
        TextGap=25; BoxSize=CSize(10,10);
        TextColor=RGB(100,100,100);
       for (int k=1;k<=nPE;k++)
       {
            PE[k].aTS[0]=0;
            PE[k].prt=0;
            PE[k].pel=0;
            PE[k].Home=CPoint(20,360+(k-1)*30);
       }
}

void CCode9::OnPaint()
{
        CPaintDC dc(this);
        CRect rc;
        CString s;
        CPen penBlue(PS_SOLID,5,RGB(0,0,255));
        CBrush BgBrush(BgColor);
        GetClientRect(&rc);
        dc.FillRect(&rc,&BgBrush);

        CPen penDrawingBox(PS_SOLID,4,RGB(100,100,100));
        dc.SelectObject(penDrawingBox);
        dc.SelectStockObject(HOLLOW_BRUSH);
        rc=CRect(TopLeft,BottomRight);
        dc.Rectangle(rc);

        dc.SelectObject(&penBlue);
        dc.SetTextColor(RGB(255,255,255));
        for (int k=1;k<=nPE;k++)
        {
            rc=CRect(PE[k].Home,CSize(25,25));
            dc.FillSolidRect(&rc,RGB(100,100,100));
```

```
          s.Format("%d",k); dc.TextOut(PE[k].Home.x+8,PE[k].Home.y+5,s);
     }
}

void CCode9::OnLButtonDown(UINT nFlags,CPoint pt)
{
     CClientDC dc(this);
     CString s;
     CPen penGray(PS_SOLID,2,TextColor);
     dc.SelectObject(penGray);

     dc.SelectObject(FontCourier);
     dc.SetTextColor(TextColor); dc.SetBkColor(BgColor);
     if (CRect(TopLeft,BottomRight).PtInRect(pt))
          if (nv<=N)
          {
               nv++;
               v[nv].Home=pt;
               v[nv].pre[0]=0;
               v[nv].ast=0;
               v[nv].act=0;
               v[nv].sta=0;            // set Node status to inactive
               v[nv].Box=CRect(CPoint(pt),CSize(BoxSize));
               dc.Rectangle(v[nv].Box);
               s.Format("%d",nv);
               dc.TextOut(v[nv].Home.x,v[nv].Home.y-15,s);
               v[nv].len=1+rand()%5;
               s.Format("%d",v[nv].len);
               dc.TextOut(v[nv].Home.x,v[nv].Home.y+10,s);
          }
}

void CCode9::OnRButtonDown(UINT nFlags,CPoint pt)
{
     CClientDC dc(this);
     CString s;

     int i,u,w,r;
     CPen penGray(PS_SOLID,1,TextColor);
     dc.SelectObject(penGray);
     time_t seed=time(NULL);
     srand((unsigned)seed);
     dc.SelectObject(FontCourier);
     dc.SetTextColor(TextColor);
     dc.SetBkColor(BgColor);
     for (i=1;i<=nv;i++)
          if (v[i].Box.PtInRect(pt))
          {
               RButtonFlag++;
               if (RButtonFlag==1)
                    Pt1=i;
               if (RButtonFlag==2)
               {
                    Pt2=i;
```

```
                        RButtonFlag=0;
                        r=++v[Pt2].pre[0];
                        v[Pt2].pre[r]=Pt1;
                        v[Pt2].preCom[r]=1+rand()%5;
                        dc.MoveTo(v[Pt1].Home);
                        dc.LineTo(v[Pt2].Home);
                        u=(v[Pt1].Home.x+v[Pt2].Home.x)/2;
                        w=(v[Pt1].Home.y+v[Pt2].Home.y)/2;
                        s.Format("%d",v[Pt2].preCom[r]);
                        dc.TextOut(u,w,s);
                    }
            }
}

void CCode9::OnClickCalc()
{
     PMM();          // sort the nodes according to their colevel values
     PreScheduler(); // initialize each PE with 1st Node
    Scheduler();     // the scheduler
     TaskInfo();     // display task information
}

void CCode9:: TaskInfo()
{
     CString s;
     TaskInfoView.DeleteAllItems();
    LV_ITEM lvItem;
    lvItem.mask = LVIF_TEXT | LVIF_STATE;
    lvItem.state = 0;
    lvItem.stateMask = 0;
     for (int i=0;i<=nv;i++)
     {
          lvItem.iItem=i;
          lvItem.iSubItem=0;
          lvItem.pszText="";
          TaskInfoView.InsertItem(&lvItem);
          if (i+1<=nv && i+1<=N)
          {
               s.Format("%d",i+1); TaskInfoView.SetItemText(i,0,s);
               s.Format("%d",v[i+1].aPE); TaskInfoView.SetItemText(i,1,s);
               s.Format("%d",v[i+1].len); TaskInfoView.SetItemText(i,2,s);
               s.Format("%d",v[i+1].ast); TaskInfoView.SetItemText(i,3,s);
               s.Format("%d",v[i+1].act); TaskInfoView.SetItemText(i,4,s);
               s.Format("%d",v[i+1].pre[0]); TaskInfoView.SetItemText(i,5,s);
               s.Format("%d",v[i+1].pmm); TaskInfoView.SetItemText(i,6,s);
               s.Format("%d",v[i+1].sort); TaskInfoView.SetItemText(i,7,s);
          }
     }
}

void CCode9::PMM()  // Sort the tasks according to levels
{
    int i,j,k,r,tmp;
    for (i=1;i<=nv;i++)
```

```
    {
           v[i].sort=i;
           if (v[i].pre[0]==0)
              v[i].pmm=0;
           else
           {
              r=v[i].pre[1];
              tmp=v[r].len+v[i].preCom[1]+v[r].pmm;
              for (j=1;j<=v[i].pre[0];j++)
              {
                   r=v[i].pre[j];
                   tmp=(tmp<v[r].len+v[i].preCom[j]+v[r].pmm)?
                        v[r].len+v[i].preCom[j]+v[r].pmm:tmp;
              }
              v[i].pmm=tmp;
           }
           for (k=1;k<=i-1;k++)
           {
                r=v[k].sort;
                if (v[i].pmm<v[r].pmm)
                {
                   for (j=i;j>=k+1;j—)
                        v[j].sort=v[j-1].sort;
                   v[k].sort=i;
                   break;
                }
           }
    }
}

void CCode9::PreScheduler()
{
    int j,i,u,k,AsPE,r,tmp;
    for (i=1;i<=nv;i++)
    {
           u=v[i].sort;
           if (v[u].pre[0]==0)
           {
              if (i<=nPE)
              {
                   AsPE=i;
                   v[u].aPE=AsPE;
                   v[u].ast=0;
                   v[u].act=v[u].ast+v[u].len;
                   v[u].sta=1;
                   PE[AsPE].aTS[0]++;
                   PE[AsPE].aTS[1]=u;
                   PE[AsPE].prt=v[u].act;
                   PE[AsPE].pel += v[u].len;
              }
              else
              {
                   tmp=PE[1].prt;
                   AsPE=1;
```

```
                for (k=1;k<=nPE;k++)
                    if (tmp>PE[k].prt)
                    {
                                tmp=PE[k].prt;
                                AsPE=k;
                    }
                v[u].aPE=AsPE;
                v[u].ast=tmp;
                v[u].act=v[u].ast+v[u].len;
                v[u].sta=1;
                r=++PE[AsPE].aTS[0];
                PE[AsPE].aTS[r]=u;
                PE[AsPE].prt=v[u].act;
                PE[AsPE].pel += v[u].len;
            }
        }
        else
            if (i<=nPE)
            {
                AsPE=i;
                v[u].aPE=AsPE;
                r=v[u].pre[1];
                tmp=v[r].act+v[u].preCom[1];
                for (j=1;j<=v[u].pre[0];j++)
                {
                    r=v[u].pre[j];
                    if (tmp<v[r].act+v[u].preCom[j])
                        tmp=v[r].act+v[u].preCom[j];
                }
                v[u].ast=tmp;
                v[u].act=v[u].ast+v[u].len;
                v[u].sta=1;
                PE[AsPE].aTS[0]++;
                PE[AsPE].aTS[1]=u;
                PE[AsPE].prt=v[u].act;
                PE[AsPE].pel += v[u].len;
            }
    }
}

void CCode9::Scheduler()
{
    CClientDC dc(this);
    CString s;
    int u,i,j,k,r,m,AsPE;
    int HiDom,HiDomIndex,HiDomPE,LoDom,LoDomIndex,tmp;
    CPen penGray(PS_SOLID,1,TextColor);

    dc.SelectObject(penGray);
    for (j=1;j<=nv;j++)
    {
        u=v[j].sort;
        if (!v[u].sta)
        {
```

```
// calculate hrt
HiDomIndex=1;
HiDom=v[u].pre[HiDomIndex];
v[u].hrt=v[HiDom].act+v[u].preCom[HiDomIndex];
for (i=1;i<=v[u].pre[0];i++)
{
     r=v[u].pre[i];
     if (v[u].hrt<v[r].act+v[u].preCom[i])
     {
          v[u].hrt=v[r].act+v[u].preCom[i];
          HiDom=r;HiDomIndex=i;
     }
}
// calculate lrt
if (v[u].pre[0]==1)
     v[u].lrt=v[u].hrt;
else
{
     LoDomIndex=((HiDomIndex==1)?2:1);
     LoDom=v[u].pre[LoDomIndex];
     v[u].lrt=v[LoDom].act+v[u].preCom[LoDomIndex];
     for (i=1;i<=v[u].pre[0];i++)
          if (i!=HiDomIndex)
          {
             r=v[u].pre[i];
             if (v[u].lrt<v[r].act+v[u].preCom[i])
                  v[u].lrt=v[r].act+v[u].preCom[i];
          }
}
HiDomPE=v[HiDom].aPE;
if (PE[HiDomPE].prt<=v[u].hrt)
{
     AsPE=HiDomPE;
     v[u].aPE=AsPE;
     if (v[u].pre[0]==1)
          v[u].ast=PE[HiDomPE].prt;
     else
          v[u].ast=((PE[HiDomPE].prt>=v[u].lrt)?
               PE[HiDomPE].prt:v[u].lrt);
     v[u].act=v[u].ast+v[u].len;
     v[u].sta=1;
     PE[AsPE].prt=v[u].act;
     r=++PE[AsPE].aTS[0];
     PE[AsPE].aTS[r]=u;
     PE[AsPE].pel+=v[u].len;
}
else
{
     AsPE=((HiDomPE==1)?2:1);
     tmp=PE[AsPE].pel;
     for (k=1;k<=nPE;k++)
          if (k!=HiDomPE)
               if (tmp>PE[k].pel)
```

```
                              {
                                   AsPE=k;
                                   tmp=PE[k].pel;
                              }
                    v[u].aPE=AsPE;
                    v[u].ast=((PE[AsPE].prt>=v[u].hrt)?PE[AsPE].prt:v[u].hrt);
                    v[u].act=v[u].ast+v[u].len;
                    v[u].sta=1;
                    PE[AsPE].prt=v[u].act;
                    r=++PE[AsPE].aTS[0];
                    PE[AsPE].aTS[r]=u;
                    PE[AsPE].pel += v[u].len;
               }
          }
     }
     dc.SelectObject(FontCourier);
     dc.SetTextColor(TextColor);
     for (i=1;i<=nv;i++)
     {
          m=v[i].aPE;

v[i].GHome=CPoint(30+PE[m].Home.x+20*v[i].ast,PE[m].Home.y);
          v[i].GBox=CRect(v[i].GHome.x,v[i].GHome.y,
                              v[i].GHome.x+20*v[i].len,v[i].GHome.y+25);
          dc.Rectangle(&v[i].GBox);
          s.Format("%d",i); dc.TextOut(2+v[i].GHome.x,5+v[i].GHome.y,s);
     }
}
```

CHAPTER 10

DISCRETE-EVENT SIMULATION

10.1 CONCEPTS OF SIMULATION

Many real-life events are difficult to implement directly due to their high initial startup costs or because they are too risky or too dangerous in terms of safety. A thorough study has to be performed well ahead of its scheduled implementation. In many cases, the study includes the development of a simulation model that will give a correct indication of the success or failure of the project. *Simulation* is defined as a small-scale imitation of a real-life event over a measuring quantity such as time. This small-scale imitation represents the event by having the controlling parameters or variables that scale upward linearly or nonlinearly. In order to achieve this upward scalability objective, a simulation model is carefully developed for producing the correct parameters for this representation. The parameters must be able to withstand the upward scalability requirements of the problems associated with this transformation. A simulation model is said to be successful in its implementation if this important factor is tackled correctly.

Simulation models are produced to give an understanding and the correct picture of a real-life event. The simulation approach of analyzing a model is opposed to the analytical approach, in which the method of analyzing the system is purely theoretical. One main objective of simulation is to produce an optimum model that meets all the requirements of the real-life problem. The real-life problem may be bugged by uncertainties and a string of constraints that may pose a big overhead in its implementation. Hence, understanding the system goes all the way to optimizing the system performance, verifying its correctness, and testing its reliability over several sets of different data. For example, a proposal for a new engine in a car involves a comprehensive study of the performance of the engine and factors such as the production costs, comparative performances of the engine against other models in the market, reliability against technical glitches, and the market demand. On the technical side, the engine performance is simulated in the laboratory for things like fuel consumption, durability, maintenance issues, and effectiveness. Several parameters and variables in the design are supplied to the simulation model to see changes in the engine performance, and the best model produced leads to a decision as to whether the model is accepted or rejected.

Simulations also represent the virtual picture of a real-life situation. For example, in the Universal Studio's "Back to the Future" virtual ride, the audience can feel the virtual environment of riding in a futuristic car. The computer has been successful in creating special effects combining audio, visual, and mechanical resources in the studio to allow interaction between the user and the environment. Such simulations are used extensively today to train military personnel for battlefield situations, at a fraction of the cost of running field exercises involving real tanks, aircrafts, and other expensive equipment.

A simulation may be done manually or using a computer. A manual simulation involves some tedious steps that could take months to complete. In many cases, a manual simulation is necessary and cannot be replaced by a computer in cases in which its parameters are manually controlled. For example, a plane crash may have been caused by some mechanical failures in the aircraft engine during a flight. To study the cause of these failures, all the debris found at the site of the crash are collected and assembled to build a model of the original aircraft. The assembled model is checked vigorously for some possible faults in the wiring or failures of certain parts of the engine. The results from this simulated study contribute toward tracing the faults and failures that caused the accident.

A more effective simulation model is based on the computer. A computer-simulated model always performs faster and provides a more convenient approach by allowing a higher degree of flexibility of the model. A computer simulation is the execution of a model, represented by a computer program that gives information about the system being investigated. Simulations through a computer can be done using several software packages as simple as Microsoft Excel and a more specific tool such as Arena. There are also some general simulation packages such as Simscript and GPSS which provide a process-based approach to writing a simulation program. With this approach, the components of the program consist of entities, which combine several related events into one process. A more flexible and challenging approach is to use a general-purpose programming language such as C++ or Java. Programming languages provide greater flexibility in the development of a simulation at the expense of a longer and more difficult learning curve.

10.2 SIMULATION MODEL DEVELOPMENT

A simulation model is necessary as the first step toward implementing an idea that has been found to be too large and costly to model without simulation. A good simulation model will eventually lead to good decision making about whether to continue or abandon the project. The analysis of the results of the simulation often gives the correct number and magnitude of the parameters or variables associated with the proposed project. It is at this point that a decision is made on the implementation of the project. For example, a city plans to embark on a light transit system based on the rail network linking various points in the city, in order to overcome traffic congestion. This is a big project that will cost billions of dollars of the taxpayers' money. One main question that becomes the responsibility of the city planners is, will the huge investment really solve the traffic congestion problem? This is a difficult question to answer as reducing traffic congestion is an objective that can be successful or end in failure. One way to predict the success or failure of this project is to perform a comprehensive simulation study. It is through this simulation that a decision can be made on whether to embark on the project or not.

The development of a simulation model takes several steps. The first is the clear formulation of the problem and its objectives. It is also necessary to break down the main problem into several smaller ones if the scope is too big. In this case, it may be necessary to state the long-term objectives as well as the short-term or immediate objectives. With the problem and its objectives well laid out, a feasibility study follows to determine whether the problem can be solved with the given timeframe and within the available resources. It is here that simulation is performed, using several variables and parameters of the "what-if" scenario. In most cases, a computer is often involved to perform the simulation. Some reliable data are fed into the simulation model for testing and validation. The results are then analyzed for correctness using some statistical and numerical tools. A good simulation often ends with a steady-state model with the desired output. The final step in the simulation involves testing and validating the model for upward scalability and robustness using several different data sets.

A simulation model can be developed as a static or dynamic model. In a static model, sometimes called a Monte Carlo simulation, a system is represented in such a way that the state of the system over time is known beforehand. In many cases, a static simulation model is suitable for developing or improving an existing model based on the available data. A dynamic simulation, on the other hand, represents a system whose state changes over time. In this case, the variables or parameters are only known *on-the-fly* or as the execution is in progress. This simulation model is often associated with the real-time systems, such as in telephone circuit switching networks.

10.3 DISCRETE-EVENT SYSTEM SIMULATIONS

A *discrete event* is an event that occurs at an instance of time. For example, pushing a stopwatch for the start of a sprint event and pushing the button of an elevator are discrete events because there is an instant of time at which each occurs. Activities such as moving a crane from one place to another are not discrete events because they have time duration. However, the event becomes discrete if time is recorded at the start of the activity and continues by monitoring the crane movement at some time marks.

A simulation that is based on a number of finite events over time is called a *discrete-event* simulation. In this case, the events are variables that change according to fixed or irregular time slots. In a similar manner, events that change continuously over time can be modeled as *continuous-event* simulation models. A good example of a continuous-event simulation model is the modeling of gas emissions in a factory. In this example, gas emission is read as a quantity that changes continuously according to the activities inside the factory.

Discrete-event simulation is an analytical technique directed toward understanding the behavior of a system or process. Some common applications include modeling of manufacturing processes, transportation systems, and human service systems. One main objective of discrete-event simulation is to provide a working model of the real event. The model should be able to display information on resource utilization, efficiency, and cost. It is also necessary to optimize the procedures and resource allocation by analyzing the relationship between the components in the system. This, in turn, contributes to producing the best model for implementation based on several "what-if" analyses.

Discrete-event simulations using computers have several advantages. First, simulations provide a visualization picture of both the problem and its solution, at a fraction of the cost of an actual model. Visualization makes the problem and its solution easier to understand, especially to people with nontechnical and management backgrounds. It is through this understanding that these people can respond positively to the project before a decision can be made. Another advantage lies in the fact that computer simulations provide some flexibility, making it easy to perform changes, and a series of "what-if" scenarios by modifying the parameters or variables in the system. By changing the values of the variables, for example, the cost of the whole operation is recalculated to produce a possible working model.

One of the most exciting applications of discrete-event simulation is the process of assigning counters to customers whose arrival is random over an interval of time. This problem is known as *scheduling.* A *scheduler* is a program that maps the customers to the counters. A *customer* here refers to anything that needs to be serviced at a *counter*. From the last chapter, a task in a task graph is a customer, whereas the processor is the counter. Another example is customers who arrive at a bank waiting for their turn to be served at a counter. One other example is the telephone calls (customers) that arrive at a public switched telephone network (PSTN) exchange waiting to be assigned with a channel (counter). In these three examples, the number of counters is limited and they may not be able to cater to all the needs of the arriving customers simultaneously. Hence, only a handful of the customers will be assigned to the counters immediately upon arrival. The rest either have to wait or are blocked from being served, depending on the type of system implemented. If the system allows waiting, then a queue is formed in which the unsuccessful customers still have the chance to be served. Otherwise, in a system that doesn't allow waiting, the customers are simply turned away or blocked.

One of the most important objectives of simulation is the development of a scheduling system for a dynamic system. A dynamic system is often associated with time, and this parameter plays a critical role as a measure of performance in the simulation. The main objective in scheduling is to obtain an *optimum completion time,* which is the earliest time for the completion of the whole project. An optimum schedule has many advantages. The most obvious is the savings to the host organization in terms of time, cost, labor, and other resources.

However, in any dynamic system an optimum schedule is not easy to obtain. Many unpredictable factors affect the performance of a dynamic system, which may cause some delay to its execution. In this case, a *feasible schedule* is good enough for most systems. A feasible schedule is one in which the completion time may not be optimum but it does not cause any discomfort to the system. In this case, a slight delay in its completion is still tolerable and the system may function normally.

A dynamic system may have a *deadline* in its implementation. A well-planned scheduling system needs to be designed so that all jobs or modules in the project will be completed before or on the deadline. The deadline is said to be a *soft deadline* if a slight delay in its completion will not cause any harmful effect to the system. A soft deadline is tolerable to the system, although it may cause some discomfort. On the other hand, if a slight delay causes a disastrous effect to the system, such as a collapse or a breakdown, then this deadline is called a *hard deadline.* The period between the soft deadline and hard deadline is considered critical to the system, and the scheduler must respond quickly to rectify the problem in order to avoid the disaster.

Figure 10.1 illustrates scheduling with deadlines. A job starts at the earliest possible time at time $t = t_0$. With this starting time, the job may be completed optimal at t_m. An earlier completion time than t_m may result in a premature ending, which is not good for the system as the job may have not been executed properly. Some slight delay in the starting time may force the job to complete at a later time. Com-

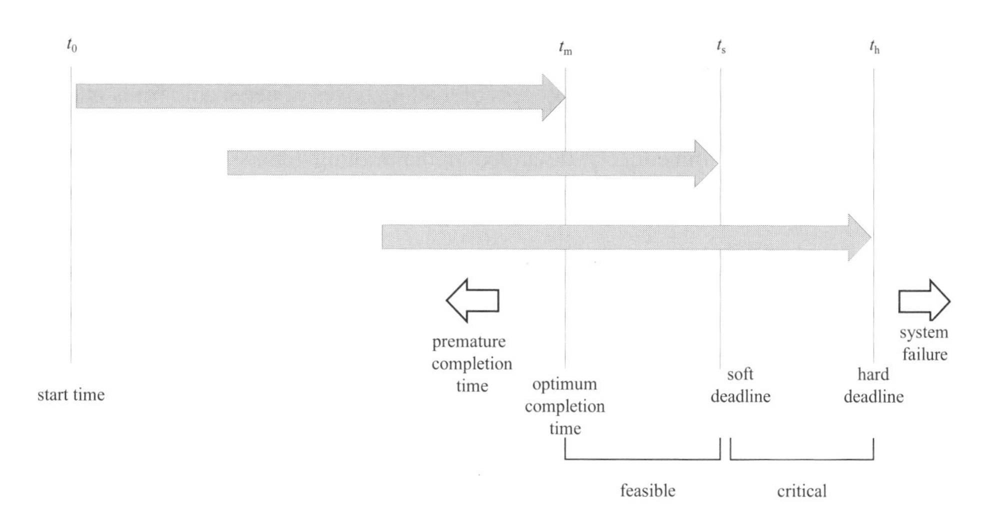

Figure 10.1 Scheduling with hard and soft deadlines.

pletion before the soft deadline t_s indicates a feasible schedule, which is tolerable to the system. A hard deadline is indicated by t_h. Completion between t_s and t_h is considered critical to the system, although the system may still function normally if that doesn't happen. A completion time above t_h is not tolerable as it may have a disastrous effect on the system.

In this chapter, we present two models of dynamic scheduling for two different types of simulation. The first is a simulation of a multicounter system with no waiting. This system does not allow waiting, and an arriving customer who fails to find a counter will be blocked and dropped from the system. The second is the simulation of a multicounter system that allows waiting. In this system, an arriving customer who cannot find a counter is given the chance to wait in a single queue. The customer will wait for his turn for assignment to a counter according to the first-come-first-served principle until it becomes available. This second system guarantees each customer service at the counter, although some of them will have to wait for quite some time. The second example is more realistic in real life than the first. It is a practice for organizations not to turn away their customers as part of their quality of service commitment. Our strategy in designing the two schedulers is to start with the first model and try to maintain the same code for the second. This implies that most of the code in the first model can be reused in the second in order to show its relevance and usefulness.

10.4 MULTICOUNTER SYSTEM WITH BLOCKING

In this section, we discuss a simple system with blocking. In this system, customers arrive at random at time t to make up a discrete event and they are assigned to a limited pool of counters on a first-come-first-serve basis. We assume a system with no waiting in which a customer who fails to find a counter at time t will be blocked. Therefore, this simple system has no queue. There is no tolerance in the time slot, and an arriving customer who fails to find a counter is immediately blocked. An example of this type of system is the process of loading boxes of perishable fruits into a number of trucks for shipment. Since the fruits cannot wait for the next few trucks to arrive, they are immediately discarded if the waiting trucks do not provide enough space for them for shipment.

Our Scheduling Model

The objective in this application is to design a *dynamic scheduler* for mapping the randomly arrived customers to the counters. A scheduler is a component of the system that maps the customers to the available counters. In doing this, the scheduler first checks the state of each counter and determines whether it is busy or available. A customer is assigned to a counter if the counter is available. Otherwise, the customer is blocked as no waiting is allowed. The progress of the services at the counter is shown in the form of Gantt charts that clearly show the start and ending time of execution for each customer.

We refer to Figure 10.2 to illustrate our scheduling model. The figure shows an instance of time at $t = t_j$ at which five customers arrive for mapping on 10 counters. The two main players in this model are the customers and the counters. Customer i is referred to as v_i and counter k as c_k. At $t = t_j$, a total of $niv = 5$ customers arrive, denoted as v_i, v_{i+1}, v_{i+2}, v_{i+3}, and v_{i+4}. The scheduler checks the state of the counters to determine whether they are available or busy. A bar in the figure represents a customer being serviced at the counter. A busy counter is indicated as having a bar crossing the time mark t_j. From the figure, counters 1, 3, 6, and 9 are available and they are immediately assigned with v_i, v_{i+1}, v_{i+2}, and v_{i+3}, respectively. Counters 2, 4, 5, 7, 8, and 10 are busy servicing customers from the previous arrivals. The last customer, v_{i+4}, is not lucky as all the counters are now occupied. Therefore, v_{i+4} is blocked.

The scheduler in this model is a C++ program that manages the assignment of customers to the counters according to the requirements of the system. The scheduler responds to the arrival of several customers generated randomly at time t. The full duties and responsibilities of the scheduler at each timeslot are outlined as follows:

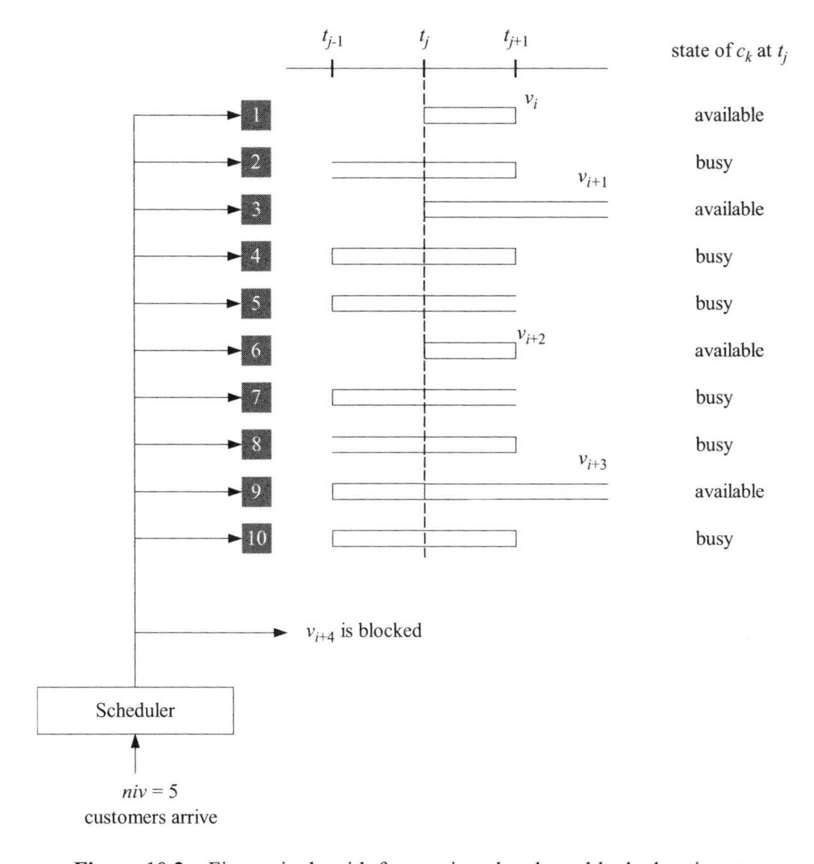

Figure 10.2 Five arrivals with four assigned and one blocked at time t_j.

1. Determine the state of the counters
2. Map the arriving customers to the counters.
3. Block the customers who fail to get counters
4. Produce Gantt charts for the assigned customers
5. Record and display the blocked customers

Code10A: Simulating Multicounter Systems

Our scheduling model is called project Code10A. Figure 10.3 shows a sample run from this project. The project simulates the assignment of 100 customers on a host of 10 counters. In the figure, the counters at which the customers are mapped are represented as shaded rectangles. The incoming customers are assumed to arrive in random numbers at the time slots, starting at $t = 0$. The output consists of Gantt charts representing the start and completion time of the customers in clear rectangles. The display also includes a list of blocked customers shown in the *Blocked Customers* area. In this application, the macros C and N represent the total number of customers (100) and counters (10), respectively.

The scheduler is the heart of the whole operation involving the mapping of the customers to the counters. The scheduler performs the discrete simulation by iteratively producing customers at discrete time t. This is done when the user presses the

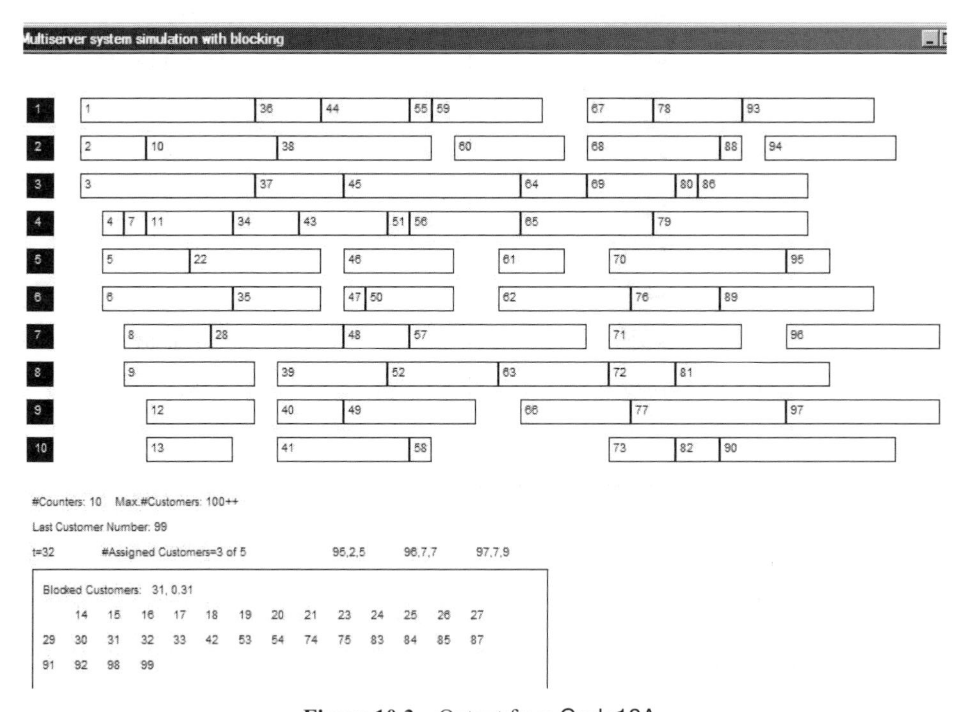

Figure 10.3 Output from Code10A.

space bar key. The initial time slot is set as $t = 0$. A press on the space bar causes t to increase by one, which means the next time slot. At each time slot, a randomly determined number of incoming customers, represented by the variable niv, arrive. The arriving customers are indexed beginning with number 1, denoted by the variable nv, in increasing order according to their time of arrival. At each arrival of customers, the scheduler checks for the available counters. Customers are assigned according to the availability of the counters. The scheduler maintains the progress of all customers by displaying all the successfully mapped customers as Gantt charts, whereas those that are blocked are listed in the *Blocked Customers* area.

Figure 10.4 shows the flowchart of our model for simulating the multicounter discrete-event assignments. In this model, an event is represented by the arrival of customers at the given timeslot t. The discrete events are controlled by time slots beginning at $t = 0$, which are activated by pressing the space bar key. We assume that a randomly generated number of customers arrive at each timeslot. In this model, the random number is represented as niv, and is assumed to have values from 0 to 4, where 0 means no customer arrival is recorded at the time slot. The arriving customers are indexed from $t = 0$, and their current number is represented by nv.

For each arriving customer v_{nv}, the scheduler checks for the state of the counters, looking for the first counter that is available for assignment. Here, the state of each counter can be either available (0) or busy (1). In this case, the scheduler checks for a counter whose state is 0, starting from the first counter. A counter found to have this value is immediately assigned with v_{nv}. The state of the counter is updated to 1. If none of the counters are available, then v_{nv} is blocked. A counter that has completed servicing a customer has its state updated to 0.

The same practice is applied to the next customer in the time slot until all the arriving customers in the timeslot have either been assigned or blocked. Another press on the space bar key produces another set of arriving customers, and the same procedure is repeated until the total number of customers reaches N.

Code10A consists of the files Code10A.h and Code10A.cpp. Table 10.1 lists the constants and global variables used in the application.

The information about the customers is stored as the structure `CUSTOMER`, as listed in Table 10.2. The information includes the length or magnitude (`len`) of the customer, actual start time (`ast`) of servicing at the counter and its completion time (`act`), the assigned counter number (`aC`), and the home screen coordinates (`Home`) of the bar in the Gantt charts.

A counter has attributes represented in the structure as `COUNTER`. Table 10.3 summarizes the list of elements of this structure. Each counter has elements including its state at time t (`sta`), the number of assigned customers (`av[0]`), the ith assigned customer (`av[i]`), the last customer number (`lv`) and its home coordinates (`Home`).

The constructor `CCode10A()` in Code10A allocates memory for the class and the structures, initializes several variables and objects, and sets the text font displaying text. The initializations include the starting time, t; customer number, `nv`; and the number of blocked customers, `nbv`. Also, each counter `c[k]`, for `k=1,2,...,C`, has no customers initially, therefore, `c[k].av[0]=0`. Each

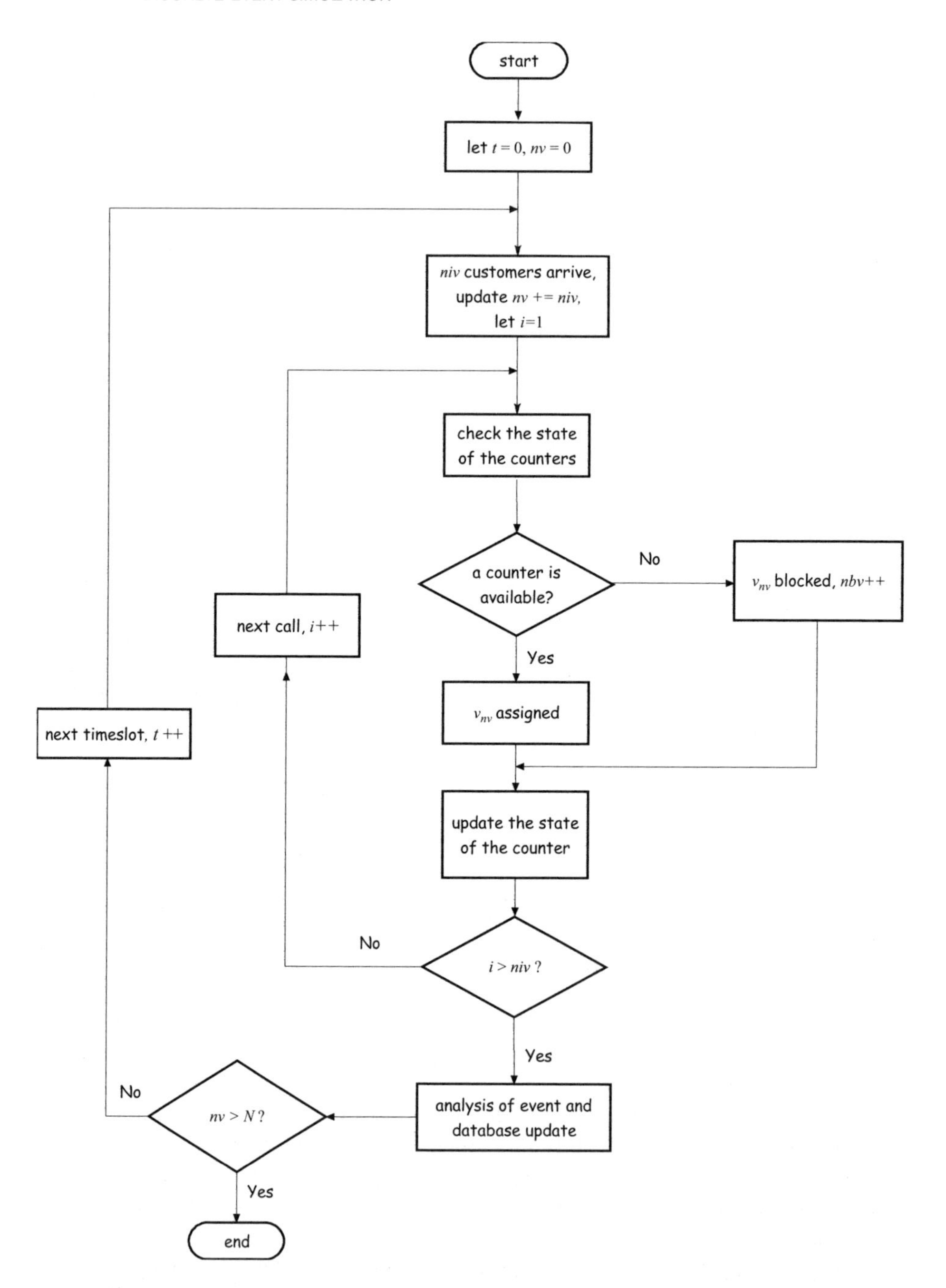

Figure 10.4 The scheduler Code10A for the multicounter system with no waiting.

Table 10.1 Constants and global variables in Code10A

Constants/variables	Description
C	The number of counters
N	The total number of customers
t	Current time slot
mLen	Maximum length (magnitude) of a customer
mniv	Maximum number of incoming customers
nv	Current customer number at time t
nbv	Total number of blocked customers at time t
wBar, hBar	Width and height of the Gantt charts bar
TopLeft, BottomRight	Top left and bottom-right coordinates of the Gantt charts area
fontTimes, lfTimes	Font object for text and its attribute

Table 10.2 Information about a customer in the structure CUSTOMER

Element	Description
v[i]	Customer i, or v_i
v[i].len	Length of v_i.
v[i].ast	Actual start time of v_i.
v[i].act	Actual completion time of v_i.
v[i].aC	The counter to which v_i is assigned
v[i].Home	Home coordinates of v_i on the screen

Table 10.3 Information about a counter in the structure COUNTER

Element	Description
c[k]	Counter k, or c_k
c[k].sta	Status of c_k at time t
c[k].av[0]	Number of assigned customers in c_k
c[k].av[i]	ith assigned customer in c_k, where $i \neq 0$
c[k].lv	Last customer in c_k
c[k].Home	Home coordinates of c_k on the screen

counter is initially idle and is ready to accept new customers. Therefore, its initial state c[k].sta is set to 0. The following code fragment performs the operation:

```
CCode10A::CCode10A()
{
	v=new CUSTOMER [N+1];
	c=new COUNTER [C+1];
	TopLeft=CPoint(0,0); BottomRight=CPoint(900,620);
	Create(NULL,"Multi-counter system simulation with blocking",
		WS_OVERLAPPEDWINDOW,CRect(TopLeft,BottomRight));
```

```
        t=0; nbv=0; nv=0;
        for (int k=1;k<=C;k++)
        {
               c[k].sta=0;
               c[k].av[0]=0;
        }
        wBar=20; hBar=20;
        ZeroMemory (&lfTimes,sizeof(lfTimes));
        lfTimes.lfHeight=80;
        fontTimes.CreatePointFontIndirect(&lfTimes);
}
```

Project Code10A has two events, namely, the initial display and the spacebar key press, as shown in Figure 10.5. The two events are detected by `WM_PAINT` and `WM_KEYDOWN`, and handled by `OnPaint()` and `OnKeyDown()`, respectively.

The function `OnPaint()` divides the window into two regions: the Gantt charts area in the upper half and the Blocked Customers area in the lower half. In the Gantt charts area, the counters are drawn as shaded rectangles. As described before, the number of counters is represented by `C` and this number can be changed by putting a new value in its `#define` statement.

The space bar key-press event is handled by the function `OnKeyDown()`. This function activates the scheduler whose operations are outlined in the flowchart in Figure 10.4. The function `OnKeyDown()` responds to the spacebar key press only, as indicated by the conditional statement:

```
if (nChar==VK_SPACE && nv<=N)
```

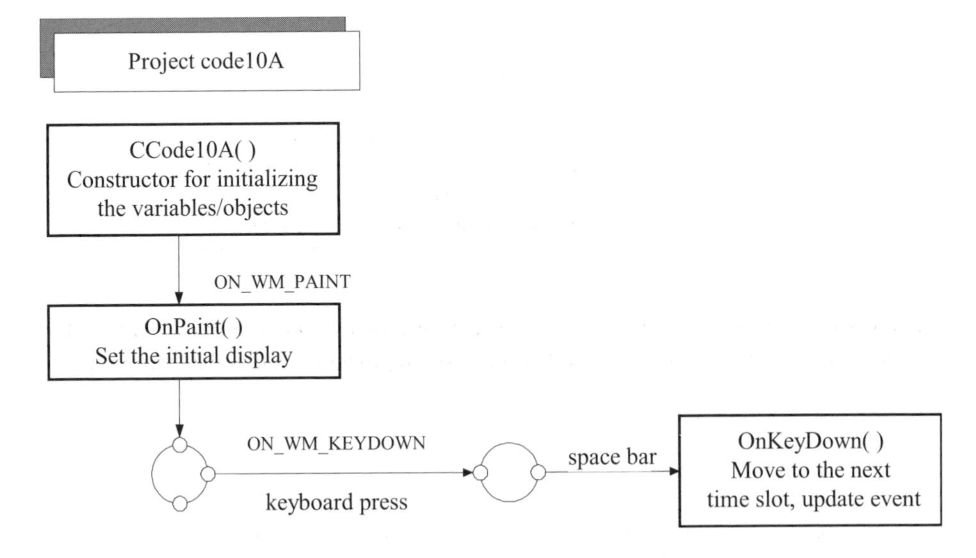

Figure 10.5 Events in Project Code10A.

Table 10.4 Some macros for the keyboard

Macro	Representation
VK_SPACE	Space bar key
VK_RETURN	Enter key
VK_F1	F1 key
VK_PGDN	Page down key
VK_PGUP	Page up key
MVK_ESC	Escape key

In the above statement, `nChar` is a local variable representing the key pressed by the user. The macro `VK_SPACE` represents the space bar key. Some commonly used macros are listed in Table 10.4.

The second condition imposed is `nv<=N` which makes sure that the number of arriving customers does not exceed the maximum number, defined to be 100 in this application. It is important to include this statement as a value `nv>N` will cause an overflow in the array `v[nv]`.

A press on the space bar key causes the time t to increase by one. At every time slot, the number of available customers, `nav`, is initialized to 0 and a random value is assigned to the number of arriving customers, `niv`. Immediately after that, the program checks for the available counters and mark them with status 0. Initially, all the counters are idle. At a later time, the counters need to undergo a test to determine if they are busy or not. The following code fragment implements this idea:

```
nav=0;
niv=rand()%mniv;
for (int k=1;k<=C;k++)
     if (c[k].av[0]>=1)
     {
          int r=c[k].lv;
          if (v[r].act<=t)
               c[k].sta=0;
     }
```

In the above code, the expression `if (c[k].av[0]>=1)` declares a counter that has more than one assigned customer to be idle if the completion time of its last customer is lower than the current time, t. This is a way of telling if the counter is busy or not at the current processing time. An idle counter is marked by `c[k].sta=0` (FALSE) and a busy one with `c[k].sta=1` (TRUE).

The next step is to assign the successful customers to the idle counters and block the unsuccessful ones. The following routine in `OnKeyDown()` implements this idea:

```
for (int i=1;i<=niv;i++)
{
      nv++;
      if (nv>N)
            return;
      for (int k=1;k<=C;k++)
      {
            v[nv].aC=0;
            if (!c[k].sta)
            {
                  aC=k; nav++;
                  v[nv].aC=aC;
                  v[nv].len=1+rand()%mLen;
                  v[nv].ast=t;
                  v[nv].act=t+v[nv].len;
                  c[aC].sta=1;
                  c[aC].lv=nv;
                  int r=++c[aC].av[0];
                  c[aC].av[r]=nv;
                  s.Format("%d,%d,%d",nv,v[nv].len,v[nv].aC);
                  dc.TextOut(240+65*i,390,s);
                  v[nv].Home=CPoint(55+t*wBar,5+30*aC);
                  rc=CRect(v[nv].Home,CSize(wBar*v[nv].len,hBar));
                  dc.Rectangle(rc);
                  s.Format("%d",nv);
                  dc.TextOut(55+t*wBar+4,5+30*aC+2,s);
                  break;
            }
      }
      if (v[nv].aC==0)
      {
            nbv++;
            int v=nbv/14, w=nbv%14;
            s.Format("%d",nv); dc.TextOut(40+30*w,440+v*20,s);
      }
      dc.SetBkColor(RGB(255,255,255));
}
```

Once the states of the counters are known, the next step is to assign the arriving customers to the idle counters. A customer is initially assigned the value of 0 to indicate it is still looking for a counter, through `v[nv].aC=0`. A test is then performed on each counter. The first found idle is immediately assigned with a customer. This is followed by the assignments of the starting time of service, which is the current time, length, and completion time. The length of service is again determined by a random number.

An update is also performed on the state of the counter. The host counter has its status changed to busy, through `c[aC].sta=1`. The number of customers in the assigned counter is increased by one with the newly assigned customer marked as its lastest customer. Finally, a bar showing the starting time, completion time, and length of the latest customer is displayed in the charts.

An unsuccessful customer is detected if its assignment value doesn't change, that is, `v[nv].aC=0`. In this case, the customer is declared as a failure and is immediately blocked. Every blocked customer is listed in the *Blocked Customers* area.

10.5 QUEUEING SYSTEMS

A discrete-event simulation model often involves a *queueing system.* A queueing system is a stochastic event that relates a customer from the time a queue is formed to its service completion. In our discussion, the word *system* used here includes both the *queue* and *service*. A queue is formed when *customers* arrive and waiting to be served through one or more *counters*. Just like in the previous application, the term "customers" here refers generally to items like people waiting to be served in a bank, packets of data for transmission in a network and customers waiting for assignment to their counters. Counters, in this context, refer to the service providers such as bank cashiers, bank teller machines, and airport immigration counters.

A queueing model is a dynamic system consisting of three main components, namely, the arrivals, service, and their queueing discipline. In the model, the states of the counters, queues, and customers depend very much on the past states. Since the arrivals and service are random processes, they are normally Markovian processes. Customers may arrive at the queue either singly or in bulk. The first case is typified by a queue at the bank teller machine, where a customer's arrival is independent of the arrivals before and after him. In the second case, a bulk arrival involves a group of customers that arrive in a queue, such as a group of people arriving at a restaurant waiting to be served.

The queue discipline is the way customers form the queue in the system. Most common is the *first-come-first-served* (FCFS) discipline wherein customers who arrive first have higher priority for being served. Another way is the *first-come-last-served* (FCLS) discipline wherein customers who arrive early will be given lower priority. FCLS discipline applies in a restaurant where a stack of trays are arranged in a vertical column.

Queueing systems are normally modeled as Markovian processes written in brief form as $a/b/C$. In this notation, a indicates the arrival pattern that forms the queue, b is the service distribution, and C is the number of counters. For example, in the $M/G/1$ system, customer arrivals form a Poisson process in which M means memoryless, with a general (G) holding time and using one counter.

M/M/1 Queueing System

A common form of queueing system is the $M/M/1$ model having only one counter to serve customers. Let $N(t)$ be the number of customers and $p_n(t)$ be the probability of having n customers in the system at time t. For a steady-state system, the expected number of customers in the system is given as

$$E[N(t)] = \sum_{n=0}^{\infty} np_n(t) \tag{10.1}$$

where $P_n = P[L(t) = n] = P_n(t)$ is the *steady-state probability* of the system. In this system, customers arrive at random and their interarrival times are assumed to form a probability distribution function. The time taken by a customer is assumed to be independent of other customers in the system. The *arrival rate* of the customers is defined as follows:

$$\text{Arrival rate, } \lambda = \frac{\text{total number of customers}}{\text{time duration}} \tag{10.2}$$

The customer arrivals can be modeled according to some established functions, such as the Poisson and Bernoulli distribution functions. The customers are serviced at the counter with mean denoted by μ. The *counter utilization rate,* sometimes called the *offered load* ρ, is given by

$$\rho = \frac{\lambda}{\mu} \tag{10.3}$$

and relates the arrivals to service at the counter. For example, in the telephone exchange application, the intensity of calls that arrive at a trunk can be measured through the counter-utilization rate. In this case, ρ may indicate the traffic intensity at the trunk. A counter is said to be stable if $\lambda < \mu$ or $\rho < 1$. The mean number of customers in the system is given by

$$N_s = \frac{\lambda}{\mu - \lambda} = \frac{\rho}{1 - \rho} \tag{10.4}$$

A useful equation called Little's equation provides a very fundamental relationship between the arrival rate of customers λ, the mean number of customers in the system N_s, and the *mean sojourn time* of customers T_s (or the *mean holding time* of customers). The equation is given as follows:

$$N_s = \lambda T_s \tag{10.5}$$

Little's equation is applied in determining, for example, the offered load of calls that arrive at a telephone exchange. In this case, the offered load at a trunk (or counter) is given simply as $a = \lambda T$.

The mean sojourn time of the customers, or the average holding time of a customer in the system, is obtained from Equations (10.4) and (10.5):

$$T_s = \frac{N_s}{\lambda} = \frac{1}{\mu - \lambda} \tag{10.6}$$

In addition, the *mean waiting time* of the customers in the queue is given by

$$w_q = T_s - \frac{1}{\mu} = \frac{\rho}{\mu - \lambda} \tag{10.7}$$

and the *mean number of customers* in the queue is given by

$$N_q = \lambda w_q = \frac{\rho^2}{1 - \rho} \tag{10.8}$$

M/M/C Queueing System

A single queue system with *c* counters is modeled as the *M*/*M*/*C* system. In this system, the arrivals are assumed to be a Poisson distribution function with mean λ. Each counter has an independent and identical exponential service–time distribution with mean given by $\mu = 1/\lambda$. Figure 10.6 shows a typical *M*/*M*/*C* queueing system having *C* counters. The first parameter is the arrival process, which is Markovian, and the second is the service process, which is also Markovian. The third parameter is the number of counters. In this illustration, an arriving customer in the queue is assigned to the first available counter.

In order for the *M*/*M*/*C* system to have a statistical equilibrium, the offered load must satisfy $\lambda/\mu < C$. The counter-utilization rate, or the traffic load ρ, is then obtained as follows:

$$\rho = \frac{\lambda}{C\mu} \tag{10.9}$$

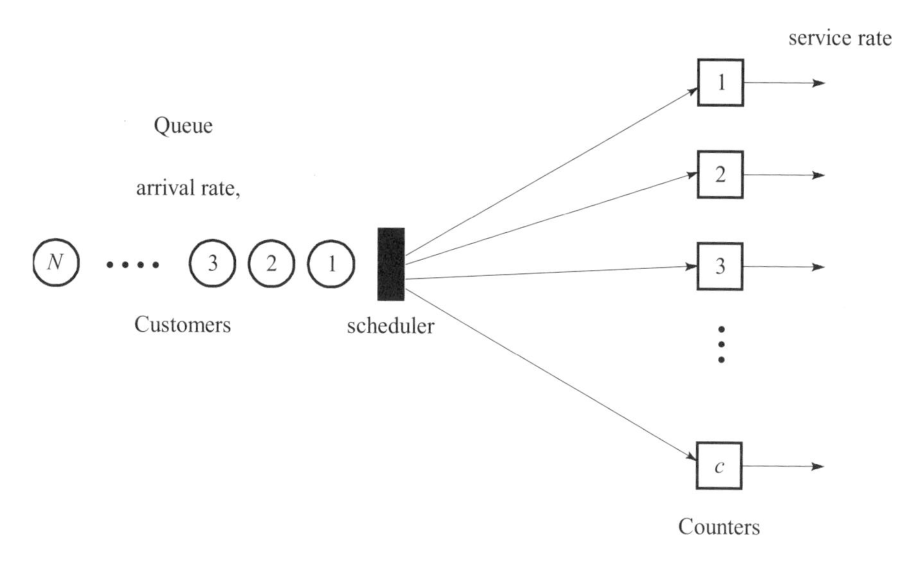

Figure 10.6 The Markovian *M*/*M*/*C* queueing system having *c* counters.

If $\lambda \geq C\mu$, then the queue is growing faster than what the counters can offer and this creates instability in the system. The system is said to have no statistical equilibrium as the arrival rate is greater than or equal to the maximum service rate of the system. The counter-utilization rate gives a good indication of the traffic pattern at the counters. At steady state, the probability of having no customers is given by

$$p_0 = 1 - \sum_{n=1}^{\infty} p_n = \left[\sum_{k=0}^{C} \frac{(C\rho)^k}{k!} + \frac{(C\rho)^C}{C!(1-\rho)} \right]^{-1} \tag{10.10}$$

The arriving customers are assigned straight to the counters if their number is less than the number of counters, or $1 \leq n \leq C$. The probability of having n customers in the system at time t in this case is given by

$$p_n = \frac{\lambda}{n\mu} p_{n-1} = \frac{(C\rho)^n}{n!} p_0 \tag{10.11}$$

Otherwise, if $n > C$, the probability is given by

$$p_n = \left(\frac{\lambda}{C\mu} \right)^{n-C} p_c = \frac{C^C \rho^n}{C!} p_0 \tag{10.12}$$

The probability that an arriving customer finds all the counters busy (so that, he or she has to queue) is given as follows:

$$p_q = \sum_{n=c}^{\infty} p_n = \frac{(C\rho)^C}{C!} \frac{p_0}{1-\rho} \tag{10.13}$$

Equation (10.13) is also known as the Erlang-C formula, which indicates that there is a waiting time in the queue before the arriving customer is assigned to a counter. The expected number of customers waiting in the queue is given by

$$N_q = \sum_{n=c}^{\infty} (n-C)p_n = p_0 \frac{(C\rho)^C}{C!} \frac{\rho}{(1-\rho)^2} = p_q \frac{\rho}{(1-\rho)} \tag{10.14}$$

The mean waiting time in the queue for a customer is

$$w_q = \frac{N_q}{\lambda} = p_q \frac{\rho}{(1-\rho)} \tag{10.15}$$

and the mean time a customer spends in the system is given by

$$T_s = w_q + \frac{1}{\mu} = p_q \frac{\rho}{\lambda(1-\rho)} + \frac{1}{\mu} \tag{10.16}$$

From Little's law in Equation (10.5), the mean number of customers in the system is

$$N_s = \lambda T_s = C\rho + \frac{(C\rho)^{c+1} p_0}{C(C!)(1-\rho)^2} = p_q \frac{\rho}{1-\rho} + C\rho \tag{10.17}$$

The Markovian *M*/*M*/*C* system has many applications in real life. Some of these examples include the arrival of customers waiting to be served by a host of counters in a bank, jobs waiting in a queue to be printed in a local area network, and telephone calls waiting to be assigned to channels in a wireless cellular network. In this section, we discuss the design of the *M*/*M*/*C* system that allows an infinite time for waiting. Unlike the previous model, this model does not block or turn away any customer. Each customer is given the chance to be served at a counter, no matter how long he or she is queueing in the system.

Code10B: Simulating the M/M/C System

The *M*/*M*/*C* system with no blocking means it allows the arriving customers an infinite time for waiting in the queue. This is illustrated in the project Code10B. The project is an extension of the earlier project Code10A and most of the basic code is retained. Figure 10.7 shows the output of Code10B. It consists of the Gantt charts

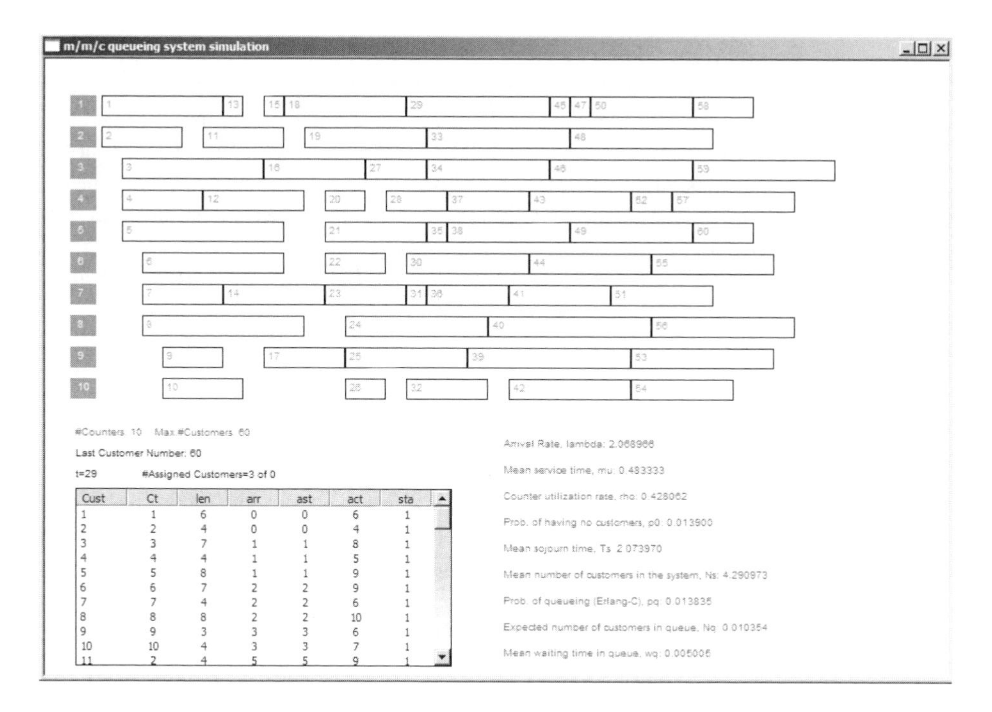

Figure 10.7 Output from Code10B.

area, the customer information in the form of a list view table, and the output analysis area. The list view table is a child window that displays the detailed information about the customers in the system. This facility is useful for viewing and validating the results from the program.

The discrete time simulation in Code10B is controlled by pressing the space bar key. At each press, the time slot t increases by one unit. A random number of customers arrive and they form a queue on a first-come-first-served basis, in which customers who arrive early will stand in front of the queue. It is assumed that the arriving customers are independent of each other, with the arrivals based on the Poisson distribution with rate λ. There are C counters for serving a total of N customers in the system. The service rate is assumed to have a probability distribution function given by the exponential function with mean $\mu = 1/\lambda$.

We discuss the design of the scheduler for the $M/M/C$ queueing system in Code10B. Figure 10.8 shows the flowchart of the scheduler. Initially, the time t and the number of customers in the system nv are set to 0. The discrete event begins when the space bar key is pressed. The key press marks the arrival of a random number of customers, represented by the variable niv. The scheduler assigns the customers with the number nv and place them in the central queue Q. The scheduler then checks the state of each counter. The first C customers in front of Q are assigned to the counters, whereas the rest are pushed C places forward in Q. When a customer v_i is assigned to counter c_k, it is removed from Q, its state is updated and the counter is marked as "busy." Once the service at the counter is completed, the counter changes its state to "available." The customer information is updated and it is now no longer in the system.

The process continues with another space bar key press, which moves the process to the next time slot. Another batch of customers arrive and they are placed behind the previous customers, who are still in Q. The same scheduling procedure is applied when the assigned customers are removed from Q and unsuccessful ones are pushed forward in the queue. The process terminates when all the customers have been assigned and complete their service at the counters.

Code10B has two files: Code10B.h and Code10B.cpp. Table 10.5 lists all the macros (constants) used throughout the program that are defined using `#define`. The constants, especially `mniv` and `mLen` are the stabilizing factors for the system, as they determine whether the system will have a statistical equilibrium or not, according to Equation (10.9). A low value of `mniv` (for example, `mniv=4`) is favorable for achieving a short queue, and this contributes to better stability. A high value, on the other hand, means a high rate of customer arrivals. This results in a long queue, which implies more counters are needed to service the incoming customers.

There are two structures used in the program, namely, `CUSTOMER` and `COUNTER`. No changes are made in `COUNTER`. The structure `CUSTOMER` stores all the information about the customers in the system, as shown in Table 10.6. Most of the elements of `CUSTOMER` are the same as in the previous model. New elements for a customer in this model are the arrival time (`art`) and the waiting time in the queue (`wait`).

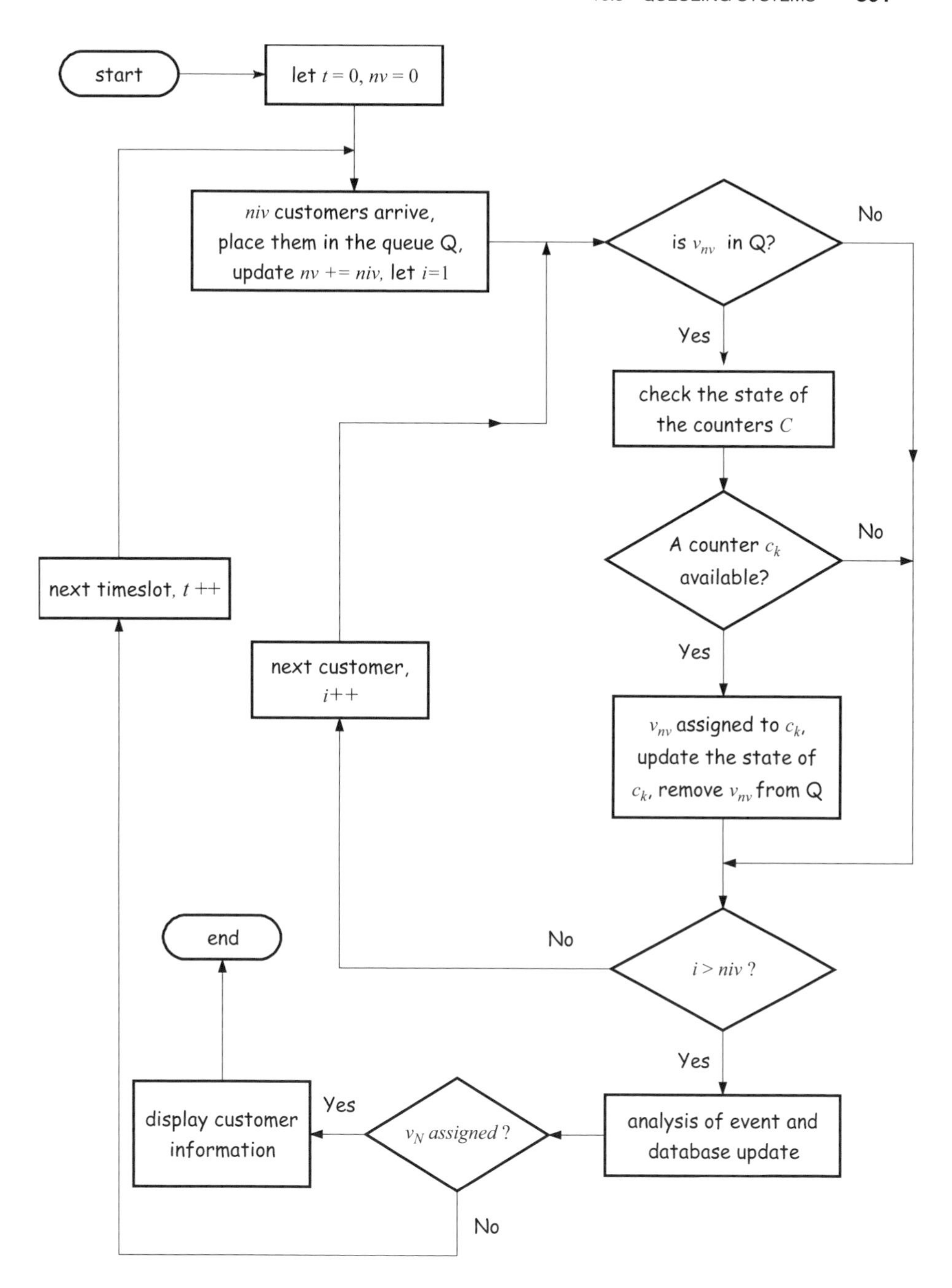

Figure 10.8 Flowchart of the scheduler in Code10B.

Table 10.5 Constants in Code10B

Macro name	Description
C	Number of counters, set as 10 in this model
N	Maximum number of customers, set as 60 in this model
mniv	Maximum number of incoming customers, set as 10 in this model
mLen	Maximum length of a customer, set as 8 in this model

Table 10.6 Elements of the structure CUSTOMER in Code10B

Element	Description
v[i]	Customer i, or v_i
v[i].Home	Home (top left) coordinates of v_i in the Gantt charts
v[i].len	Length of v_i
v[i].art	The arrival time of v_i
v[i].ast	Actual start time of v_i
v[i].act	Actual completion time of v_i
v[i].aC	The counter that v_i is assigned to
v[i].sta	Current status of v_i
v[i].wait	Waiting time of v_i in the queue before it is assigned to a counter

The functions in Code10B are listed in Table 10.7. The functions `OnPaint()` and `OnKeyDown()` are the methods that respond to the initial display and keystroke events, respectively. The functions `CustomerInfo()`, `Analysis()`, and `Factorial()` are new functions for displaying the customer information, analyzing the results, and computing the factorial of a number, respectively.

The constructor Code10B() allocates memory for the class and the arrays, besides initializing several global variables in the application. The time mark t is initially set to –1. This value will increase by one at each space bar keystroke. Therefore, with the first space bar key press t becomes 0, and this signals the start of the simulation. In addition, the constructor initializes the state of the counters to 0 to indicate they are all idle and ready to accept the assignments. The number of assigned

Table 10.7 Functions in Code10B

Function	Description
CCode10B()	The constructor
~CCode10B()	The destructor
OnPaint()	Initial display in the main window
OnKeyDown()	Method to respond to the key-press event
CustomerInfo()	Customer information tabulated as a list view window
Analysis()	Statistical analysis of the scheduling results
Factorial()	Computes the factorial of a number

customers in each counter is also initialized to 0. The code fragment for these activities is given as follows:

```
t=-1; nv=0; tnav=0;
for (int k=1;k<=C;k++)
{
      c[k].sta=0;
      c[k].av[0]=0;
}
```

The state of each customer v_i at time t is a flag indicated by `v[i].sta`. The state may change with one of the values listed in Table 10.8. Initially, the flag is set to 0, which means the customer does not exist yet, as follows:

```
for (int i=1;i<=N;i++)
        v[i].sta=0;
```

Once a customer arrives at time t, its state changes to 1, which means it is waiting in the central queue for assignment. The state changes to 2 once it is assigned and serviced at a counter. The state again changes to 3 when the service at the counter is completed. This flag value means the customer is no longer in the system and is terminated.

Information about the customers are displayed as a table using the list view window. The window is created from the the `CListCtrl` object called `vInfoView`, and it consists of seven fields. The following code segment creates the table with the field titles that describe the customers:

```
vInfoView.Create(WS_VISIBLE | WS_CHILD | WS_BORDER | LVS_REPORT
    | LVS_NOSORTHEADER,CRect(30,410,400,580), this, IDC_vINFO);
char* column[nFIELDS+1]=
     {"Cust","Ct","len","arr","ast","act","sta"};
int columnWidth[nFIELDS+1]={50,50,50,50,50,50,50};
LV_COLUMN lvColumn;
lvColumn.mask = LVCF_WIDTH | LVCF_TEXT | LVCF_FMT | LVCF_SUBITEM;
lvColumn.fmt = LVCFMT_CENTER;
lvColumn.cx = 85;
for (i=0;i<=nFIELDS;i++)
{
     lvColumn.iSubItem = 0;
     lvColumn.pszText = column[i];
     vInfoView.InsertColumn(i,&lvColumn);
     vInfoView.SetColumnWidth(i,columnWidth[i]);
}
```

The function `OnKeyDown()` responds to the key press that simulates the discrete event. At time t, the scheduler checks the state of each counter, `c[k].sta`, to determine whether it is busy or available. One way to tell this is to check the last

Table 10.8 Status of a customer, `v[i].sta`, at time t

v[i].sta	Description
0	v_i does not exist yet
1	v_i has arrived and is waiting in the queue for assignment (the customer is in Q)
2	v_i has arrived and been assigned to a counter but has not completed its execution at the counter yet
3	v_i has arrived, been assigned, and terminated

customer assigned to the counter, `v[i].lv`. If the completion time of the last customer in the counter is less than t then the counter is idle, and its status is set to 0. Otherwise, the status is 1 to show that it is busy. The following code fragment performs these tasks:

```
for (k=1;k<=C;k++)
     if (c[k].av[0]>=1)
     {
          int r=c[k].lv;
          if (v[r].act<=t)
               c[k].sta=0;
     }
```

At each timeslot t, the number of assigned customers `nav` in the timeslot is initially set to 0. The total number of customers `nv` is updated to include the new arrival of customers. Initially, the state of a customer, `v[nv].sta`, is set to 0 in the constructor, which means it does not exist yet. At time t, a random number of `niv` customers arrive and they are placed in the central queue. The number is generated using the C++ function `rand()`, and its value ranges from 0 to the maximum number allowed, `mniv` (a macro defined in the header file). The state of each of these customers changes to 1. The current time t is marked and this value is immediately assigned as the arrival time, `v[nv].art`, of the customers. The following code fragment shows this process:

```
niv=rand()%mniv; nav=0;
for (i=1;i<=niv;i++)
{
     nv++;
     if (nv<=N)
     {
          v[nv].art=t;
          v[nv].sta=1;
     }
}
```

Once the initializations have been performed at time t, the next step is to assign the customers to the available counters. The scheduler obtains the queueing list Q to check the current state of all the customers. A customer is in Q if its flag status is 1, and this is a prerequisite before it can be considered for assignment. This strategy eliminates others that have been assigned before. Customers who arrive early are placed in front of the queue for a higher priority of assignment.

Initially, each customer in Q is assigned to counter 0 to denote it is not assigned to any counter yet, or `v[i].aC=0`. The scheduler's next task is to find an idle counter for the assignment. This is achieved by checking the Boolean value of each counter, `c[k].sta`. A FALSE (0) value means the counter is idle, whereas TRUE (1) means it is idle. The assignments start with the customer standing in front of Q being assigned to the first available counter, followed by the next-highest customer in Q, and so on until all the counters are occupied.

With each assignment, other information about the customer is updated. This includes the customer's length (`len`) which is randomly determined, the starting time (`ast`) of service and its completion time (`act`), and the waiting time in the queue (`wait`). With this assignment, the customer is removed from the queue by setting the flag `v[i].sta=2`. The following code fragment performs these tasks:

```
aC=k; nav++; tnav++;
v[i].aC=aC;
v[i].len=1+rand()%mLen;
v[i].ast=t;
v[i].act=t+v[i].len;
v[i].sta=2;
v[i].wait=v[i].ast-v[i].art;
```

The customer assignment is shown as a bar in the Gantt charts as follows:

```
v[i].Home=CPoint(55+t*wBar,5+30*aC);
rc=CRect(v[i].Home,CSize(wBar*v[i].len,hBar));
dc.Rectangle(rc);
```

Each assignment also requires some update to the state of the counter. This includes setting the counter status to "busy," adding the assigned customer as the latest customer in the counter, and increasing the number of assigned customers in the counter by one, as follows:

```
c[aC].sta=1;
c[aC].lv=i;
r=++c[aC].av[0];
c[aC].av[r]=i;
```

The scheduler also checks the state of the customer undergoing service at the counter. If the customer has completed its service at the counter, its state changes to "terminated," as follows:

```
if (t>=v[i].act && v[i].sta==2)
     v[i].sta=3;
```

The event at time completes with an update to the customer information table in the list view window. This is performed in the function `CustomerInfo()`, as follows:

```
void CCode10B:: CustomerInfo()
{
     CString s;
     vInfoView.DeleteAllItems();
     LV_ITEM lvItem;
     lvItem.mask = LVIF_TEXT | LVIF_STATE;
     lvItem.state = 0;
     lvItem.stateMask = 0;
     for (int i=0;i<=tnav;i++)
     {
          lvItem.iItem=i;
          lvItem.iSubItem=0;
          lvItem.pszText="";
          vInfoView.InsertItem(&lvItem);
          if (i+1<=tnav && i+1<=N)
          {
               s.Format("%d",i+1);
               vInfoView.SetItemText(i,0,s);
               s.Format("%d",v[i+1].aC);
               vInfoView.SetItemText(i,1,s);
               s.Format("%d",v[i+1].len);
               vInfoView.SetItemText(i,2,s);
               s.Format("%d",v[i+1].art);
               vInfoView.SetItemText(i,3,s);
               s.Format("%d",v[i+1].ast);
               vInfoView.SetItemText(i,4,s);
               s.Format("%d",v[i+1].act);
               vInfoView.SetItemText(i,5,s);
               s.Format("%d",v[i+1].sta);
               vInfoView.SetItemText(i,6,s);
          }
     }
}
```

The last customer in the system is `v[N]`. Once this customer has been assigned to a counter, an analysis is performed on the data gathered from the discrete events. The function `Analysis()` performs this task. The following items are included in the analysis:

1. The arrival rate of customers (using Equation 10.2), `lambda`
2. The mean service time of the customers, `mu`
3. The counter utilization rate (Equation 10.9), `rho`
4. The probability of having no customers in the system (Equation 10.10), `p0`
5. The mean sojourn time (Equation 10.16), `Ts`
6. The mean number of customers in the system (Equation 10.17), `Ns`
7. The probability of queueing (Erlang's C-formula, Equation 10.13), `pq`
8. The expected number of customers in the queue (Equation 10.14), `Nq`
9. The mean waiting time in the queue (Equation 10.15), `wq`

In computing the probability of having no customers in the system, using Equation (10.10), a recursive function for computing the factorial of a number is required and presented as follows:

```
double CCode10B::factorial(double k)
{
     if (k==0)
          return 1;
     return k*factorial(k-1);
}
```

Data analysis is performed wholly inside the function `Analysis()`. All the variables are declared locally inside the function. The function is written as follows:

```
void CCode10B::Analysis()
{
     CClientDC dc(this);
     CString s;
     int k;
     double lambda, mu, rho, temp, tC;
     double p0;               // prob. of having no customers
     double pq,Nq,wq,Ts,Ns;
     tC=(double)C;
     lambda=(double)N/t;
     mu=1/lambda;
     rho=lambda/(tC*mu);
     temp=(pow(tC*rho,C))/(factorial(tC)*(1-rho));
     for (k=1;k<=C;k++)
          temp += pow(tC*rho,k)/factorial(k);
```

```
    p0=1/temp;
    pq=pow(tC*rho,C)*p0/(factorial(tC)*(1-rho));
    Nq=pq*rho/(1-rho);
    wq=Nq/lambda;
    Ts=wq+1/mu;
    Ns=lambda*Ts;
    dc.SelectObject(fontTimes);
    dc.SetBkColor(RGB(255,255,255));
    dc.SetTextColor(RGB(100,100,100));
    s.Format("Arrival Rate, lambda: %lf",lambda);
    dc.TextOut(450,360,s);
    s.Format("Mean service time, mu: %lf",mu);
    dc.TextOut(450,385,s);
    s.Format("Counter utilization rate, rho: %lf",rho);
    dc.TextOut(450,410,s);

    s.Format("Prob. of having no customers, p0: %lf",p0);
    dc.TextOut(450,435,s);
    s.Format("Mean sojourn time, Ts: %lf",Ts);
    dc.TextOut(450,460,s);
    s.Format("Mean number of customers in the system, Ns: %lf",Ns);
    dc.TextOut(450,485,s);
    s.Format("Prob. of queueing (Erlang-C), pq: %lf",pq);
    dc.TextOut(450,510,s);
    s.Format("Expected number of customers in queue, Nq: %lf",Nq);
    dc.TextOut(450,535,s);
    s.Format("Mean waiting time in queue, wq: %lf",wq);
    dc.TextOut(450,560,s);
}
```

10.6 SUMMARY AND CONCLUSION

Computer simulation is an important step in evaluating the performance of a discrete event. Two models of simulations are discussed in this chapter. We model a discrete event based on time ticks—at every instance of time, a number of customers arrive. In the first model, customers are not allowed to wait in the queue. An arriving customer at time t is either assigned to a counter immediately or blocked, depending on whether a counter is available or not. There is no buffering in this system as no waiting is allowed. As a result, this model produces stable results in a system with many counters and low arrival rate.

Due to its high dependence on the number of counters, the first model is not very useful for implementation in real life. However, its mechanism contributes to the design of a more practical system involving waiting in a queue. The second model presented in this chapter makes use of the idea of the first system by retaining most of the code and adding new ones for supporting a queue. The second model is the $M/M/C$ queueing system that allows infinite waiting and no blocking. In this system, no customer is turned away as he will be staying in the central queue until a counter becomes available. This provides some buffering in the form of a queue that tends to stabilize the system. An arriving customer at time t is either assigned directly to a counter (if the counter is available), or placed in a central queue. The assignment of customers in this system is based on the first-come-first-served priority list. The performance of this system is measured using some statistical tools, such as the mean waiting time in the queue, mean sojourn time, mean number of customers in the system, and the counter utilization rate.

A queueing system itself has many forms. The simulation models presented here can be extended to include several other applications in real life. Most relevant to the second model, for example, is the $M/M/C$ system that allows finite waiting time. In this system, a customer is allowed to wait in a queue up to a preset time period. If a counter is still not available, then the customer is blocked and terminated from the system. One such application is the telephone calls that arrive at a local switching center, whereby each call has a waiting time limit. The blocking probability in this case is measured using the Erlang's B-formula. A lot of ideas from the second model can easily be extended to develop this system.

BIBLIOGRAPHY

1. L. Kleinrock, *Queueing Systems Theory, Volume 1,* Wiley-Interscience, 1975.
2. J. Banks, J. S. Carson, B. L. Nelson, and D. M. Nicol, *Discrete-Event System Simulation,* Prentice-Hall, 1991.
3. F. Cottet, J. Delacroix, C. Kaiser, and Z. Mammeri, *Scheduling in Real-Time Systems,* Wiley, 2002.
4. J. Medhi and J. Medhi, *Stochastic Models in Queueing Theory,* Academic Press, 2002.
5. L. Kleinrock, *Queueing Systems Theory Volume 1,* Wiley-Interscience, 1975.
6. J. Banks (Ed.), *Handbook of Simulation: Simulations, Principles, Advances, Application and Practice,* Wiley-Interscience, 1998.

CODE LISTINGS

Code10A: Discrete-Event Simulation with Blocking

```
// code10A.h
#include <afxwin.h>
#define C 10                        // number of counters
#define N 100                       // total number of customers
#define mLen 8                      // max length of a customer
#define mniv 7                      // max #incoming customers

class CCode10A : public CFrameWnd
{
private:
       LOGFONT lfTimes; CFont fontTimes;
```

```
        CPoint TopLeft, BottomRight;
        int wBar, hBar, t;           // width,height,time
        int nv, nbv;                 // customer number, #blocked customers
        typedef struct
        {
              CPoint Home;
              bool sta;                 // status
              int av[N+1],lv;        // ith customer, last customer
        } COUNTER;
        COUNTER *c;
        typedef struct
        {
              CPoint Home;           // coordinates of customer
              int len;                  // length
              int ast,act;           // actual start,completion time
              int aC;                   // assigned counter no.
        } CUSTOMER;
        CUSTOMER *v;
public:
        CCode10A();
        ~CCode10A();
        afx_msg void OnPaint();
        afx_msg void OnKeyDown (UINT nChar, UINT nRep, UINT nFlags);
        DECLARE_MESSAGE_MAP();
};

class CMyWinApp : public CWinApp
{
public:
  virtual BOOL InitInstance();
};

// code10A.cpp
#include "code10A.h"

CMyWinApp  MyApplication;

BOOL CMyWinApp::InitInstance()
{
        CCode10A* pFrame = new CCode10A;
        m_pMainWnd = pFrame;
        pFrame->ShowWindow(SW_SHOW);
        pFrame->UpdateWindow();
        return TRUE;
}

BEGIN_MESSAGE_MAP(CCode10A, CFrameWnd)
        ON_WM_PAINT()
        ON_WM_KEYDOWN()
END_MESSAGE_MAP()

CCode10A::CCode10A()
{
        v=new CUSTOMER [N+1];
        c=new COUNTER [C+1];
```

```
	TopLeft=CPoint(0,0); BottomRight=CPoint(900,620);
	Create(NULL,"Multiserver system simulation with blocking",
		WS_OVERLAPPEDWINDOW,CRect(TopLeft,BottomRight));
	t=0; nbv=0; nv=0;
	for (int k=1;k<=C;k++)
	{
		c[k].sta=0;
		c[k].av[0]=0;
	}
	wBar=20; hBar=20;
	ZeroMemory (&lfTimes,sizeof(lfTimes));
	lfTimes.lfHeight=80;
	fontTimes.CreatePointFontIndirect(&lfTimes);
}

CCode10A::~CCode10A()
{
	delete c,v;
}

void CCode10A::OnPaint()
{
	CPaintDC dc(this);
	CString s;
	CRect rc;

	dc.SelectObject(fontTimes);
	dc.SetTextColor(RGB(255,255,255));
	for (int k=1;k<=C;k++)
	{
		c[k].Home.x=25; c[k].Home.y=35+(k-1)*30;
		rc= CRect(c[k].Home,CSize(wBar+5,hBar));
		dc.FillSolidRect(&rc,RGB(150,150,150));
		s.Format("%d",k);
		dc.TextOut(c[k].Home.x+7,7+30*k,s);
	}
	dc.SetBkColor(RGB(255,255,255));
	dc.SetTextColor(RGB(100,100,100));
	s.Format("#Counters: %d    Max.#Customers: %d++",C,N);
	dc.TextOut(30,350,s);
	rc=CRect(30,410,500,560);
	dc.Rectangle(rc);
	s.Format("Blocked Customers:");
	dc.TextOut(40,420,s);
}

void CCode10A::OnKeyDown(UINT nChar, UINT nRep, UINT nFlags)
{
	CClientDC dc(this);
	CString s;
	CRect rc;
	CBrush* pWhite=new CBrush(RGB(255,255,255));
	int aC,niv,nav; // assigned,no.of incoming,no.of available customers
```

```
dc.SelectObject(fontTimes);
dc.SetTextColor(RGB(0,0,0));

time_t seed=time(NULL);
srand((unsigned)seed);
if (nChar==VK_SPACE && nv<=N)
{
        dc.FillRect(CRect(30,390,750,410),pWhite);
        s.Format("t=%d",t++);
        dc.TextOut(30,390,s);
        nav=0;
        niv=rand()%mniv;
        for (int k=1;k<=C;k++)
                if (c[k].av[0]>=1)
                {
                        int r=c[k].lv;
                        if (v[r].act<=t)
                                c[k].sta=0;
                }

        dc.SetTextColor(RGB(150,150,150));
        for (int i=1;i<=niv;i++)
        {
                nv++;
                if (nv>N)
                        return;
                for (int k=1;k<=C;k++)
                {
                        v[nv].aC=0;
                        if (!c[k].sta)
                        {
                                aC=k; nav++;
                                v[nv].aC=aC;
                                v[nv].len=1+rand()%mLen;
                                v[nv].ast=t;
                                v[nv].act=t+v[nv].len;
                                c[aC].sta=1;
                                c[aC].lv=nv;
                                int r=++c[aC].av[0];
                                c[aC].av[r]=nv;
                                s.Format("%d,%d,%d",nv,
                                        v[nv].len,v[nv].aC);
                                dc.TextOut(240+65*i,390,s);
                                v[nv].Home=CPoint(55+t*wBar,5+30*aC);
                                rc=CRect(v[nv].Home,
                                        CSize(wBar*v[nv].len,hBar));
                                dc.Rectangle(rc);
                                s.Format("%d",nv);
                                dc.TextOut(55+t*wBar+4,5+30*aC+2,s);
                                break;
                        }
                }
                if (v[nv].aC==0)
                {
```

```
                        nbv++;
                        int v=nbv/14, w=nbv%14;
                        s.Format("%d",nv);
                        dc.TextOut(40+30*w,440+v*20,s);
                    }
                    dc.SetBkColor(RGB(255,255,255));
                }
                s.Format("#Assigned Customers=%d of %d",nav,niv);
                dc.TextOut(95,390,s);
                s.Format("Last Customer Number: %d",nv);
                dc.TextOut(30,370,s);
                s.Format("%d, %.2lf",nbv,(double)nbv/nv);
                dc.TextOut(140,420,s);
        }
}
```

Code10B: M/M/C Queueing without Blocking

```
// code10B.h
#include <afxwin.h>
#include <afxcmn.h>
#include <math.h>
#define IDC_vINFO 501
#define nFIELDS 6
#define C 10                    // number of counters
#define N 60                    // total number of customers
#define mniv 5                  // max. #incoming customers per slot
#define mLen 8                  // max. length of customer

class CCode10B : public CFrameWnd
{
private:
        LOGFONT lfTimes; CFont fontTimes;
        CPoint TopLeft, BottomRight;
        CListCtrl vInfoView;
        int wBar, hBar, t;      // width,height,time
        int nv,tnav;            // customer number,tot #assigned customers
        typedef struct
        {
              CPoint Home;
              bool sta;         // status
              int av[N+50],lv;// ith, last customer
        } COUNTER;
        COUNTER *c;
        typedef struct
        {
              CPoint Home;      // coordinates of customer
              int len;          // length
              int art;          // arrival time
              int ast,act;      // actual start,completion time
              int aC;                       // assigned counter no.
              int sta;          // status, 0=not arr., 1=arr.assg.,
                                   2=arr.not assg., 99=dead
              int wait;           // waiting time in the queue
```

```
        } CUSTOMER;
        CUSTOMER *v;
public:
        CCode10B();
        ~CCode10B();
        afx_msg void OnPaint();
        afx_msg void OnKeyDown (UINT nChar, UINT nRep, UINT nFlags);
        DECLARE_MESSAGE_MAP();
        void CustomerInfo(),Analysis();
        double factorial(double);
};

class CMyWinApp : public CWinApp
{
public:
  virtual BOOL InitInstance();
};


// code10B.cpp: M/M/C queueing simulation with no blocking
#include "code10B.h"

CMyWinApp  MyApplication;

BOOL CMyWinApp::InitInstance()
{
        CCode10B* pFrame = new CCode10B;
        m_pMainWnd = pFrame;
        pFrame->ShowWindow(SW_SHOW);
        pFrame->UpdateWindow();
        return TRUE;
}

BEGIN_MESSAGE_MAP(CCode10B, CFrameWnd)
        ON_WM_PAINT()
        ON_WM_KEYDOWN()
END_MESSAGE_MAP()

CCode10B::CCode10B()
{
        v=new CUSTOMER [N+1];
        c=new COUNTER [C+1];
        TopLeft=CPoint(0,0);
        BottomRight=CPoint(900,620);
        t=-1; nv=0; tnav=0;
        for (int k=1;k<=C;k++)
        {
               c[k].sta=0;
               c[k].av[0]=0;
        }
        for (int i=1;i<=N;i++)
               v[i].sta=0;
        wBar=20; hBar=20;
        ZeroMemory (&lfTimes,sizeof(lfTimes));
```

```
        lfTimes.lfHeight=80;
        fontTimes.CreatePointFontIndirect(&lfTimes);

        Create(NULL,"m/m/c queueing system simulation",
               WS_OVERLAPPEDWINDOW,CRect(TopLeft,BottomRight));
        vInfoView.Create(WS_VISIBLE | WS_CHILD | WS_BORDER | LVS_REPORT
               | LVS_NOSORTHEADER,CRect(30,410,400,580), this, IDC_vINFO);
        char* column[nFIELDS+1]
               ={"Cust","Ct","len","arr","ast","act","sta"};
        int columnWidth[nFIELDS+1]={50,50,50,50,50,50,50};
        LV_COLUMN lvColumn;
        lvColumn.mask = LVCF_WIDTH | LVCF_TEXT | LVCF_FMT | LVCF_SUBITEM;
        lvColumn.fmt = LVCFMT_CENTER;
        lvColumn.cx = 85;
        for (i=0;i<=nFIELDS;i++)
        {
               lvColumn.iSubItem = 0;
               lvColumn.pszText = column[i];
               vInfoView.InsertColumn(i,&lvColumn);
               vInfoView.SetColumnWidth(i,columnWidth[i]);
        }
}

CCode10B::~CCode10B()
{
        delete c,v;
}

void CCode10B::OnPaint()
{
        CPaintDC dc(this);
        CString s;
        CRect rc;

        dc.SelectObject(fontTimes);
        dc.SetTextColor(RGB(255,255,255));
        for (int i=1;i<=C;i++)
        {
               c[i].Home.x=25; c[i].Home.y=35+(i-1)*30;
               rc= CRect(c[i].Home,CSize(wBar+5,hBar));
               dc.FillSolidRect(&rc,RGB(150,150,150));
               s.Format("%d",i);
               dc.TextOut(c[i].Home.x+7,7+30*i,s);
        }
        dc.SetBkColor(RGB(255,255,255));
        dc.SetTextColor(RGB(100,100,100));
        s.Format("#Counters: %d    Max.#Customers: %d",C,N);
        dc.TextOut(30,350,s);
}

void CCode10B::OnKeyDown(UINT nChar,UINT nRep,UINT nFlags)
{
        CClientDC dc(this);
        CString s;
```

```
CRect rc;
CBrush* pWhite=new CBrush(RGB(255,255,255));
int aC, niv, nav;  // assigned, #incoming customers, #available customers
int i,k,r;

time_t seed=time(NULL);
srand((unsigned)seed);
if (nChar==VK_SPACE && v[N].sta!=3)
{
      dc.FillRect(CRect(30,390,750,410),pWhite);
      dc.SelectObject(fontTimes);
      dc.SetBkColor(RGB(255,255,255));
      dc.SetTextColor(RGB(0,0,0));
      s.Format("t=%d",++t); dc.TextOut(30,390,s);
      for (k=1;k<=C;k++)
            if (c[k].av[0]>=1)
            {
                  int r=c[k].lv;
                  if (v[r].act<=t)
                        c[k].sta=0;
            }

      niv=rand()%mniv; nav=0;
      dc.SetTextColor(RGB(150,150,150));

      for (i=1;i<=niv;i++)
      {
            nv++;
            if (nv<=N)
            {
                  v[nv].art=t;
                  v[nv].sta=1;
            }
      }

      for (i=1;i<=nv;i++)
      {
            if (v[i].sta==1)
            {
                  v[i].aC=0;
                  for (k=1;k<=C;k++)
                        if (!c[k].sta)
                        {
                              aC=k; nav++; tnav++;
                              v[i].aC=aC;
                              v[i].len=1+rand()%mLen;
                              v[i].ast=t;
                              v[i].act=t+v[i].len;
                              v[i].sta=2;
                              v[i].wait=v[i].ast-v[i].art;
                              v[i].Home
                                    =CPoint(55+t*wBar,5+30*aC);
                              c[aC].sta=1;
                              c[aC].lv=i;
```

```
                                    r=++c[aC].av[0];
                                    c[aC].av[r]=i;
                                    rc=CRect(v[i].Home,
                                          CSize(wBar*v[i].len,hBar));
                                    dc.Rectangle(rc);
                                    s.Format("%d",i);
                                    dc.TextOut(55+t*wBar+4,
                                          5+30*aC+2,s);
                                    break;
                              }
                  }
                  if (t>=v[i].act && v[i].sta==2)
                        v[i].sta=3;
            }
            dc.SetTextColor(RGB(0,0,0));
            s.Format("#Assigned Customers=%d of %d",nav,niv);
            dc.TextOut(95,390,s);
            s.Format("Last Customer Number: %d",tnav);
            dc.TextOut(30,370,s);
            CustomerInfo();
      }
      if (v[N].sta==2)
      {
            v[N].sta=3;
            Analysis();
            return;
      }
}

void CCode10B:: CustomerInfo()
{
      CString s;
      vInfoView.DeleteAllItems();
      LV_ITEM lvItem;
      lvItem.mask = LVIF_TEXT | LVIF_STATE;
      lvItem.state = 0;
      lvItem.stateMask = 0;
      for (int i=0;i<=tnav;i++)
      {
            lvItem.iItem=i;
            lvItem.iSubItem=0;
            lvItem.pszText="";
            vInfoView.InsertItem(&lvItem);
            if (i+1<=tnav && i+1<=N)
            {
                  s.Format("%d",i+1);
                  vInfoView.SetItemText(i,0,s);
                  s.Format("%d",v[i+1].aC);
                  vInfoView.SetItemText(i,1,s);
                  s.Format("%d",v[i+1].len);
                  vInfoView.SetItemText(i,2,s);
                  s.Format("%d",v[i+1].art);
                  vInfoView.SetItemText(i,3,s);
                  s.Format("%d",v[i+1].ast);
```

```
                        vInfoView.SetItemText(i,4,s);
                        s.Format("%d",v[i+1].act);
                        vInfoView.SetItemText(i,5,s);
                        s.Format("%d",v[i+1].sta);
                        vInfoView.SetItemText(i,6,s);
                }
        }
}

void CCode10B::Analysis()
{
        CClientDC dc(this);
        CString s;
        int k;
        double lambda, mu, rho, temp, tC;
        double p0;                    // prob. of having no customers
        double pq,Nq,wq,Ts,Ns;
        tC=(double)C;
        lambda=(double)N/t;
        mu=1/lambda;
        rho=lambda/(tC*mu);
        temp=(pow(tC*rho,C))/(factorial(tC)*(1-rho));
        for (k=1;k<=C;k++)
                temp += pow(tC*rho,k)/factorial(k);
        p0=1/temp;
        pq=pow(tC*rho,C)*p0/(factorial(tC)*(1-rho));
        Nq=pq*rho/(1-rho);
        wq=Nq/lambda;
        Ts=wq+1/mu;
        Ns=lambda*Ts;
        dc.SelectObject(fontTimes);
        dc.SetBkColor(RGB(255,255,255));
        dc.SetTextColor(RGB(100,100,100));
        s.Format("Arrival Rate, lambda: %lf",lambda);
        dc.TextOut(450,360,s);
        s.Format("Mean service time, mu: %lf",mu);
        dc.TextOut(450,385,s);
        s.Format("Counter utilization rate, rho: %lf",rho);
        dc.TextOut(450,410,s);

        s.Format("Prob. of having no customers, p0: %lf",p0);
        dc.TextOut(450,435,s);
        s.Format("Mean sojourn time, Ts: %lf",Ts);
        dc.TextOut(450,460,s);
        s.Format("Mean number of customers in the system, Ns: %lf",Ns);
        dc.TextOut(450,485,s);
        s.Format("Prob. of queueing (Erlang-C), pq: %lf",pq);
        dc.TextOut(450,510,s);
        s.Format("Expected number of customers in queue, Nq: %lf",Nq);
        dc.TextOut(450,535,s);
        s.Format("Mean waiting time in queue, wq: %lf",wq);
        dc.TextOut(450,560,s);
}
```

```
double CCode10B::factorial(double k)
{
      if (k==0)
            return 1;
      return k*factorial(k-1);
}
```

CHAPTER 11

MODELING WIRELESS NETWORKS

11.1 WIRELESS CELLULAR NETWORKS

Communication using radio waves was invented by an Italian physicist, Guglielmo Marconi, in 1895. Since then, rapid changes in the electronics industry have seen tremendous improvement in the way people communicate. The standard form of telephone communication is still the wired PSTN (public-switched telephone network). With the integration of the telecommunications industry with computers, several new forms of communications involving wired and wireless media have evolved. An emerging standard is the wireless digital cellular telephone network, which is a component of the *personal communications services* (PCS) system. PCS is a digital system that uses the 1900 MHz radio frequency spectrum band.

To help understand wireless cellular network communication, a few basic terminologies are discussed. Wireless communication is made possible through the transmission of energy waves in the form of electromagnetic radiation in the air, referred to as the *electromagnetic spectrum.* The length between the peaks of an energy wave is called its *wavelength.* The number of cycles per second of the energy wave is called its *frequency.* Frequency is measured in hertz (Hz), which refers to the number of cycles per second of the oscillating wave. Frequency is inversely proportional to wavelength—a high wavelength means low frequency, and vice versa. A wave with low frequency may have a value as big as 10^{23} Hz, whereas a high-frequency wave can be as short as 10^2 Hz.

Electromagnetic spectrum is a broad term for energy waves including visible light, radio frequencies, microwave, infrared, ultraviolet, X-rays and gamma rays. The frequency of visible light is around 10^{15} Hz, whereas the gamma ray has the highest frequency at around 10^{21} Hz. From this classification, *radio frequency* (RF) occupies the lowest class and it is often used to transmit all kinds of data including voice, digital data, and television signals. RF has the longest wavelengths and the lowest frequencies, ranging from 10^2 Hz to 10^{10}, and is, therefore, suitable for wireless communication. Cellular networks uses RF to transmit and receive data in all their operations. For transmitting data, the information is piggybacked to the air in the form of waves using a device called a *modulator*. When the data reaches its destination, the waves are converted back to their original form using a device called a *demodulator*.

A cellular telephone network consists of several hexagonal areas called *cells*. Figure 11.1 shows two cellular networks, denoted by H_1 and H_2, with the cells numbered starting at 1. The subscript n in the symbol H_n is the radius of the network, or the layer of cells covering the network center at cell 1. It can be seen from the figure that H_1 is made up of seven cells, whereas H_2 has 19 cells. Each cell has a *base station,* a tall tower that has antennas, amplifiers, transmitters, and receivers for serving the mobile users in the cell. The facilities at the base stations also include the modulator and demodulator, which convert data streams into RF and vice versa. A base station has the capability to transmit and receive several groups of data at any instance of time using several different nonconflicting frequencies. Each stream of data occupies a *channel* or allocated frequency. The relationship is explained as follows: channel k has a frequency given as f_k. Therefore, a base station transmits and receives data using its allocated resources in the form of channels. The number of channels allocated to a base station varies from one cell to another, depending on the strength of its antenna. Normally, stronger antennae are installed on base stations in areas where the demand is high, like in a city.

The base stations in the cells are connected by high-speed cables to the mobile switching center (MSC), which serves as the headquarters for several cells in the network. The MSC is responsible for managing the calls in their respective cells by identifying the callers/receivers, allocating bandwidths, and recording these events in its database. To achieve these objectives, the mobile switching centers are linked to the central office for access and share infrastructures such as satellites, PSTN, the Internet, and other facilities.

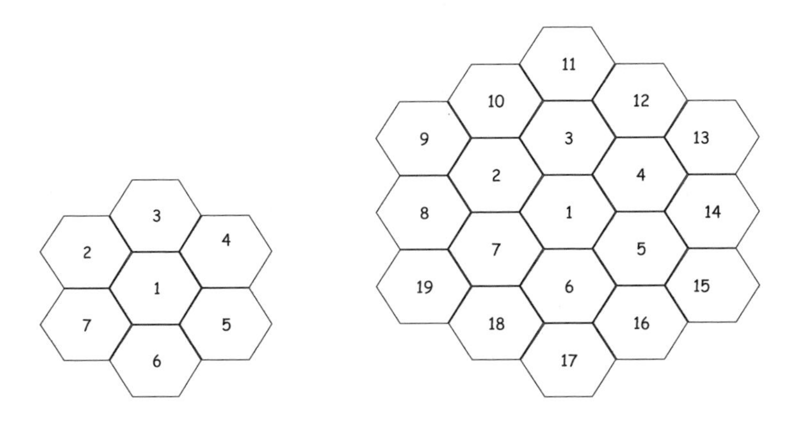

Figure 11.1 H_1 (left) and H_2 (right) cellular networks, both with their centers at cell 1.

In allocating channels to the cells, one important feature must be considered. The allocated frequencies must be separated so that they don't interfere with one another. Frequencies, such as 90.5MHz and 90.6MHz, that are too close to each other will definitely not be suitable for allocation in the same cell as they may cause severe interference. The two frequencies may still be used by placing them in two different cells separated some distance away. The way frequencies are allocated to the cells makes up the *frequency assignment problem,* or *channel assignment problem.*

Calls in a cellular network are assigned in a circuit-switching manner. *Circuit-switching* is a routing technique involving a pair of users, the caller and his receiver. In circuit-switching, the paths or channels assigned to the caller and his receiver are reserved wholly for them throughout the duration of the call. In comparison, *packet-switching* is another routing technique that allows data to be distributed into several packets over several different links in the network.

We discuss two models of channel assignments in this chapter. The first model involves the assignment of fixed channels allocated to the cells in a network to mobile users roaming in the network. The second is the problem of allocating the minimum number of channels to the cells in the network in such a way as to avoid the interference that arises from three types of electromagnetic constraints.

11.2 CHANNEL-ASSIGNMENT PROBLEM

The channel-assignment problem has its prototype in the graph-coloring problem, and is stated as follows: *Given a graph G(V, E) with the set of vertices V and a set of edges E, the objective is to find the minimum number of colors required for the vertices in such a way that no two adjacent nodes share the same color.*

In another form called the *edge-coloring* problem, the problem can be stated as finding the minimum number of colors for the edges in such a way that no two edges belonging to the same vertex have the same color. Graph coloring is one of the most researched graph theory problems. The high interest in this problem is due to the fact that many applications use the graph-coloring problem as their prototype. One common application of the graph-coloring problem is in coloring countries in a map with different colors so that no two neighboring countries share the same color. Having neighbors with different colors is a form of constraint that makes the problem difficult to solve. The problem is further complicated by adding the neighborhood constraint. In this constraint, any two vertices in a graph can share the same color if they are separated by at least one node. The channel-assignment problem has this type of constraint.

The channel-assignment problem is a constrained-optimization problem. Basically, the problem can be viewed as a type of graph-coloring problem that finds its solution by applying an effective graph-coloring algorithm. The number of colors in this problem refers to the number of frequencies for assignment. Therefore, the main objective of the channel-assignment problem is to produce an assignment that minimizes the span of the frequencies, or the use of bandwidth. This is an important objective in channel assignment as bandwidth is a very scarce, expensive, and valuable resource.

In assigning the channels, the electromagnetic interference that arises from certain constraints is to be avoided and kept to a minimum. Three common constraints are to be considered, namely, the cochannel, adjacent channel, and cosite constraints. In the *cochannel constraint,* equal frequencies may be assigned only to antennas that are separated by some minimum distance from each other, normally given as three hexagons away. This implies that two cells a and b must be separated by a minimum distance of i in order to avoid interference when both of them are using the same channel k simultaneously. This can be stated as follows:

$$|d(a, b, k)| \geq i \tag{11.1}$$

In the above equation, $d(a, b, k)$ represents the separation distance between cells a and b in order to share the same channel k. Referring to the H_2 network in Figure 11.1, if the separation distance is three hexagons, then cells 6 and 10 can share the same channel k at the same time, as they are three hexagons apart. Cells 6 and 2 cannot share the same channel as their separation is less than three hexagons.

In the *adjacent-channel constraint,* the frequencies f_a and f_b assigned to the channels in the adjacent cells a and b, respectively, must differ by at least j units, given as follows:

$$|f_a - f_b| \geq j \tag{11.2}$$

From Figure 11.1, the frequencies f_6 and f_7 allocated to cells 6 and 7, respectively, must differ by at least j units as the cells are adjacent to each other. This constraint does not apply if the two cells are not neighbors.

In the *cosite constraint,* the frequencies f_1 and f_2 of any two channels in a cell must be separated by a minimum distance of k, given as follows:

$$|f_1 - f_2| \geq k \tag{11.3}$$

This constraint is almost similar to the adjacent-channel constraint except that it applies to the same cell. If the frequencies are labeled in order from lowest to highest as $f_1, f_2, \ldots, f_n$, and $k = 3$, then a practical allocation of frequencies to cell 6 is f_6, f_9, f_{12}, and so on.

Channel assignments can be viewed as the problem of finding the right permutation of the available resources in order to keep the bandwidth to a minimum. This objective requires an effective allocation scheme that encourages channel reuse. The simultaneous use of the same channel in several cells in a nonconflicting manner will definitely minimize the utilization of resources in the network. Among the models proposed for solving the channel-assignment problem include graph theory and heuristical methods that consider one or all the constraints. Newer approaches include intelligent methods such as neural networks, simulated annealing, and genetic algorithms.

The theoretical foundation of the channel assignment problem was developed under the assumption that channel assignment is a mapping problem involving matching the prey (cell) to its predator (channel). The channels may be allocated to the cells in the network in a decentralized or centralized manner according to the schemes used. In the decentralized scheme, the channels are divided and allocated to the cells permanently. This type of assignment is called the *static allocation scheme.* Each cell is allocated with a fixed number of nonconflicting channels for servicing the mobile users in the cell. If all the allocated channels are used, then the cell makes a request to borrow a channel from one of its neighbors if the need arises. In the same manner, the cell can also lend its channels to its neighbors if at least one of its channels is available.

In the centralized scheme, all the channels are controlled directly by MSC. These channels are allocated to the cells based on the *dynamic allocation scheme,* according to demand from the mobile users in the cells. Therefore, the channels are allocated on a temporary basis and they are immediately removed from the cells upon the completion of the calls so they can be reused by other cells.

11.3 CHANNEL ASSIGNMENTS: DISCRETE MODEL

In this section, we discuss a discrete-event model involving the static allocation scheme in which a fixed number of channels are allocated to the cells. A model, shown in Figure 11.2, is used as an example to illustrate the scheme. The figure shows the H_1 network with seven cells, labeled from 1 to 7. To the right of the figure is the adjacency matrix of the cells in the network. The cells may not be equal in size, depending on the geographical location and demand. For example, in a densely populated suburban area of a city, the hexagon may be small in size as the base station in the cell needs to serve a large number of customers. We illustrate the discrete model by allocating four channels to each cell in the H_1 network. The channel number in each cell is shown below the cell number, as shown in the figure.

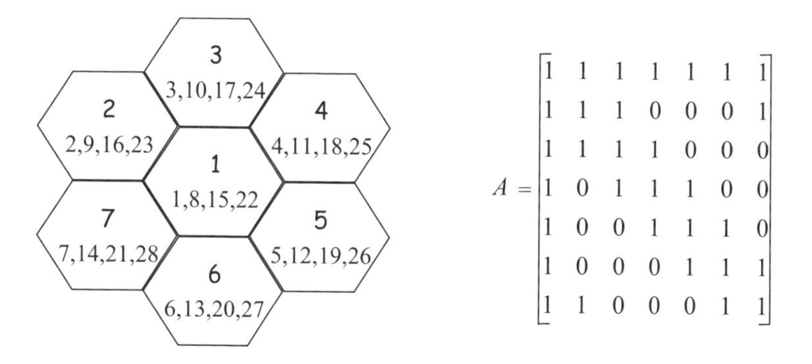

$$A = \begin{bmatrix} 1 & 1 & 1 & 1 & 1 & 1 & 1 \\ 1 & 1 & 1 & 0 & 0 & 0 & 1 \\ 1 & 1 & 1 & 1 & 0 & 0 & 0 \\ 1 & 0 & 1 & 1 & 1 & 0 & 0 \\ 1 & 0 & 0 & 1 & 1 & 1 & 0 \\ 1 & 0 & 0 & 0 & 1 & 1 & 1 \\ 1 & 1 & 0 & 0 & 0 & 1 & 1 \end{bmatrix}$$

Figure 11.2 H_1 network model (left) and its adjacency matrix (right).

Code11A: Channels for Mobile Users

Our simulation model is called Code11A. The project consists of the files Code11A.h and Code11A.cpp, plus an input file called Code11A.in. In our model, each cell in the network is allocated with four nonconflicting channels, shown as numbers below the cell number in each cell in Figure 11.2. In the figure, each cell in the network is allocated with four cells. The channels allocated to each cell are called the *primary channels* of the cell. There are 28 channels in total, and they are numbered from 1 to 28 according to the order from the lowest to highest frequencies. Furthermore, the model assumes that only one MSC controls the whole network and that it is geographically located within the network. This MSC is responsible for assigning the channels for all calls originating from the network. It is also assumed that, for a call, both the caller and the receiver are located internally within the network. Therefore, the model does not support calls from external sources or destinations.

In this model, the cochannel constraint as expressed by Equation (11.1) is not applicable as the model does not support the simultaneous deployment of the same channel in two or more cells. However, the adjacent-channel constraint is observed by setting $j = 1$ in Equation (11.2). Therefore, putting channels 6 and 7 in cells 6 and 7, respectively, will not cause any electromagnetic interference according to this assumption. The cosite constraint is also observed by setting $k = 1$ in Equation (11.3). This assumption provides flexibility, allowing a cell to borrow any channel belonging to its neighbor without causing any conflict.

A similarity with the graph-coloring problem is observed by noting that a cell in the network cannot have two successively numbered channels in its cell. Each channel in a cell must be separated by at least one frequency in order to avoid conflict with another channel. The arrangement of channels in Figure 11.2 only satisfies the minimum requirement for avoiding conflicts. In reality, several other constraints need to be observed in order to create a practical and interference-free cellular network.

Each cell in the network is allowed to borrow one or more channels from its neighboring cells to service the mobile users in its cell, provided they are available at that instance of time. Only neighboring cells can share channels in this case. At the time of borrowing, the program refers to the adjacency matrix of the network to determine if the borrowing cell and the host cell are adjacent to each other. The borrowed channels are called the *secondary channels*. In the same manner, each cell also must be kind enough to lend its unused channels to other cells whenever a request is received. One requirement that must be satisfied before the secondary channels can be applied to the mobile users in the host cell is that they must not interfere with the electromagnetic spectrum of the primary channels in the cell. To avoid this kind of problem, the channels in the network are normally assigned in such a way to satisfy the electromagnetic constraints not only in their cells but also in their neighboring cells. This is one strategy to allow the borrowing of channels between the neighboring cells in the network. Cell borrowing is an important feature of a cellular network as it makes it possible for the network to service mobile users in a dynamic manner. This is important as the network does not know the exact number of calls in each cell at any instance of time.

Table 11.1 summarizes the main characters in the model. The discrete events at time t are characterized by the random arrival of calls at several cells in the network. Cell m in the network is represented by $cell_m$. Calls are numbered in increasing order according to their arrival time. Call i is denoted by v_i. We refer to channel k as f_k.

The attributes of a call are summarized in Table 11.2. A call v_i is created at its arrival time (`art`), starts its execution (`ast`) at the assigned channel, completes its execution (`act`), and is terminated. To facilitate the assignment of calls, the scheduler monitors the state (`sta`) of the call at every time slot t where several flag values are assigned. The duration or length of the call (`len`) is the difference between the starting time (`ast`) and its completion time (`act`). The time that a call spends in the queue is its waiting time (`awt`), determined by the difference between the arrival time (`art`) and the starting time (`ast`). A call is originated from the source cell (`SCell`) with the assigned channel (`Sf`), whereas its receiver is assigned with another channel (`Df`) in the destination cell (`DCell`).

A cell in the network is a geographical region in which the mobile users are located. A mobile user has an id that correctly identifies his or her identification in order to get access to the network. To determine its presence, the cellular phone belonging to the mobile user transmits a beacon periodically and this message is picked up by the base station in the host cell. The beacon is passed to MSC, which updates the user location in its database. Through regular beacon transmission, the network knows the location of the mobile user at any given time, and this information helps in speedy circuit-switching connection of a call.

Table 11.3 lists the attributes of a cell. In this simple model, there are four allocated channels in every cell. In reality, there could be 1000 channels or more per cell. This high number of channels reflects the high demand for channels, especially in densely populated areas, like in a city. The model can easily be modified to test for some higher values by changing the macro `R` in the program. A cell is also characterized by its adjacency or separation (`sep`) status with other cells in the network, where a value of 1 means the two cells are adjacent and 0 otherwise. This value is referred to every time a cell makes a request to borrow a channel from its neighbor. Our model assumes that a cell may borrow a channel if both the host cell and the cell in which a channel is requested are neighbors.

Table 11.1 Main symbols and variables

Items	Symbol	C++ Variable
time	t	`t`
Call i	v_i	`v[i]`
Channel k	f_k	`f[k]`
Cell m	$cell_m$	`Cell[m]`

Table 11.2 Features of a call

Element	Description
`v[i]`	Call i, or v_i
`v[i].sta`	State of v_i where 0 = call created, 1 = call assigned, 2 = call arrived, and 3 = call terminated
`v[i].art`	Actual arrival time of v_i
`v[i].ast`	Actual start time for execution of v_i
`v[i].act`	Actual completion time of v_i
`v[i].len`	Length of v_i
`v[i].awt`	Actual waiting time of v_i in the queue
`v[i].Sf`	Source channel of
`v[i].Df`	Destination channel of v_i
`v[i].SCell`	Source cell of v_i
`v[i].DCell`	Destination cell of v_i

The attributes of a channel are listed in Table 11.4. Channel k is represented as f_k, where the symbol f has been selected as it can also be interpreted as frequency. The state of the channel (`sta`) is checked at every time slot t to determine whether the channel is available (0) or busy (1). Since the model is based on the static allocation scheme, each allocated channel serves as a primary channel in its host cell (`hCell`). However, this channel can be borrowed by the host cell neighbor as a secondary channel to the neighboring cell if the request for borrowing is granted. Each channel also records the calls assigned to it (`av`).

We model the arrival of calls and their assignments to the channels as discrete events. Also, the assignment of calls to the channels is a scheduling problem in which the scheduler in this case is the program, or the flowchart in the figure. The process of assignments is best described through Gantt charts, as the charts show the detailed information about the starting time, completion time, length of the calls, and the channels involved.

All the operations of the model are summarized in the flowchart in Figure 11.3. An event at time t is characterized by the arrival of calls. At each time slot, calls arrive at random and they are numbered in increasing order. When a call is made, the base station in the host cell detects the message and identifies the caller id, its location, and the call destination. The base station transmits the information to MSC and makes a request for a channel. Upon receiving this information, MSC

Table 11.3 Features of a cell

Element	Description
`Cell[m]`	Cell m, or $cell_m$
`Cell[m].f[k]`	mth channel allocated to $cell_m$
`Cell[m].sep[w]`	Adjacency status between cells m and w, where 1 = adjacent and 0 = not adjacent

Table 11.4 Attributes of a channel

Element	Description
`f[k]`	Channel k, or f_k
`f[k].sta`	State of f_k where 0 = channel is available, 1 = channel is busy
`f[k].hCell`	Host cell of f_k
`f[k].av[i]`	ith call assigned to f_k. However, `f[k].av[0]` represents the number of calls assigned to f_k
`f[k].lv`	Last call assigned to f_k

checks for the status of both the call source and its destination and the state of the channels in its database. The receiver's location and id are immediately determined from the database. Once the location is known, MSC checks for the available channels allocated to the two base stations. A channel each is required for the caller and its receiver, and the two channels must be assigned simultaneously to the call. In each cell, if a primary channel is available, then it is immediately as-

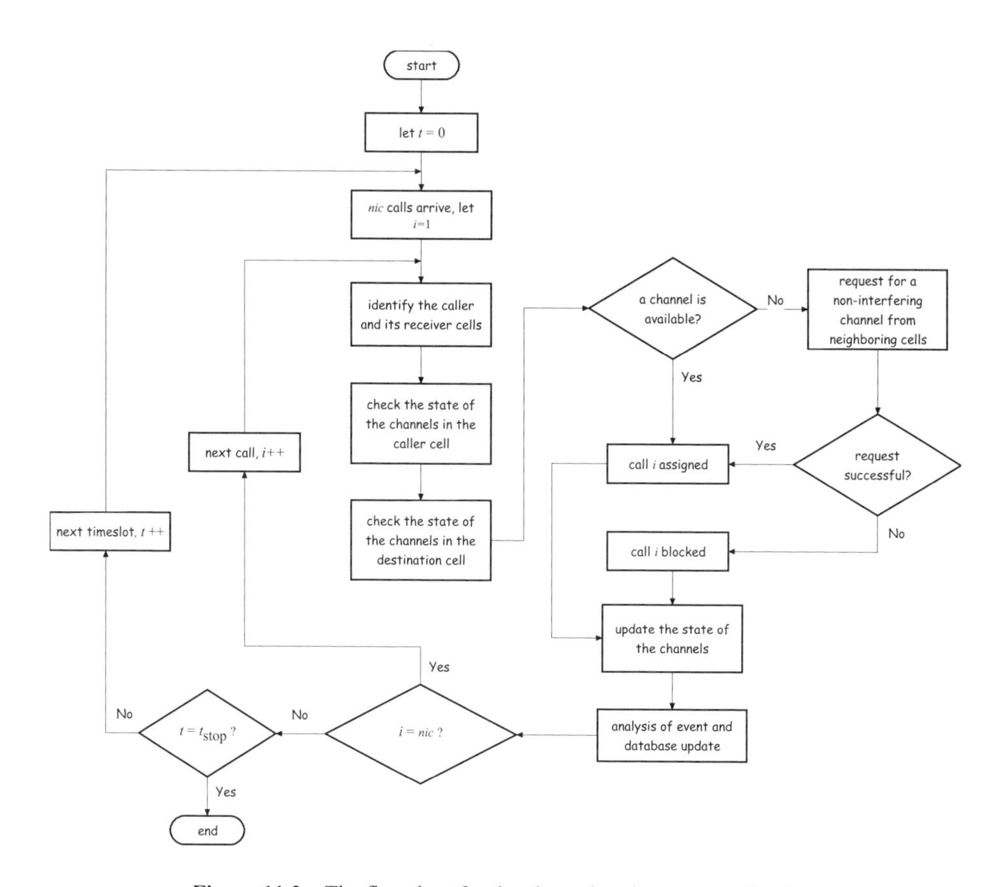

Figure 11.3 The flowchart for the channel-assignment application.

signed. Otherwise, the cell makes a request for a secondary channel from its neighbors, one by one, until the channel is found. This channel is then assigned to the user.

For a call to be established, a channel each must be allocated to both the caller and his receiver. If one or both channels are unavailable then the call is placed in the central queue waiting for the next time slot, and so on, until a pair of channels is found. A channel is available if the previous caller has completed his call before or at the current time t. The availability of the channels is indicated through a Boolean flag called `sta`, or state, where the value of 0 means available and 1 is busy. All channels start with the initial state of 0 at time $t = 0$. When a channel is assigned to a call, its flag immediately changed to 1. Once the call is completed, the flag is returned to 0 to indicate it is available for reuse. This flag is regularly checked and updated by the scheduler at every time slot.

Figure 11.4 shows the output of a sample run of the project Code11A. The calls are marked by shaded and unshaded rectangles in the left-hand part of the figure. A shaded rectangle represents the caller, whereas an unshaded one represents the re-

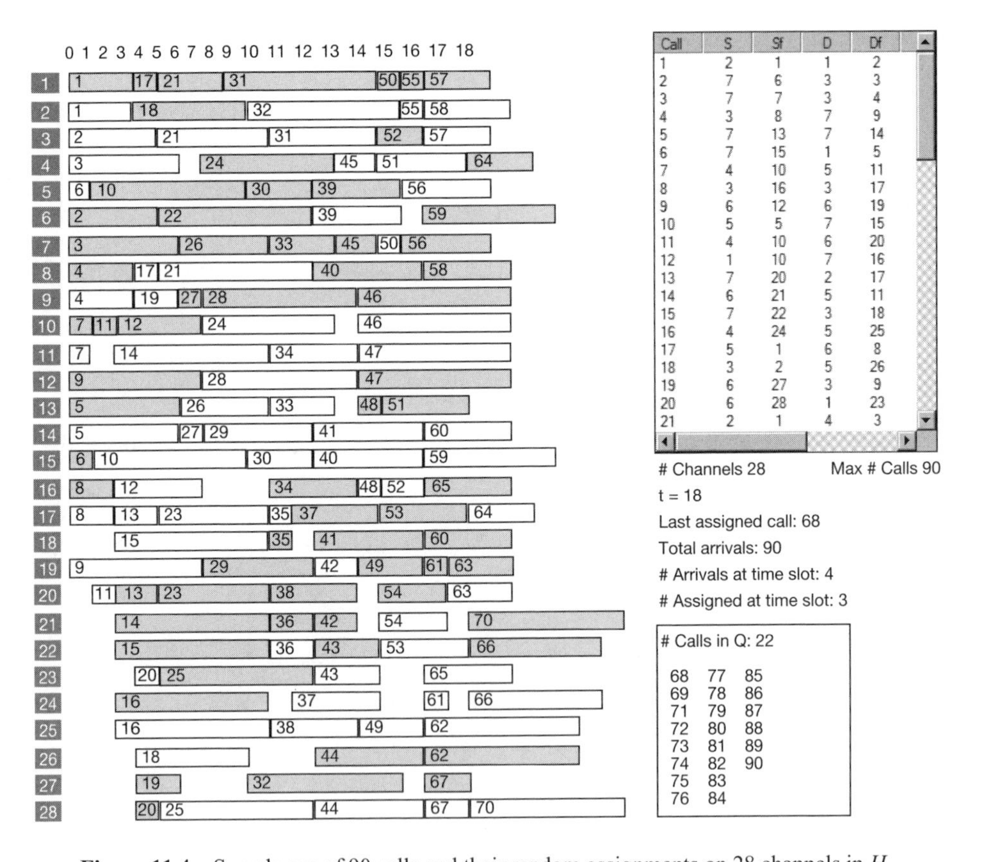

Figure 11.4 Sample run of 90 calls and their random assignments on 28 channels in H_2.

ceiver of the call. There are 28 channels distributed evenly on the seven cells and these are shown at the far left of the left-hand part of the figure as small shaded rectangles. The output displays the information of all the 90 generated calls tabulated in the list view scrollable window at the right.

The initial setup consists of the declarations and initial values for several macros and variables. The constants are declared in Code11A.h, shown in Table 11.4.

At time t, the total number of calls in the system and the total number of assigned calls are denoted by the variables `nv` and `tnav`, respectively. In the constructor, their values are set to 0 initially at $t = 0$. Other variables initialized are the state of the each channel and the number of assigned calls in each channel. This is followed by the state of each call. The following code fragments in the constructor perform the initializations:

```
nv=0; tnav=0;
for (k=1;k<=M;k++)
{
     f[k].sta=0;
     f[k].av[0]=0;
}
for (i=1;i<=N;i++)
     v[i].sta=0;
```

The constructor also reads the information about the connectivity of the cells in the network. The connectivity is determined through the adjacency matrix as displayed in Figure 11.2. The values are read from an input file called Code11A.in. The next step in the constructor is to allocate four primary channels each to the cells according to channel numbers shown in Figure 11.2. The numbers are allocated using a simple formula given by:

```
Cell[i].f[r]=i+7*(r-1)
```

The above formula gives the rth channel in `Cell[i]`. For example, the first and second channels of `Cell[5]` are given by 5 and 12 from this formula. Once the channels have been allocated to the cells, an update on the channel host cell is also

Table 11.4 Macros for the initial setup

Macro	Value	Description
R	4	Number of channels per cell
L	7	Total number of cells
M	28	Total number of channels
N	90	Total number of calls
mniv	10	Maximum number of calls per time slot
mLen	7	Maximum length of a call

necessary and is accomplished by letting `f[w].hCell=i`, where `w` is the allocated channel number. The following routine performs all these steps:

```
for (i=1;i<=L;i++)
{
     for (j=1;j<=i;j++)
     {
          fscanf(InFile,"%d",&Cell[i].sep[j]); // read cell adjacency info
          Cell[j].sep[i]=Cell[i].sep[j];
     }
     for (r=1;r<=R;r++)
     {
          w=Cell[i].f[r]=i+7*(r-1);   // allocate channels to cells
          f[w].hCell=i;                    // update channel information
     }
}
```

The constructor also creates the list view window for displaying the information on the arriving calls. The window is conveniently located on the right side of the main window. A list view window is necessary as it provides all the detailed information about the calls. The following routine creates the list view window:

```
vInfoView.Create(WS_VISIBLE | WS_CHILD | WS_BORDER | LVS_REPORT
   | LVS_NOSORTHEADER,CRect(BottomRight.x-250,TopLeft.y+10,
     BottomRight.x-20,TopLeft.y+350), this, IDC_vINFO);
char* column[nFIELDS+1]=
     {"Call","S","Sf","D","Df","arr","ast","act","sta"};
int columnWidth[nFIELDS+1]={40,40,40,40,40,40,40,40,40};
LV_COLUMN lvColumn;
lvColumn.mask = LVCF_WIDTH | LVCF_TEXT | LVCF_FMT | LVCF_SUBITEM;
lvColumn.fmt = LVCFMT_CENTER;
lvColumn.cx = 85;
for (i=0;i<=nFIELDS;i++)
{
     lvColumn.iSubItem = 0;
     lvColumn.pszText = column[i];
     vInfoView.InsertColumn(i,&lvColumn);
     vInfoView.SetColumnWidth(i,columnWidth[i]);
}
```

The discrete event is simulated at time t by a press on the space bar key. The event is detected by `WM_KEYDOWN` in the message map and handled by the function `OnKeyDown()`. At $t = 0$, all channels are available for assignment. A randomly determined number of calls arrive, denoted by the variable `niv`. The number of incoming calls per time slot ranges from 0 to the maximum value set by `mniv`, which is a constant defined as a macro in Code11A.h. In this application, `mniv` is set to 11.

The total number of calls `nv` now becomes its old value added to `niv`. The scheduler then checks the state of the channels. The channels whose last assigned calls complete before the current time are marked as available, using the following routine:

```
for (k=1;k<=M;k++)
     if (f[k].av[0]>=1)
     {
          r=f[k].lv;
          if (v[r].act<=t)
               f[k].sta=0;
     }
```

For each incoming call, the scheduler records its arrival time and assigns its assignment state to 2, to denote that it has not been assigned yet. The origin of the call and its destination is also detected. In our model, both the source and destination cells are determined randomly from the seven cells in the network. The following routine performs these tasks:

```
for (i=1;i<=niv;i++)
{
     nv++;
     if (nv<=N)
     {
          v[nv].art=t;
          v[nv].sta=2;
          v[nv].SCell=1+rand()%L;
          v[nv].DCell=1+rand()%L;
     }
}
```

A call between a caller and receiver can only be connected if a channel each is found for both of them. The scheduler performs a check on the available channels in the source cell first. A primary channel in the cell is available if its current state is 0, that is, `!f[k].sta`, where `k` is the channel number. If this channel is not available, then the scheduler looks for a channel in the neighboring cell. To do this, the scheduler refers to a separation cell between the source cell and its neighbor. A channel from a neighbor is available through the conditional expression, `!f[k].sta && Cell[p].sep[z]`. Once a channel is obtained, it is reserved for the caller by setting the temporary variable `u` to the channel number. The state of the channel is set to busy (1) to prevent the channel from being used by another call. If a channel is not obtained despite the attempt to borrow from its neighbors, the variable `u` is set to 0 and the call is put in the queue. The following routine performs channel assignment on the caller:

```
z=f[k].hCell;
if (!f[k].sta && Cell[p].sep[z])
```

```
{
      u=k;
      v[i].Sf=u;
      f[u].sta=1;
      break;
}
if (k==M && !f[M].sta)
      u=0;
```

The search for a channel for the receiver follows only if a channel for the caller is obtained. This is determined if the value of u is not 0. A similar procedure is performed to search for a channel in the destination cell. If a primary channel is available, it is immediately assigned, otherwise a request for borrowing is made on its neighbors. The first available channel found is assigned to the receiver by setting the temporary variable w equal to the channel number. With the assignment of a channel to the receiver in the destination cell, the call is removed from the queue and confirmed with the assigned channels. The state of the call is set to 1 to denote it has been assigned with the two required channels. The call starts at the current time and the length of the call is determined by a random number. The flags at the assigned channels are set to 1 to indicate they are now busy. Some update is made by adding the call to the list of calls assigned to both channels. The following code fragment shows this update:

```
r=f[k].hCell;
if (!f[k].sta && Cell[q].sep[r])
{
      w=k; nav++; tnav++;
      v[i].Df=w;
      v[i].len=1+rand()%mLen;
      v[i].ast=t;
      v[i].act=t+v[i].len;
      v[i].sta=1;
      v[i].awt=v[i].ast-v[i].art;
      f[w].sta=1;
      f[w].lv=f[u].lv=i;
      z=++f[w].av[0]; f[w].av[z]=i;
}
```

Once the states of the call and their assigned channels have been updated, the next step is to display the assigned call as Gantt charts in the window. The display consists of blue and red rectangles which indicate the call as the source and destination, respectively. The following routine performs the assignments of the calls and their display as Gantt charts:

```
r=f[k].hCell;
if (!f[k].sta && Cell[q].sep[r])
```

```
{
    v[i].Home=CPoint(f[w].Home.x+30+
                t*wBar,f[w].Home.y);
    rc=CRect(v[i].Home,
         CSize(wBar*v[i].len,hBar));
    dc.SelectObject(&DPen); dc.Rectangle(rc);
    s.Format("%d",i);

    dc.TextOut(2+v[i].Home.x,1+v[i].Home.y,s);

    z=++f[u].av[0]; f[u].av[z]=i;
    v[i].Home=CPoint(f[u].Home.x+30+
                t*wBar,f[u].Home.y);

    rc=CRect(v[i].Home,CSize(wBar*v[i].len,hBar));
    dc.SelectObject(&SPen); dc.Rectangle(rc);
    s.Format("%d",i);
    dc.TextOut(2+v[i].Home.x,1+v[i].Home.y,s);
    break;
}
```

The last step in each time slot is to display the information about all the information about the assigned calls in the list view table. This step requires calling the function `CallInfo()`, given as follows:

```
void CCode11A:: CallInfo()
{
     CString s;
     vInfoView.DeleteAllItems();
     LV_ITEM lvItem;
     lvItem.mask = LVIF_TEXT | LVIF_STATE;
     lvItem.state = 0;
     lvItem.stateMask = 0;
     for (int i=0;i<=tnav;i++)
     {
          lvItem.iItem=i;
          lvItem.iSubItem=0;
          lvItem.pszText="";
          vInfoView.InsertItem(&lvItem);
          if (i+1<=tnav && i+1<=N)
          {
               s.Format("%d",i+1); vInfoView.SetItemText(i,0,s);
               s.Format("%d",v[i+1].SCell); vInfoView.SetItemText(i,1,s);
               s.Format("%d",v[i+1].Sf); vInfoView.SetItemText(i,2,s);
               s.Format("%d",v[i+1].DCell); vInfoView.SetItemText(i,3,s);
               s.Format("%d",v[i+1].Df); vInfoView.SetItemText(i,4,s);
               s.Format("%d",v[i+1].art); vInfoView.SetItemText(i,5,s);
```

```
                    s.Format("%d",v[i+1].ast); vInfoView.SetItemText(i,6,s);
                    s.Format("%d",v[i+1].act); vInfoView.SetItemText(i,7,s);
                    s.Format("%d",v[i+1].awt); vInfoView.SetItemText(i,8,s);
            }
        }
}
```

The same procedure is repeated in the next time slot t. The scheduler performs the same strategy for assigning channels to the arriving calls. If a pair channels are available, they are immediately assigned to the call, otherwise the call is placed in the central queue. At each time slot, the information about the assigned call is displayed in the list view window. The simulation is terminated when the last call is completed, by changing its state to 3.

11.4 SOLVING THE CHANNEL-ASSIGNMENT PROBLEM

Channel assignment in a cellular network is a constrained graph-coloring problem. As discussed in Section 11.2, three constraints need to be overcome in order to allocate radio channels to the cells in a wireless cellular network: the adjacent channel, cochannel, and cosite constraints. A violation of any of these constraints will cause electromagnetic interference on calls originating from or destined to the respective cells. We discuss this problem and develop its solution based on a greedy allocation method.

In this section, a simulation model using 19 cells on the H_2 network has been developed for allocating one channel per cell by taking into consideration the adjacent channel and the cochannel constraints. The cosite constraint does not apply in this model as it can make an impact only if two or more channels need to be allocated. Some assumptions are made in the simulation model to realize the idea of channel assignments. This includes limiting the channel separation for the adjacency cell constraint in Equation (11.2) to two channels, or $j = 2$. Another assumption is limiting the cell separation for the cochannel constraint in Equation (11.1) to three cells, or $i = 3$.

Code11B: Solving the Channel-Assignment Problem

The model is developed under the project Code11B, and it includes the files Code11B.h, Code11B.cpp, and Code11B.rc. Figure 11.5 shows the output from this project. It consists of a diagram showing the H_2 network on the left, and the adjacent channel and cochannel matrices on the right. The number at the center of each cell is the cell number, whereas the one below it is the output in the form of the channel assigned to this cell. Both the adjacency and cochannel matrices have been created based on the H_2 network diagram. Due to their nature and properties, both matrices are assumed to be symmetric, so that a constraint from cells i to j is the same as from cells j to i.

The cell-adjacency matrix is symmetric and has entries of either 0 or 1. An entry of 1 means the cells represented by its row and column numbers are adjacent, where-

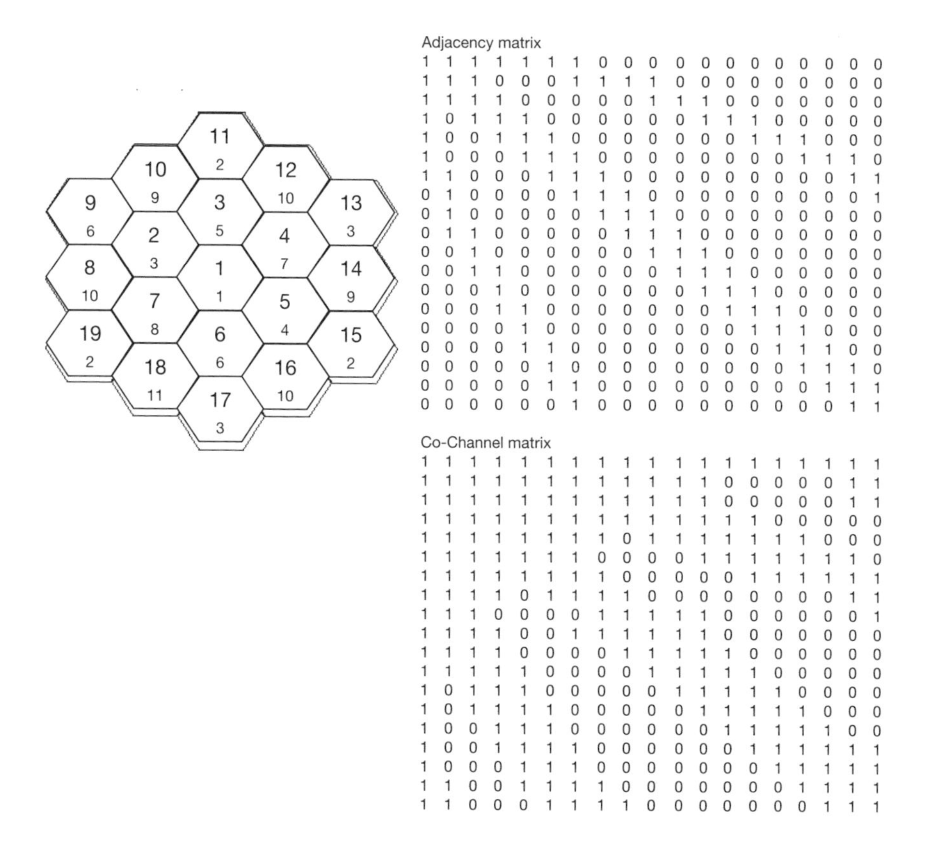

Figure 11.5 Output from Code11B.

as 0 indicates they are not. The cell-adjacency constraint applies to the cells that are adjacent by assuming a channel separation distance of 2 in Equation (11.2), or $j = 2$. For example, cell number 2 is adjacent to cells 1, 3, 7, 8, 9, and 11. Therefore, row number 2 has entries of 1's with columns 1, 3, 7, 8, 9, and 10, and 0's with the rest.

The cochannel constraint applies on every pair of cells using Equation (11.1) by assuming a separation distance of 3, that is, $i = 3$. Similarly, in the cochannel matrix, an entry of 1 means the two cells are affected by the cochannel constraint and 0 means they are not. For example, referring to the network diagram, cell number 2 has the cochannel limitation with all cells in the network except with cells 13, 14, 15, 16, and 17. Therefore, row 2 of the matrix has 0's on columns 13, 14, 15, 16, and 17, and 1's on the rest.

The global variables, objects, and functions in the project are declared in Code11B.h. The application class in this project is Code11B and it consists of several member variables and functions. The number of cells in the H_2 network is 19, and this number is represented by the macro `L`. The class also includes the structure `CELL`, whose elements are summarized in Table 11.5.

Table 11.5 Elements of the structure CELL

Element	Description
Cell[i].f	Channel assigned to cell i
Cell[i].home	Home coordinates of cell i
Cell[i].adj[j]	Adjacency status between cells i and j
Cell[i].coc[j]	Cochannel status between cells s i and j

The cell-adjacency constraint between the cells i and j is checked through Cell[i].adj[j]. This variable has a Boolean value where TRUE (1) indicates the cells i and j are adjacent, and FALSE (0) means they are not. This variable gets its values defined by reading the input data from the file Code11BAdj.in.

The cochannel constraint between the cells i and j is represented by the variable Cell[i].coc[j], which obtains its values from the diagram by reading the input data from the file Code11BCoc.in. The variable type is declared as Boolean, where a value of 1 means the cells i and j are affected by the cochannel constraint, and 0 means they are not.

Each cell in the network has home coordinates located at its center in the network diagram. This location is represented by the element home, which is declared as a CPoint object. The home coordinates of these cells are read from the file Code11BHome.in. Below the cell number of each cell is the channel number assigned to the cell. This number is the ultimate output produced from the assignment policy used in the model in such a way that the assigned channel abides by the adjacent-channel and cochannel constraints.

Another variable declared in the class is the CBitmap object h2. This variable represents the bitmap image h2.bmp, which is the cellular network diagram displayed in the window. In order to display this image, the resource file Code11B.rc is included in the project that has this bitmap file attached as a resource item. The image file h2.bmp has its id assigned as IDB_BITMAP1 in the resource file. It is loaded into the memory from the constructor as follows:

```
h2.LoadBitmap(IDB_BITMAP1);
```

The input files Code11BAdj.in, Code11BCoc.in, and Code11BHome.in are referred to in the constructor as the pointers InFile1, InFile2, and InFile3, respectively. The files are opened and read from the constructor using the C file input operations, as follows:

```
FILE *InFile1,*InFile2,*InFile3;
InFile1=fopen("Code11BAdj.in","r");
InFile2=fopen("Code11BCoc.in","r");
InFile3=fopen("Code11BHome.in","r");
for (i=1;i<=L;i++)
{
```

```
    for (j=1;j<=i;j++)
    {
        fscanf(InFile1,"%d",&Cell[i].adj[j]); // read cell adjacency info
        Cell[j].adj[i]=Cell[i].adj[j];
    }
    for (j=1;j<=i;j++)
    {
        fscanf(InFile2,"%d",&Cell[i].coc[j]); // read cell co-chan info
        Cell[j].coc[i]=Cell[i].coc[j];
    }
    fscanf(InFile3,"%d %d",
        &Cell[i].home.x,&Cell[i].home.y);      // read cell home coordinates
}
```

The input files refer to Code11BAdj.in, Code11BCoc.in, and Code11BHome.in, which represent the adjacency matrix, the cochannel matrix, and the home coordinates of the cells, respectively. Each of the adjacency and cochannel matrices is symmetric. Therefore, the statements:

```
Cell[j].adj[i]=Cell[i].adj[j];
Cell[j].coc[i]=Cell[i].coc[j];
```

are added as input statements for completing the entries in the arrays.

To facilitate the channel-assignment operations, each cell in the network is initially assigned with channel 0, as follows:

```
for (i=1;i<=L;i++)
    Cell[i].f=0;   // initialize all the cells with channel 0
```

These initial values are necessary as the assignment policy is based on an iterative process in which the current value refers to some previous values. An initial value of 0 means no channel has been assigned, and this value can act as a reminder so that the real channel will be assigned to the cell.

The output display on the window using `WM_PAINT` is the only event in this application. The event gets its response from the function `OnPaint()`. The output begins by displaying the image file `h2.bmp`. In order to display this image, a device context object called `memDC` is created from the MFC class `CDC`. This object is made compatible with the Windows device context object `dc` using the function `CreateCompatibleDC()`. Once the compatibility has been established, the image is displayed using the function `BitBlt()`. The image output routine is described as follows:

```
CDC memDC;
memDC.CreateCompatibleDC(&dc);
memDC.SelectObject(&h2);
dc.BitBlt(0,50,400,320,&memDC,0,0,SRCCOPY);
```

Figure 11.6 shows the channel-assignment policy for cell *i* in the network. A scheduler is first established for the job of assigning the channels to the cells in the network. The scheduler is based on a channel-assignment policy that considers the adjacent channel and cochannel constraints. The assignment policy in the scheduler is a greedy method that assigns channels by giving priority to the lowest number, in ascending order. The concept used here is very simple: a cell checks for channels starting from the lowest, and compares them to other cells in the network for the adjacent-channel and cochannel compliance. The first available channel that complies to these two constraints is immediately assigned to the cell.

In assigning the channel, a temporary variable `k` is used as the channel number. A temporary array called `w[k]` is a flag in which 0 denotes that channel `k` complies with the constraint and 1 is a violation. Initially, `w[k]` is 0 and this value changes to 1 if a constraint is violated. The variable `k` starts from 1, and increases its value whenever any of the adjacency or cochannel constraints are violated.

The assignment starts by assigning cell 1 with channel 1, or `Cell[1].f=1`, as the initial value. The next cell, or cell 2, then checks the channel in the cell before this, or cell 1. This channel is used to compare with the generated channels, `k, starting with k=1`. The generated channel `k` is checked for compliance to the adjacent-channel constraint using the following test:

```
if (Cell[i].adj[j] && abs(k-Cell[j].f)<2)
{
      Cell[i].f=k+1;
      w[k]=1;
}
```

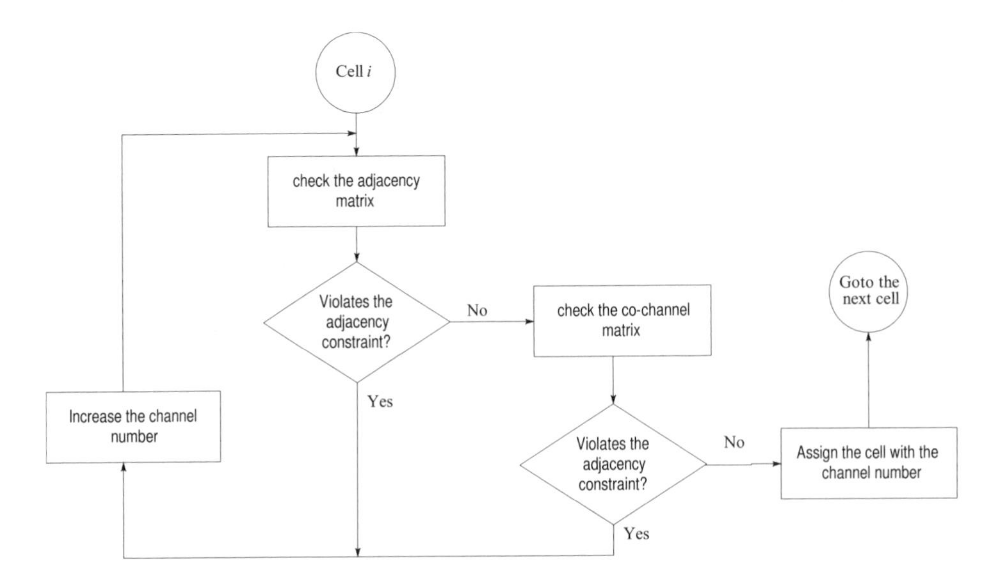

Figure 11.6 Channel-assignment policy for cell *i* in the network.

The cochannel constraint is checked using the following code fragments:

```
if (Cell[i].coc[j] && Cell[i].f==Cell[j].f)
{
      Cell[i].f=k+1;
      r++;
}
```

The scheduling policy used in this model produces an optimum result in which only 11 channels are required. The following routine shows the scheduling policy that successfully assigns the channels to the cells according to this priority policy:

```
Cell[1].f=1;
for (i=2;i<=L;i++)
{
      k=1;
      while (k<=L)
      {
            w[k]=0;
            for (j=1;j<=i-1;j++)
                  if (Cell[i].adj[j] && abs(k-Cell[j].f)<2)
                  {
                        Cell[i].f=k+1;
                        w[k]=1;
                  }
            if (!w[k])
            {
                  r=0;
                  for (j=1;j<=i-1;j++)
                  {
                        Cell[i].f=k;
                        if (Cell[i].coc[j] && Cell[i].f==Cell[j].f)
                        {
                              Cell[i].f=k+1;
                              r++;
                        }
                  }
                  if (r==0)
                        break;
            }
            k++;
      }
}
```

Once the channels have been assigned, the next step is to display the assigned number in the cells. At the same time, the adjacency and cochannel matrices are also displayed in order to see their relevance to the network diagram. This is done as follows:

```
dc.SelectObject(fontTimes);
for (i=1;i<=L;i++)
{
     dc.TextOut(380,5,"Adjacency matrix");
     dc.TextOut(380,315,"Co-Channel matrix");
     for (j=1;j<=L;j++)
     {
          s.Format("%d",Cell[i].adj[j]);
          dc.TextOut(380+(i-1)*20,20+(j-1)*15,s);
          s.Format("%d",Cell[i].coc[j]);
          dc.TextOut(380+(i-1)*20,330+(j-1)*15,s);
     }
     s.Format("%d",Cell[i].f);
     dc.TextOut(Cell[i].home.x,Cell[i].home.y,s);
}
```

11.5 SUMMARY AND CONCLUSION

Channel assignments have been illustrated by two models in this chapter. We tackle two problems here, one dealing with the assignment of channels to mobile users and the other with the optimization problem of minimizing channels subject to two common electromagnetic constraints. The first simulation model presented here is about a discrete-event simulation model for managing the requirements of mobile users, which makes use of the Markovian queueing system. The discrete-event model has been well illustrated by making the assumption that the arriving customers and their call duration are randomly determined. The model makes use of an unlimited buffer facility in the queue to avoid blocking. Therefore, the waiting time is infinite and every call involving a caller and a receiver will be entertained.

The second model is the channel-assignment problem, which considers the adjacent channel and cochannel constraints. The problem has its root in the well-known graph-coloring problem. We developed a priority-scheduling policy that assigns the first lowest available channels that comply with the two constraints to the cells. This simple assignment method is a greedy method that produces optimum results in the form of the minimum number of channels in the H_2 network model.

The two models presented here can easily be expanded to include other major factors for achieving interference-free environments in cellular networks. In real life, many other factors affect the quality of service provided by a network. One obvious factor seen in the two models is the cosite constraint, which evolves from the assignment of multiple channels in a given cell. Any two channels used in a cell must have some separation distance in order to avoid electromagnetic interference. In a real application, a cell may be allocated with hundreds or thousands of channels to serve the mobile users in its area.

At an instance of time, a cell may find that its allocated channels will not be enough to meet the demand. In this case, the cell needs to borrow some channels from its neighbors. Channel borrowing optimizes the channels in the network as it prevents some channels from being idle. Therefore, channel borrowing and reuse are two important components in the system that allow the network to respond to the requirements quickly. However, channel borrowing needs to be handled carefully as it may cause a violation of one of the constraints. A simulation model can be developed to test this problem by extending the work presented in this chapter.

BIBLIOGRAPHY

1. R. Leese and S. Hurley, *Methods and Algorithms for Radio Channel Assignments,* Oxford University Press, 2002.
2. T. S. Rappaport and T. Rappaport, *Wireless Communications: Theory and Practice,* 2nd ed., Pearson Education, 2001.
3. K. Pahlavan and P. Krishnamurthy, *Principles of Wireless Networks: A Unified Approach,* Prentice-Hall, 2001.
4. I. Katzela and M. Naghshineh, Channel Assignment Schemes for Cellular Telecommunications Systems, *IEEE Personal Communications, 3,* 3, 10–31, 1996.
5. M. Zhang and P. Y. Tak-Shing, Comparisons of Channel Assignment Strategies in Cellular Mobile Telephone Systems, *IEEE Transactions on Vehicular Technology,* vol. 38, no. 4, pp. 211–215, 1989.

CODE LISTINGS

Code11A: Channel Assignments to Mobile Users

```
// Code11A.h
#include <afxwin.h>
#include <afxcmn.h>
#define IDC_vINFO 501
#define nFIELDS 8
#define R 4                          // #channels per cell
#define L 7                          // #cells
#define C R*L                               // #channels
#define N 90                         // total #calls
#define mniv 10                      // max. #incoming calls/slot
#define mLen 7                       // max. length of call

class CCode11A : public CFrameWnd
{
private:
     LOGFONT lfTimes; CFont fontTimes;
     CPoint TopLeft,TextArea;
     CListCtrl vInfoView;
     int wBar, hBar, t;          // width,height,time
     int nv, tnav;               // call#, total #assg calls
     typedef struct
     {
```

```
        CPoint Home;
        int hCell;            // host cell
        bool sta;             // status
        int av[N+50],lv;   // ith, last call
    } CHANNEL;
    CHANNEL *f;
    typedef struct
    {
        CPoint Home;          // coordinates of call
        int art;              // arrival time
        int len;              // length
        int ast,act;          // actual start,completion time
        int Sf,Df;            // assigned channel no.
        int sta;              // state of call,
                              // 0=doesn't exist yet, 1=in queue,
                              // 2=assigned, 3=terminated
        int awt;              // actual waiting time in the queue
        int SCell,DCell;      // host cell: S=source,D=dest
    } CALL;
    CALL *v;
    typedef struct
    {
        int f[R+1];           // allocated channels
        bool sep[L+1];        // adjacency: 0=no, 1=yes
    } CELL;
    CELL *Cell;
public:
    CCode11A();
    ~CCode11A();
    afx_msg void OnPaint();
    afx_msg void OnKeyDown (UINT nChar, UINT nRep, UINT nFlags);
    DECLARE_MESSAGE_MAP();
    void CallInfo();
};

class CMyWinApp : public CWinApp
{
public:
  virtual BOOL InitInstance();
};

// Code11A.cpp
#include "Code11A.h"

CMyWinApp  MyApplication;
BOOL CMyWinApp::InitInstance()
{
      CCode11A* pFrame = new CCode11A;
      m_pMainWnd = pFrame;
      pFrame->ShowWindow(SW_SHOW);
      pFrame->UpdateWindow();
      return TRUE;
}
```

```
BEGIN_MESSAGE_MAP(CCode11A, CFrameWnd)
      ON_WM_PAINT()
      ON_WM_KEYDOWN()
END_MESSAGE_MAP()

CCode11A::CCode11A()
{
      int i,j,k,r,w;
      TopLeft=CPoint(0,0);
      TextArea=CPoint(750,10);
      f=new CHANNEL [C+1];
      v=new CALL [N+1];
      Cell=new CELL [L+1];
      Create(NULL,"Cellular telephone system simulation",
            WS_OVERLAPPEDWINDOW,CRect(0,0,1000,660));

      // channel and call initialization
      t=-1; nv=0; tnav=0;
      for (k=1;k<=C;k++)
      {
            f[k].sta=0;
            f[k].av[0]=0;
      }
      for (i=1;i<=N;i++)
            v[i].sta=0;

      // call assignment table
      vInfoView.Create(WS_VISIBLE | WS_CHILD | WS_BORDER | LVS_REPORT
            | LVS_NOSORTHEADER,CRect(TextArea.x,TextArea.y,
            TextArea.x+230,TextArea.y+330),this, IDC_vINFO);
      char* column[nFIELDS+1]=
            {"Call","S","Sf","D","Df","arr","ast","act","sta"};
      int columnWidth[nFIELDS+1]={40,40,40,40,40,40,40,40,40};
      LV_COLUMN lvColumn;
      lvColumn.mask = LVCF_WIDTH | LVCF_TEXT | LVCF_FMT | LVCF_SUBITEM;
      lvColumn.fmt = LVCFMT_CENTER;
      lvColumn.cx = 85;
      for (i=0;i<=nFIELDS;i++)
      {
            lvColumn.iSubItem = 0;
            lvColumn.pszText = column[i];
            vInfoView.InsertColumn(i,&lvColumn);
            vInfoView.SetColumnWidth(i,columnWidth[i]);
      }
      FILE *InFile;
      InFile=fopen("Code11A.in","r");
      for (i=1;i<=L;i++)
      {
            for (j=1;j<=i;j++)
            {
                  fscanf(InFile,"%d",&Cell[i].sep[j]); // read cell adjacency info
                  Cell[j].sep[i]=Cell[i].sep[j];
            }
```

```
                for (r=1;r<=R;r++)
                {
                        w=Cell[i].f[r]=i+7*(r-1); // allocate channels to cells
                        f[w].hCell=i;
                }
        }
        fclose(InFile);
        wBar=18; hBar=15;
        ZeroMemory(&lfTimes,sizeof(lfTimes));
        lfTimes.lfHeight=80;
        fontTimes.CreatePointFontIndirect(&lfTimes);
}

CCode11A::~CCode11A()
{
        delete f,v,Cell;
}

void CCode11A::OnPaint()
{
        CPaintDC dc(this);
        CString s;
        CRect rc;

        dc.SelectObject(fontTimes);
        dc.SetTextColor(RGB(255,255,255));
        for (int i=1;i<=C;i++)
        {
                f[i].Home.x=15; f[i].Home.y=35+(i-1)*(hBar+6);
                rc=CRect(f[i].Home,CSize(wBar+5,hBar));
                dc.FillSolidRect(&rc,RGB(150,150,150));
                s.Format("%d",i);
                dc.TextOut(f[i].Home.x+7,f[i].Home.y+2,s);
        }
        dc.SetBkColor(RGB(255,255,255));
        dc.SetTextColor(RGB(100,100,100));
        s.Format("#Channels: %d       Max.#Calls: %d",C,N);
        dc.TextOut(TextArea.x,TextArea.y+330,s);

rc=CRect(TextArea.x,TextArea.y+480,TextArea.x+230,TextArea.y+600);
        dc.Rectangle(&rc);
}

void CCode11A::OnKeyDown(UINT nChar,UINT nRep,UINT nFlags)
{
        CClientDC dc(this);
        CString s;
        CRect rc;
        CBrush* pWhite=new CBrush(RGB(255,255,255));
        CPen SPen(PS_SOLID,1,RGB(0,0,200));
        CPen DPen(PS_SOLID,1,RGB(200,0,0));
        int niv, nav;               // #inc calls, #avail calls
        int i,j,k,p,q,r,u,w,z;
```

```
j=0;                        // initialize #calls in Q
time_t seed=time(NULL);    srand((unsigned)seed);
if (nChar==VK_SPACE && v[N].sta!=3)
{
       dc.SelectObject(fontTimes);
       dc.SetBkColor(RGB(255,255,255));
       dc.SetTextColor(RGB(0,0,0));
       s.Format("t=%d",++t);
       dc.TextOut(TextArea.x,TextArea.y+350,s);
       s.Format("%d",t);
       dc.TextOut(25+f[1].Home.x+t*wBar,f[1].Home.y-20,s);
       rc=CRect(TextArea.x+5,TextArea.y+490,
              TextArea.x+225,TextArea.y+590);
       dc.FillRect(&rc,pWhite);
       for (k=1;k<=C;k++)
              if (f[k].av[0]>=1)
              {
                     r=f[k].lv;
                     if (v[r].act<=t)
                            f[k].sta=0;
              }

       niv=rand()%mniv; nav=0;
       dc.SetTextColor(RGB(150,150,150));

       for (i=1;i<=niv;i++)
       {
              nv++;
              if (nv>N)
                     nv=N;
              v[nv].art=t;
              v[nv].sta=1;
              v[nv].SCell=1+rand()%L;
              v[nv].DCell=1+rand()%L;
       }
       for (i=1;i<=nv;i++)
       {
              u=0; w=0;
              if (v[i].sta==1)
              {
                     v[i].Sf=0; p=v[i].SCell;
                     for (k=1;k<=C;k++)
                     {
                            z=f[k].hCell;
                            if (!f[k].sta && Cell[p].sep[z])
                            {
                                   u=k;
                                   v[i].Sf=u;
                                   f[u].sta=1;
                                   break;
                            }
                            if (k==C && !f[C].sta)
                                   u=0;
                     }
```

```
            v[i].Df=0; q=v[i].DCell;
            if (f[u].sta && u>0)
                for (k=1;k<=C;k++)
                {
                    r=f[k].hCell;
                    if (!f[k].sta && Cell[q].sep[r])
                    {
                        w=k; nav++; tnav++;
                        v[i].Df=w;
                        v[i].len=1+rand()%mLen;
                        v[i].ast=t;
                        v[i].act=t+v[i].len;
                        v[i].sta=2;
                        v[i].awt=v[i].ast-v[i].art;
                        f[w].sta=1;
                        f[w].lv=f[u].lv=i;

                        z=++f[w].av[0]; f[w].av[z]=i;
                        v[i].Home=CPoint(f[w].Home.x+30
                            +t*wBar,f[w].Home.y);
                        rc=CRect(v[i].Home,
                            CSize(wBar*v[i].len,hBar));
                        dc.SelectObject(&DPen);
                        dc.Rectangle(rc);
                        s.Format("%d",i);
                        dc.TextOut(2+v[i].Home.x,
                            1+v[i].Home.y,s);

                        z=++f[u].av[0]; f[u].av[z]=i;
                        v[i].Home=CPoint(f[u].Home.x
                            +30+t*wBar,f[u].Home.y);
                        rc=CRect(v[i].Home,
                            CSize(wBar*v[i].len,hBar));
                        dc.SelectObject(&SPen);
                        dc.Rectangle(rc);
                        s.Format("%d",i);
                        dc.TextOut(2+v[i].Home.x,
                            1+v[i].Home.y,s);
                        break;
                    }
                    if (k==C && !f[C].sta)
                        w=0;
                }
            if (u==0 || w==0)
            {
                j++;
                v[i].Sf=v[i].Df=0;
                v[i].sta=1;
                s.Format("%d",i);
                dc.TextOut(TextArea.x+20+((j-1)/8)*30,
                    TextArea.y+510+((j-1)%8)*10,s);
            }
        }
        if (v[i].sta==2 && t>=v[i].act)
            v[i].sta=3;
```

```
            }
            dc.SetTextColor(RGB(0,0,0));
            s.Format("Last Assigned Call: %d",tnav);
            dc.TextOut(TextArea.x,TextArea.y+370,s);
            s.Format("Total Arrivals: %d",nv);
            dc.TextOut(TextArea.x,TextArea.y+390,s);
            s.Format("#Arrivals at timeslot: %d",niv);
            dc.TextOut(TextArea.x,TextArea.y+410,s);
            s.Format("#Assigned at timeslot: %d",nav);
            dc.TextOut(TextArea.x,TextArea.y+430,s);
            s.Format("#Calls in Q: %d",j);
            dc.TextOut(TextArea.x+10,TextArea.y+490,s);
            CallInfo();
      }
      if (v[N].sta==2)
      {
            v[N].sta=3;
            return;
      }
}

void CCode11A:: CallInfo()
{
      CString s;
      vInfoView.DeleteAllItems();
      LV_ITEM lvItem;
      lvItem.mask = LVIF_TEXT | LVIF_STATE;
      lvItem.state = 0;
      lvItem.stateMask = 0;
      for (int i=0;i<=tnav;i++)
      {
            lvItem.iItem=i;
            lvItem.iSubItem=0;
            lvItem.pszText="";
            vInfoView.InsertItem(&lvItem);
            if (i+1<=tnav && i+1<=N)
            {
                  s.Format("%d",i+1); vInfoView.SetItemText(i,0,s);
                  s.Format("%d",v[i+1].SCell); vInfoView.SetItemText(i,1,s);
                  s.Format("%d",v[i+1].Sf); vInfoView.SetItemText(i,2,s);
                  s.Format("%d",v[i+1].DCell); vInfoView.SetItemText(i,3,s);
                  s.Format("%d",v[i+1].Df); vInfoView.SetItemText(i,4,s);
                  s.Format("%d",v[i+1].art); vInfoView.SetItemText(i,5,s);
                  s.Format("%d",v[i+1].ast); vInfoView.SetItemText(i,6,s);
                  s.Format("%d",v[i+1].act); vInfoView.SetItemText(i,7,s);
                  s.Format("%d",v[i+1].sta); vInfoView.SetItemText(i,8,s);
            }
      }
}
```

Code11B: Channel-Assignment Problem

```
// Code11B.h
#include <afxwin.h>
```

```
#include <math.h>
#include "resource.h"
#define L 19                    // #cells

class CCode11B : public CFrameWnd
{
private:
      LOGFONT lfTimes; CFont fontTimes;
      CPoint TopLeft,TextArea;
      CBitmap h2;
      typedef struct
      {
            int f;              // allocated channels
            bool adj[L+1];      // adjacency: 0=no, 1=yes
            bool coc[L+1];      // co-channel: 0=no, 1=yes
            CPoint home;
      } CELL;
      CELL *Cell;
public:
      CCode11B();
      ~CCode11B();
      afx_msg void OnPaint();
      DECLARE_MESSAGE_MAP();
};

class CMyWinApp : public CWinApp
{
public:
  virtual BOOL InitInstance();
};

// Code11B.cpp: Channel Assignment Problem
#include "Code11B.h"

CMyWinApp  MyApplication;
BOOL CMyWinApp::InitInstance()
{
      CCode11B* pFrame = new CCode11B;
      m_pMainWnd = pFrame;
      pFrame->ShowWindow(SW_SHOW);
      pFrame->UpdateWindow();
      return TRUE;
}

BEGIN_MESSAGE_MAP(CCode11B, CFrameWnd)
      ON_WM_PAINT()
END_MESSAGE_MAP()

CCode11B::CCode11B()
{
      int i,j;
      ZeroMemory(&lfTimes,sizeof(lfTimes));
      lfTimes.lfHeight=80; fontTimes.CreatePointFontIndirect(&lfTimes);
      Cell=new CELL [L+1];
```

```
        Create(NULL,"Channel Assignment Problem: One channel per cell",
                WS_OVERLAPPEDWINDOW,CRect(0,0,800,660));
        h2.LoadBitmap(IDB_BITMAP1);

        for (i=1;i<=L;i++)
                Cell[i].f=0;  // initialize the cells with channel 0

        // call assignment tables
        FILE *InFile1,*InFile2,*InFile3;
        InFile1=fopen("Code11BAdj.in","r");
        InFile2=fopen("Code11BCoc.in","r");
        InFile3=fopen("Code11BHome.in","r");
        for (i=1;i<=L;i++)
        {
                for (j=1;j<=i;j++)   // read cell adjacency info
                {
                        fscanf(InFile1,"%d",&Cell[i].adj[j]);
                Cell[j].adj[i]=Cell[i].adj[j];
                }
                for (j=1;j<=i;j++)   // read cell co-chan info
                {
                        fscanf(InFile2,"%d",&Cell[i].coc[j]);
                        Cell[j].coc[i]=Cell[i].coc[j];
                }
                // read cell home
                fscanf(InFile3,"%d %d",&Cell[i].home.x,&Cell[i].home.y);
}
        fclose(InFile1);
        fclose(InFile2);
        fclose(InFile3);
}

CCode11B::~CCode11B()
{
        delete Cell;
}

afx_msg void CCode11B::OnPaint()
{
        CPaintDC dc(this);
        CString s;
        int i,j,k,r;
        bool w[L+1];

        // Draw the h2 hexagon
        CDC memDC;
        memDC.CreateCompatibleDC(&dc);
        memDC.SelectObject(&h2);
        dc.BitBlt(0,50,400,320,&memDC,0,0,SRCCOPY);

        // assign the channels
        Cell[1].f=1;
        for (i=2;i<=L;i++)
```

```
	{
		k=1;
		while (k<=L)
		{
			w[k]=0;
			for (j=1;j<=i-1;j++)
				if (Cell[i].adj[j] && abs(k-Cell[j].f)<2)
				{
					Cell[i].f=k+1;
					w[k]=1;
				}
			if (!w[k])
			{
				r=0;
				for (j=1;j<=i-1;j++)
				{
					Cell[i].f=k;
					if (Cell[i].coc[j]
						&& Cell[i].f==Cell[j].f)
					{
						Cell[i].f=k+1;
						r++;
					}
				}
				if (r==0)
					break;
			}
			k++;
		}
	}

	// display the output
	dc.SelectObject(fontTimes);
	for (i=1;i<=L;i++)
	{
		dc.TextOut(380,5,"Adjacency matrix");
		dc.TextOut(380,315,"Co-Channel matrix");
		for (j=1;j<=L;j++)
		{
			s.Format("%d",Cell[i].adj[j]);
			dc.TextOut(380+(i-1)*20,20+(j-1)*15,s);
			s.Format("%d",Cell[i].coc[j]);
			dc.TextOut(380+(i-1)*20,330+(j-1)*15,s);
		}
		s.Format("%d",Cell[i].f);
		dc.TextOut(Cell[i].home.x,Cell[i].home.y,s);
	}
}
```

Code11BAdj.in

```
1
1 1
```

```
1 1 1
1 0 1 1
1 0 0 1 1
1 0 0 0 1 1
1 1 0 0 0 1 1
0 1 0 0 0 0 1 1
0 1 0 0 0 0 0 1 1
0 1 1 0 0 0 0 0 1 1
0 0 1 0 0 0 0 0 0 1 1
0 0 1 1 0 0 0 0 0 0 1 1
0 0 0 1 0 0 0 0 0 0 0 1 1
0 0 0 1 1 0 0 0 0 0 0 0 1 1
0 0 0 0 1 0 0 0 0 0 0 0 0 1 1
0 0 0 0 1 1 0 0 0 0 0 0 0 0 1 1
0 0 0 0 0 1 0 0 0 0 0 0 0 0 0 1 1
0 0 0 0 0 1 1 0 0 0 0 0 0 0 0 0 1 1
0 0 0 0 0 0 1 1 0 0 0 0 0 0 0 0 0 1 1
```

Code11BCoc.in

```
1
1 1
1 1 1
1 1 1 1
1 1 1 1 1
1 1 1 1 1 1
1 1 1 1 1 1 1
1 1 1 1 0 1 1 1
1 1 1 0 0 0 1 1 1
1 1 1 1 0 0 1 1 1 1
1 1 1 1 0 0 0 0 1 1 1
1 1 1 1 1 0 0 0 0 1 1 1
1 0 1 1 1 0 0 0 0 0 1 1 1
1 0 1 1 1 1 0 0 0 0 0 1 1 1
1 0 0 1 1 1 0 0 0 0 0 0 1 1 1
1 0 0 1 1 1 1 0 0 0 0 0 0 1 1 1
1 0 0 0 1 1 1 0 0 0 0 0 0 0 1 1 1
1 1 0 0 1 1 1 1 0 0 0 0 0 0 0 1 1 1
1 1 0 0 0 1 1 1 1 0 0 0 0 0 0 0 1 1 1
```

Code11BHome.in

```
200 220
145 190
200 160
255 190
255 250
200 275
145 250
85 220
85 160
145 135
200 110
255 135
```

```
310 160
310 220
310 275
255 300
200 325
145 300
85 275
```

INDEX